Interactive Textbook

HOLT, RINEHART AND WINSTON
A Harcourt Education Company
Orlando • **Austin** • New York • San Diego • London

Printed in the United States of America

ISBN 0-03-079082-4

10 11 2266 18 17 16

4500635053

Contents

CHAPTER 17 Birds and Mammals

CHAPTER 18 Interactions of Living Things

CHAPTER 19 Cycles in Nature

CHAPTER 20 The Earth's Ecosystems

CHAPTER 21 Environmental Problems and Solutions

CHAPTER 22 Body Organization and Structure

CHAPTER 23 Circulation and Respiration

CHAPTER 24 The Digestive and Urinary Systems

CHAPTER 25 Communication and Control

CHAPTER 26 Reproduction and Development

Name ______________________ Class ______________ Date ______________

CHAPTER 1 The World of Life Science

SECTION 1

Asking About Life

BEFORE YOU READ

After you read this section, you should be able to answer these questions:

- What is life science?
- Why is life science important for everyday life?

What Is Life Science?

Imagine that it is summer. You are lying on the grass in a park watching dogs play and bees visiting flowers. An ant carries away a crumb from your lunch. Suddenly, questions pop into your head: How do ants find food? Why don't bees visit every flower? Why do dogs play? By asking these question, you are thinking like a life scientist.

STUDY TIP

Predict As you read this section, write a list of 10 questions that a life scientist might ask.

Life science is the study of living things. Asking questions about the world around you is the first step in any scientific investigation. ☑

1. Identify What is the first step in a scientific investigation?

____

____

Part of science is asking questions about the world around you.

What Questions Do Life Scientists Ask?

Take a look around your home or neighborhood. Just about anywhere you go, you will find some kind of living thing. The world around us is full of an amazing diversity of life. Single-celled algae, giant redwood trees, and 40-ton whales are all living things. For any living thing, you can ask: How does it get its food? Where does it live? How does it behave? Life scientists ask questions like these to learn about the world.

Critical Thinking

2. Predict Aside from studying the environment, how can life scientists affect your life? Give two ways.

____

____

____

____

____

____

What Do Life Scientists Do?

Life scientists can study many different topics. Many of these topics can affect your life. As you study life science, you will begin to see how important life science is in your life. Answering questions can help life scientists learn how to fight disease, produce food, and protect the environment.

FIGHTING DISEASE

Scientists have been successful at getting rid of some diseases. For example, *polio* is a disease that affects the brain and nerves. Polio can make it hard for a person to breathe or walk. Polio used to be very common, but today it is a very rare disease. This is because life scientists studied polio and learned how to keep it from spreading.

Today, scientists are looking for ways to stop the spread of the virus that causes *acquired immune deficiency syndrome* (AIDS). By studying how this virus affects the body and causes AIDS, scientists hope to find a cure.

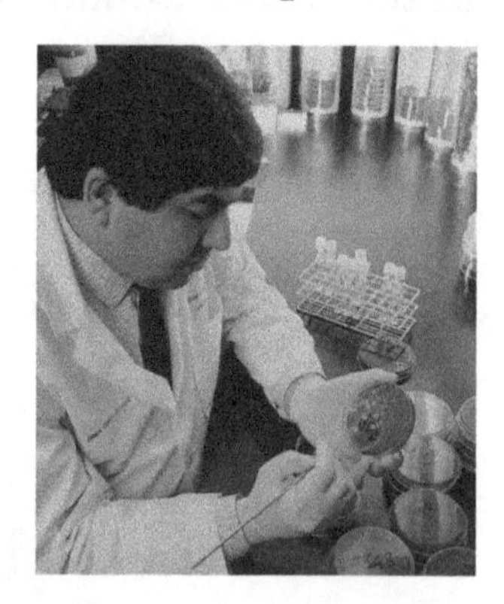

Abdul Lakhani is a life scientist who studies the AIDS virus. He is trying to find a cure for the disease.

PRODUCING FOOD

How can we produce enough food to feed everyone? How can we make sure that foods are safe to eat? To answer these questions, some scientists design experiments to learn what makes plants grow larger or faster. Other scientists look for ways to preserve foods better so that they will last longer. ☑

3. Identify Give one question about producing food that life scientists are trying to answer.

____

____

____

PROTECTING THE ENVIRONMENT

Many environmental problems are caused by people misusing natural resources. Life scientists try to understand how we affect the world around us. We can use this information to find solutions to environmental problems.

Who Is a Life Scientist?

A *life scientist* is anyone who studies *organisms*, or living things. The women and men who are life scientists can live and work anywhere in the world. Some life scientists work on farms. Others study organisms in forests or in oceans. Some even work in space! ☑

Life scientists can study many different features of organisms. They may study how organisms behave and how organisms affect their environments. Some life scientists study how organisms change with time and how they pass on their features to their young. Some life scientists even study organisms that lived millions of years ago.

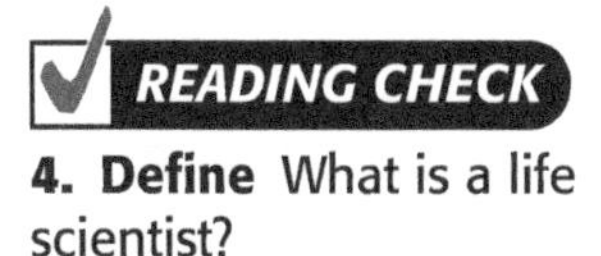

4. Define What is a life scientist?

____

____

Irene Duhart Long studies how space travel affects the human body.

Geerat Vermeij studies how the shells of certain animals have changed with time.

Irene Pepperberg studies whether parrots can learn human language.

TAKE A LOOK

5. Infer What questions are these scientists probably trying to answer?

Irene Duhart Long:

____

____

____

Geerat Vermeij:

____

____

____

Irene Pepperberg:

____

____

____

Name ______________________ Class ______________ Date ______________

Section 1 Review

SECTION VOCABULARY

life science the study of living things	

1. **Explain** How do scientists gather information about the world?

____

____

2. **Describe** What is a life scientist?

____

____

3. **List** Give four places that a life scientist can work.

____

____

____

____

4. **Identify** A life scientist has just discovered an organism that no one has ever seen before. Give four questions that the scientist may ask about the organism.

____

____

____

____

5. **Explain** Why is polio a very rare disease today?

____

____

6. **Describe** How can life scientists help people protect the environment?

____

____

____

____

Name ______________________ Class ______________ Date ______________

CHAPTER 1 The World of Life Science

SECTION 2

Scientific Methods

BEFORE YOU READ

After you read this section, you should be able to answer these questions:

- What are scientific methods?
- What is a hypothesis?
- How do scientists test a hypothesis?

What Are Scientific Methods?

A group of students in Minnesota went on a field trip to a wildlife refuge. They noticed that some of the frogs they saw looked strange. For example, some of the frogs had too many legs or eyes. The frogs were *deformed.* The students wondered what made the frogs deformed. They decided to carry out an investigation to learn what happened to the frogs.

By making observations and asking questions about them, the students were using scientific methods. **Scientific methods** are a series of steps that scientists use to answer questions and to solve problems. The figure below shows the steps in scientific methods. ☑

STUDY TIP

Outline As you read this section, make a chart showing the different steps in scientific methods. In the chart, describe how the students in Minnesota used each step to investigate the deformed frogs.

READING CHECK

1. Define What are scientific methods?

____

____

____

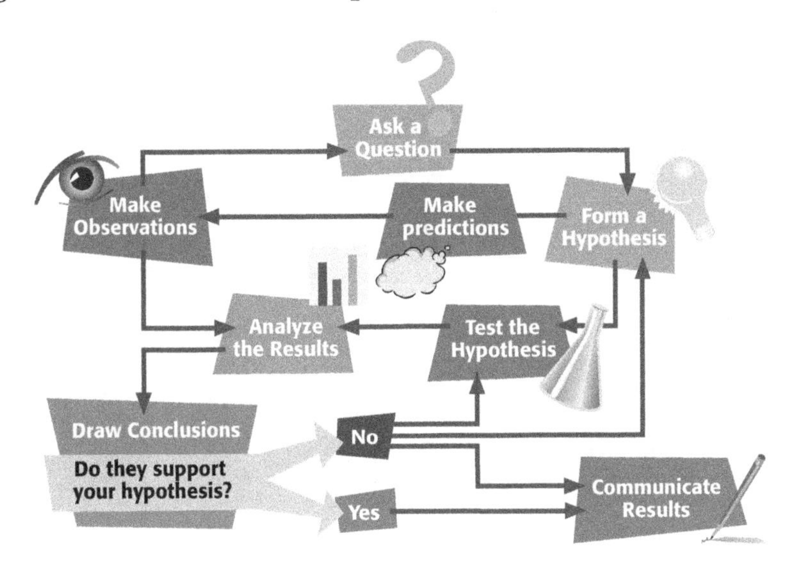

TAKE A LOOK

2. Use Models Starting with "Ask a question," trace two different paths through the figure to "Communicate results." Use a colored pen or marker to trace your paths.

As you can see, the order of steps in scientific methods can vary. Scientists may use all of the steps or just some of the steps during a certain investigation. They may even repeat some of the steps. The order depends on what works best to answer a certain question.

Why Is It Important to Ask a Question?

Asking a question helps scientists focus their research on the most important things they want to learn. In many cases, an observation leads to a question. For example, the students in Minnesota observed that some of the frogs were deformed. Then they asked the question, "Why are some of the frogs deformed?" Answering questions often involves making more observations. ☑

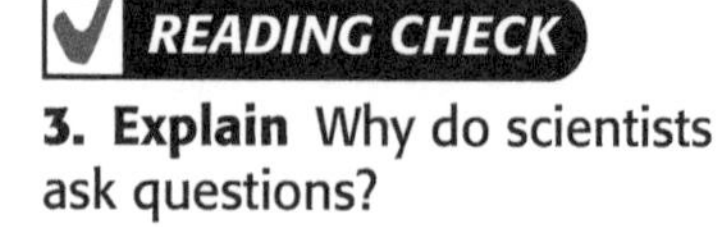

3. Explain Why do scientists ask questions?

____

____

How Do Scientists Make Observations?

The students in Minnesota made careful observations to help them answer their question. The students caught many frogs. Then they counted how many normal and deformed frogs they caught. They photographed, measured, and described each frog. They also tested the water the frogs were living in. The students were careful to record their observations accurately.

Like the students, scientists make many different kinds of observations. They may measure length, volume, time, or speed. They may describe the color or shape of an organism. They may also describe how an organism behaves. When scientists make and record their observations, they are careful to be accurate. Observations are useful only if they are accurate.

Critical Thinking

4. Explain Why is it important for observations to be accurate?

____

____

____

____

TAKE A LOOK

5. Identify Give three kinds of observations that can be made with the tools in the picture.

____

____

____

Scientists use many different tools, such as microscopes, rulers, and thermometers, to make observations.

What Is a Hypothesis?

After asking questions and making observations, scientists may form a hypothesis. A **hypothesis** (plural, *hypotheses*) is a possible answer to a question. A good hypothesis is based on observations and can be tested. When scientists form a hypothesis, they base it on all of the observations and information that they have. ☑

A single question can lead to more than one hypothesis. The students in Minnesota learned about different things that can cause frogs to be deformed. They used this information to form three hypotheses to answer their question. These hypotheses are shown in the figure below.

Hypothesis 1:
The deformities were caused by one or more chemical pollutants in the water.

Hypothesis 2:
The deformities were caused by attacks from parasites or other frogs.

Hypothesis 3:
The deformities were caused by an increase in exposure to ultraviolet light from the sun.

More than one hypothesis can be made for a single question.

READING CHECK

6. Define What is a hypothesis?

____

____

Say It

Discuss In a group, talk about some other possible hypotheses that the students could have come up with.

TAKE A LOOK

7. Describe What are two things that all of the hypotheses have in common?

____

____

____

____

Critical Thinking

8. Make Connections What is the connection between hypotheses and tests in an investigation?

____

____

____

____

____

____

PREDICTIONS

Before a scientist can test a hypothesis, the scientist must make predictions. A *prediction* is a statement that explains how something can cause an effect. A prediction can be used to set up a test of a hypothesis. Predictions are usually stated in an if-then format, as shown in the figure below. More than one prediction may be made for a hypothesis.

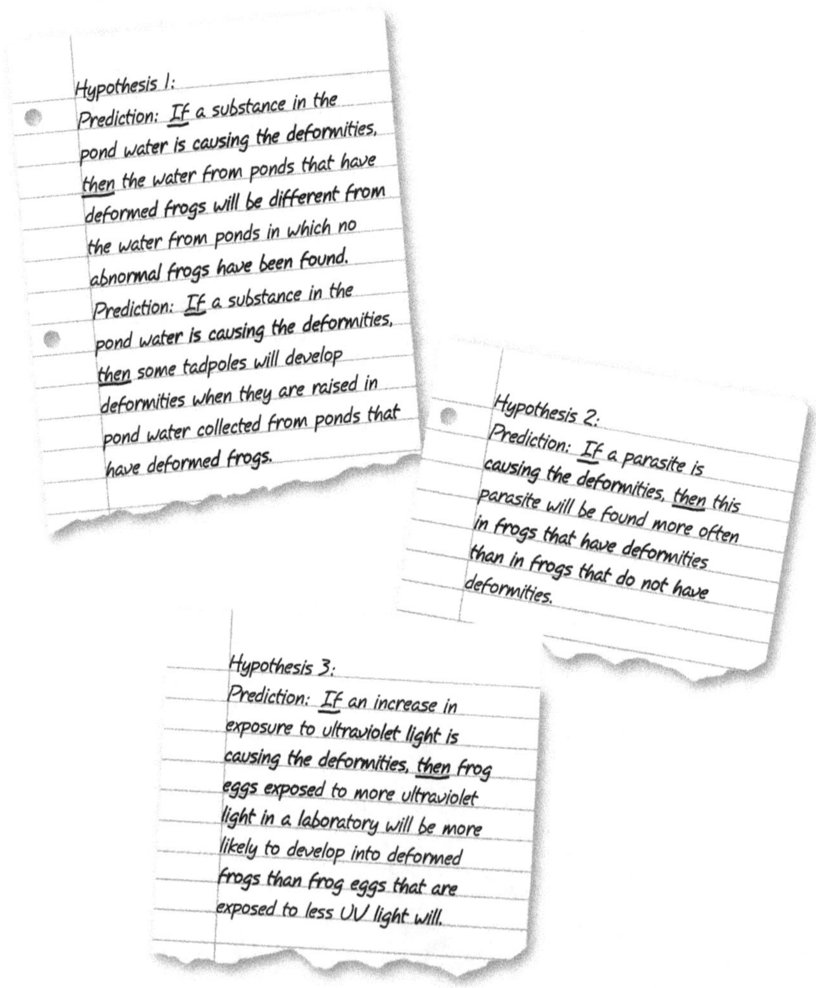

Hypothesis 1:
Prediction: If a substance in the pond water is causing the deformities, then the water from ponds that have deformed frogs will be different from the water from ponds in which no abnormal frogs have been found.
Prediction: If a substance in the pond water is causing the deformities, then some tadpoles will develop deformities when they are raised in pond water collected from ponds that have deformed frogs.

Hypothesis 2:
Prediction: If a parasite is causing the deformities, then this parasite will be found more often in frogs that have deformities than in frogs that do not have deformities.

Hypothesis 3:
Prediction: If an increase in exposure to ultraviolet light is causing the deformities, then frog eggs exposed to more ultraviolet light in a laboratory will be more likely to develop into deformed frogs than frog eggs that are exposed to less UV light will.

More than one prediction may be made for a single hypothesis.

TAKE A LOOK

9. Explain What kind of tests could the students do to test the prediction for Hypothesis 2?

____

____

____

____

____

Scientists can perform experiments to test their predictions. In many cases, the results from the experiments match a prediction. In other cases, the results may not match any of the predictions. When this happens, the scientist must make a new hypothesis and perform more tests.

How Do Scientists Test a Hypothesis?

Scientists perform experiments to show whether a certain factor caused an observation. A *factor* is anything in an experiment that can change the experiment's results. Some examples of factors are temperature, the type of organism being studied, and the weather in an area. ☑

To study the effect of each factor, scientists perform controlled experiments. A **controlled experiment** tests only one factor at a time. These experiments have a control group and one or more experimental groups.

The factors for the control group and the experimental groups are the same, except for the one factor being tested. This factor is called the **variable**. The variable is different in each experimental group. Any difference in the results between the control and experimental groups is probably caused by the variable. ☑

READING CHECK

10. Define What is a factor?

____

____

____

READING CHECK

11. Compare How are control groups and experimental groups different?

____

____

____

DESIGNING AN EXPERIMENT

Experiments must be carefully planned. Every factor should be considered when designing an experiment. It is also important for scientists to use ethical guidelines when they design and carry out experiments. These guidelines help to make sure that the scientists do not cause unnecessary harm to the organisms in the experiment.

The table below shows an experiment to test whether UV light can cause frogs to be deformed. This experiment has one control group and two experimental groups. All the factors between these groups are the same except the amount of UV light exposure. The control group receives no UV light. The number of days that the frog eggs are exposed to UV light is different between the experimental groups. Therefore, exposure to UV light is the variable.

		Control Factors			**Variable**
Group	**Tank**	Kind of frog	Number of eggs	Temperature of water (°C)	UV light exposure (days)
#1	A	leopard frog	50	25	0
	B	leopard frog	50	25	0
#2	C	leopard frog	50	25	15
	D	leopard frog	50	25	15
#3	E	leopard frog	50	25	24
	F	leopard frog	50	25	24

TAKE A LOOK

12. Apply Concepts Which group (1, 2, or 3) is the control group? Explain your answer.

____

____

____

____

COLLECTING DATA

Scientists often try to test many individuals. For example, in the UV light experiment, a total of 300 frogs were tested. By testing many individuals, scientists can account for the effects of normal differences between individuals in each group. They can be more certain that differences between the control and experimental groups are caused by the variable. ☑

READING CHECK

13. Explain Why do scientists try to use many individuals in their experiments?

____

____

____

Scientists will often repeat an experiment to determine if it produces the same results every time. If an experiment produces the same results again and again, scientists can be more certain that the results are true.

The figure below shows the setup of the UV light experiment. It also shows the results of the experiment.

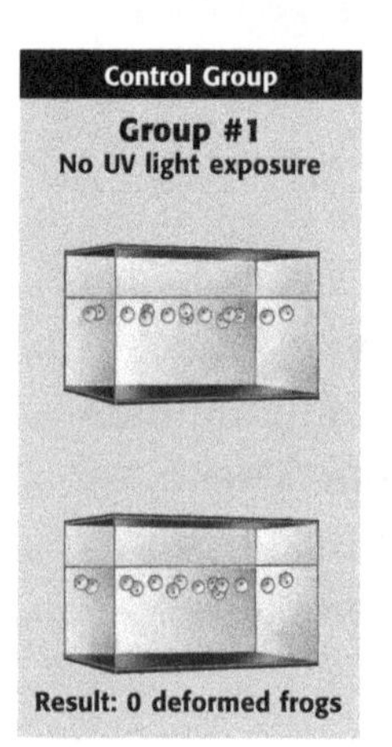

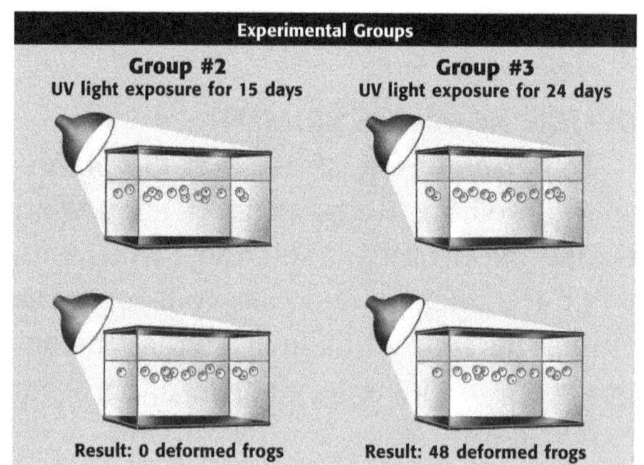

How Do Scientists Analyze Results?

When scientists finish an experiment, they must analyze the results. The information they collect during the analysis helps them explain their observations.

To organize their data, scientists often make tables and graphs. Scientists study the organized data to learn how the variable affected the experiment. The data from the UV light experiment is shown in the table below. This table shows that frogs that were exposed to 24 days of UV light developed deformities.

Number of days of UV exposure	Number of deformed frogs
0	0
15	0
24	48

Math Focus

14. Make a Graph Use the information in the table to fill in the bar graph below.

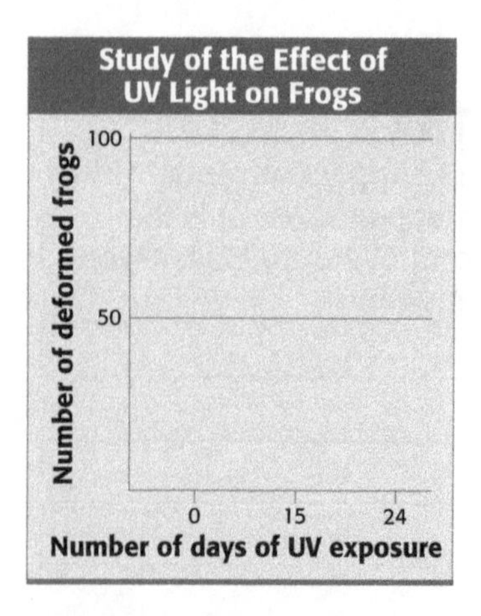

What Are Conclusions?

After analyzing results from experiments, scientists must decide if the results support the hypotheses. This is called *drawing conclusions.* Finding out that a hypothesis is not true can be as valuable as finding out that a hypothesis is true.

Sometimes, the results do not support the hypothesis. When this happens, scientists may repeat the investigation to check for mistakes. Scientists may repeat experiments hundreds of times. Another option is to ask another question and make a new hypothesis.

The UV light experiment showed that UV light can cause frogs to be deformed. However, this does not mean that UV light definitely caused the frog deformities in Minnesota. Many other factors may affect the frogs. Some of these factors may be things that scientists have not even thought of yet.

Questions as complicated as the deformed frogs are rarely solved with a single experiment. The search for a solution may continue for many years. Finding an answer doesn't always end an investigation. In many cases, the answer begins another investigation. In this way, scientists continue to build knowledge.

Critical Thinking

15. Infer How can finding out that a hypothesis is not true be useful for a scientist?

____

Why Do Scientists Share Their Results?

After finishing a study, scientists share their results with others. They write reports and give presentations. They can also put their results on the Internet.

Sharing information allows others the chance to repeat the experiments for themselves. Data from new experiments can either support the original hypothesis, or show that it needs to be changed.

Communicating the results of experiments is an important step in scientific methods.

TAKE A LOOK

16. Describe Why is it important for scientists to share their results?

____

Name ______________________ Class ______________ Date ______________

Section 2 Review

SECTION VOCABULARY

controlled experiment an experiment that tests only one factor at a time by using a comparison of a control group with an experimental group

hypothesis a testable idea or explanation that leads to scientific investigation

scientific methods a series of steps followed to solve problems

variable a factor that changes in an experiment in order to test a hypothesis

1. **Describe** In a controlled experiment, how are the control and experimental groups the same? How are they different?

____

____

____

2. **Infer** Why might a scientist need to repeat a step in scientific methods?

____

____

____

3. **Identify** What are two ways that scientists can share the results of their experiments?

____

____

4. **Define** What is a prediction?

____

____

5. **Explain** Why might a scientist repeat an experiment?

____

____

____

6. **Describe** What can scientists do if the results of an experiment do not support a hypothesis?

____

____

Name ______________________ Class ______________ Date ______________

CHAPTER 1 The World of Life Science

SECTION 3

Scientific Models

BEFORE YOU READ

After you read this section, you should be able to answer these questions:

- How do scientists use models?
- What are scientific theories and laws?

What Are Models?

You need a microscope to see inside most cells. How can you learn about the parts of a cell if you don't have a microscope? One way is to use a model. Scientists use models to learn about things that they cannot see or touch.

A **model** is something scientists use to represent an object or event to make it easier to study. Scientists study models to learn how things work or are made in the natural world. However, you cannot learn everything by studying a model, because models are not exactly like the objects they represent. Some types of scientific models are physical models, mathematical models, and conceptual models. ☑

STUDY TIP

Learn New Words As you read, underline words you don't know. When you figure out what they mean, write the words and their definitions in your notebook.

READING CHECK

1. Explain Why can't you learn everything about an object or event by studying a model?

____

____

____

PHYSICAL MODELS

A toy rocket and a plastic skeleton are examples of physical models. *Physical models* are models that you can see or touch. Many physical models look like the things they represent. The figure shows students using a model of a human body to learn how the body works. However, because the model is not alive, the students cannot learn exactly how the body functions.

TAKE A LOOK

2. Compare Give one way that the model is like a person and one way the model is not like a person.

____

____

____

____

MATHEMATICAL MODELS

A *mathematical model* is made up of mathematical equations and data. Some mathematical models are simple. For example, a Punnett square is a model of how the traits of parents can be combined in their offspring. Using this model, scientists can predict how often certain traits will appear in the offspring of certain parents.

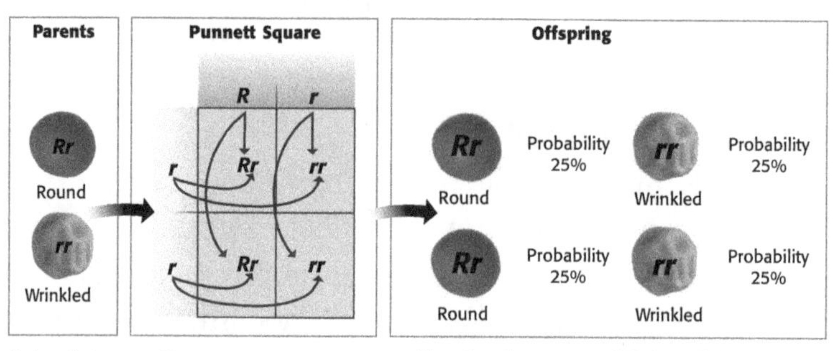

Scientists use Punnett squares to predict the features of living things.

Math Focus

3. Calculate What percentage of the offspring in the figure are round?

____

Some mathematical models are more complicated than the Punnett square. Scientists may use computers to help them interpret more complicated models. Computers work faster and make fewer mistakes than people do.

CONCEPTUAL MODELS

A *conceptual model* is a description of how something works or is put together. Some conceptual models represent ideas. Others connect things that we know to things that we are unfamiliar with. For example, scientists use conceptual models to classify the ways that animals behave. Scientists can use these models to predict how an animal will act in a certain situation.

WHY SCIENTISTS USE MODELS

Scientists use models to study things that are too difficult or dangerous to study in real life. Scientists use models to study very small things, such as atoms, or very large things, such as the Earth. Some scientists use models to predict things that haven't happened yet, or to study events that happened long ago. For example, scientists use computers to produce models of dinosaurs. These models can show how dinosaurs may have looked and moved. ☑

4. Explain Why do scientists use models?

____

____

____

How Does Scientific Knowledge Grow?

Science is always changing. Two scientists can study the same data and make different conclusions. When new technology is developed, scientists often review old data and come to new conclusions. By observing patterns in the world, scientists can create scientific theories and laws.

A scientific **theory** is a scientific explanation that connects and explains many observations. Scientific theories are based on observations. They explain all of the observations about a topic that scientists have at a certain time. Theories are conceptual models that help organize scientific thinking. They are used to explain and predict situations.

A scientific **law** is a statement or equation that can predict what will happen in certain situations. Unlike theories, scientific laws do not explain why something happens. They only predict what will happen. Many people think that scientific theories become scientific laws, but this is not true. Actually, many scientific laws provide evidence to support scientific theories.

Name	What it is
	an explanation that connects and explains evidence and observations
	a statement or equation that predicts what will happen in a certain situation

How Do Scientific Ideas Change?

Scientists are always discovering new information. This new information may show that a theory is incorrect. When this happens, the theory must be changed so that it explains the new information. Sometimes, scientists have to develop a totally new theory to explain the new and old information.

Sometimes, more than one new theory is given to explain the new information. How do scientists know that a new theory is accurate? They use scientific methods to test the new theory. They also examine all the evidence to see if it supports the new theory. Scientists accept a new theory when many tests and pieces of evidence support it.

Critical Thinking

5. Infer Why can two scientists study the same data, but come to different conclusions about it?

____

____

____

____

TAKE A LOOK

6. Identify Fill in the blank boxes in the figure with the terms *scientific law* and *scientific theory.*

Investigate Use the Internet or the library to learn about a scientific idea that interests you. Study how the idea has changed with time. Share your findings with your class.

Name ______________________ Class ______________ Date ______________

Section 3 Review

SECTION VOCABULARY

law a descriptive statement or equation that reliably predicts events under certain conditions **model** a pattern, plan, representation, or description designed to show the structure or workings of an object, system, or concept	**theory** a system of ideas that explains many related observations and is supported by a large body of evidence acquired through scientific investigation

1. Identify How are scientific theories related to observations and evidence?

____

____

____

2. Explain Why do scientists use models?

____

____

____

3. Explain How do scientists know that a new theory is accurate?

____

____

4. Describe What effect can new observations have on a scientific theory?

____

____

5. Identify Give three types of models and an example of each type.

____

____

____

6. Compare How is a scientific theory different from a scientific law?

____

____

____

____

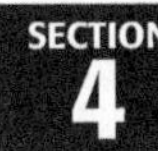

SECTION 4

Tools, Measurement, and Safety

BEFORE YOU READ

After you read this section, you should be able to answer these questions:

- How do tools help scientists?
- How do scientists measure length, area, mass, volume, and temperature?

What Tools Do Scientists Use?

Scientists can use technology to find information and to solve problems. **Technology** is the application of science for practical purposes. New technology can allow scientists to get information that was not available before.

STUDY TIP

Compare As you read this section, make a table comparing how scientists measure length, area, mass, volume, and temperature. Include the tools and units that scientists use for each type of measurement.

CALCULATORS AND COMPUTERS

Scientists analyze, or examine, data using many different tools. Computers and calculators can help scientists do calculations quickly. Computers are also very important tools for collecting, storing, and studying data.

COMPOUND LIGHT MICROSCOPES

Scientists use microscopes to see things that are very small. One kind of microscope is a compound light microscope. A **compound light microscope** is a tool that magnifies small objects. It has three main parts: a stage, a tube with two or more lenses, and a light. Items are placed on the stage. Light passes through them. The lenses help to magnify the image.

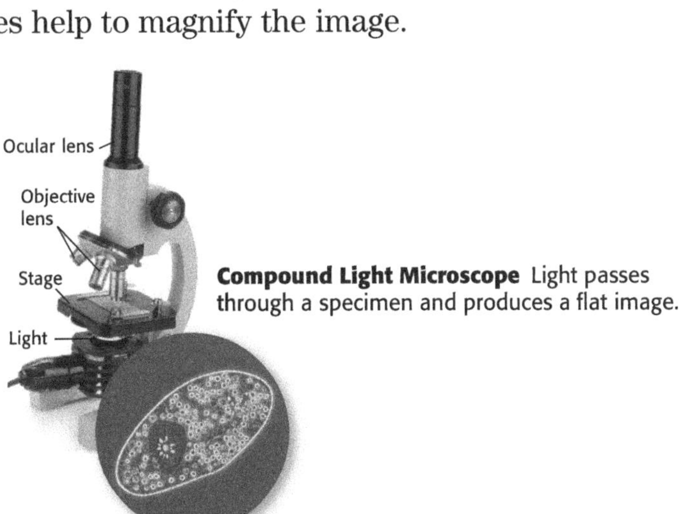

Compound Light Microscope Light passes through a specimen and produces a flat image.

TAKE A LOOK

1. Identify What are the three main parts of a compound light microscope?

____

____

____

Critical Thinking

2. Identify A scientist wants to look at a living cell. Should the scientist use a compound light microscope or an electron microscope? Explain your answer.

____

____

____

____

____

ELECTRON MICROSCOPES

Electron microscopes use tiny particles called *electrons* to produce magnified images. Electron microscopes make clearer and more detailed images than light microscopes do. However, unlike light microscopes, electron microscopes cannot be used to study things that are alive.

Transmission Electron Microscope Electrons pass through the specimen and produce a flat image.

Scanning Electron Microscope Electrons bounce off the surface of the specimen and produce a three-dimensional (3-D) image.

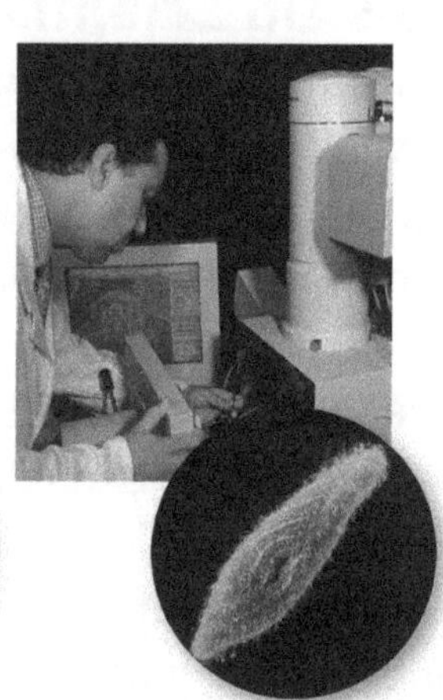

How Do Scientists Measure Objects?

Scientists make many measurements as they collect data. It is important for scientists to be able to share their data with other scientists. Therefore, scientists use units of measurement that are known to all other scientists. One system of measurement that most scientists use is called the International System of Units.

Critical Thinking

3. Predict Consequences What could happen if all scientists used different systems of measurement to record their data?

____

____

____

____

THE INTERNATIONAL SYSTEM OF UNITS

The *International System of Units*, or *SI*, is a system of measurement that scientists use when they collect data. This system of measurement has two benefits. First, scientists around the world can easily share and compare their data because all measurements are made in the same units. In addition, SI units are based on the number 10. This makes it easy to change from one unit to another.

It is important to learn the SI units that are used for different types of measurements. You will use SI units when you make measurements in the science lab.

LENGTH

Length is a measure of how long an object is. The SI unit for length is the *meter* (m). Centimeters (cm) and millimeters (mm) are used to measure small distances. There are 100 cm in 1 m. There are 1,000 mm in 1 m. Micrometers (µm) are used to measure things that are very small, such as cells. There are 1 million µm in 1 m. Rulers and metersticks are used to measure length.

Length tools: ruler or meterstick	SI Unit: meter (m) kilometer (km) centimeter (cm) millimeter (mm)	1 km = 1,000 m 1 cm = 0.01 m 1 mm = 0.001 m 1 µm = 0.000001 m

TAKE A LOOK

4. Identify What is the SI unit for length?

____

AREA

Area is a measure of how much surface an object has. For most objects, area is calculated by multiplying two lengths together. For example, you can find the area of a rectangle by multiplying its length by its width. Area is measured in square units, like square meters (m^2) or square centimeters (cm^2). There are 10,000 cm^2 in 1 m^2. ☑

There is no tool that is used to measure area directly. However, you can use a ruler to measure length and width. Multiply these measurements to find area.

Area tool: ruler (to measure lengths)	square meter (m^2) square centimeter (cm^2)	1 cm^2 = 0.0001 m^2

READING CHECK

5. Explain How can you find the area of a rectangle?

____

____

VOLUME

Volume is the amount of space an object takes up. You can find the volume of a box-shaped object by multiplying its length, width, and height together. You can find the volume of objects with many sides by measuring how much liquid they can push out of a container, as shown in the figure on the next page. You can measure the volume of a liquid using a beaker or a graduated cylinder. ☑

Volume is often measured in cubic units. For example, very large objects can be measured in cubic meters (m^3). Smaller objects can be measured in cubic centimeters (cm^3). There are 1 million cm^3 in 1 m^3. The volume of a liquid is sometimes given in units of liters (L) or milliliters (mL). One mL has the same volume as one cm^3. There are 1,000 mL in 1 L. There are 1,000 L in one m^3.

Volume tools: graduated cylinder, beaker	cubic meter (m^3) cubic centimeter (cm^3) liter (L) milliliter (mL)	1 cm^3 = 0.000001 m^3 1 L = 0.001 m^3 1 mL = 1 cm^3

READING CHECK

6. Define What is volume?

____

____

You can find the volume of this rock by measuring how much liquid it pushes out of the way. The graduated cylinder has 70 mL of liquid in it before the rock is added.

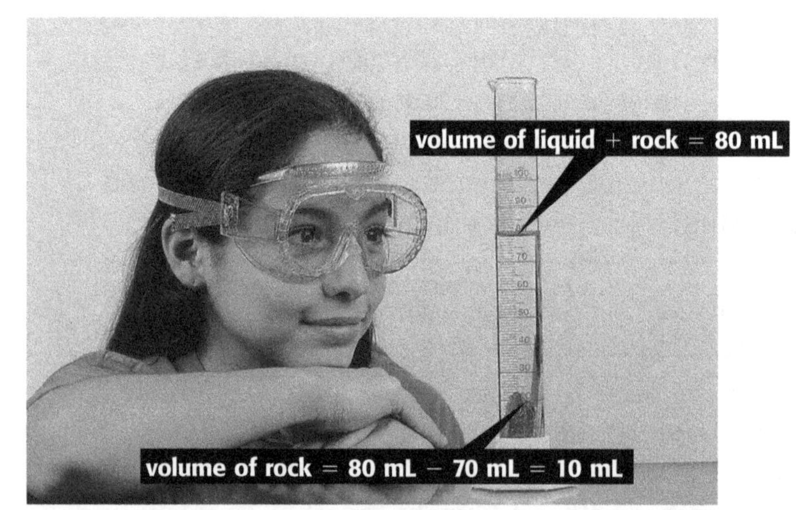

The rock made the volume of material in the cylinder go up to 80 mL. The rock pushed 10 mL of liquid out of the way. The volume of the rock is 10 mL. Because 1 mL = 1 cm^3, the volume of the rock can also be written as 10 cm^3.

TAKE A LOOK

7. Explain How do you know that the rock in the figure has a volume of 10 mL?

____

____

____

____

MASS

Mass is a measurement of the amount of matter in an object. The SI unit for mass is the kilogram (kg). The masses of large objects, such as people, are measured using kg. The masses of smaller objects, such as an apple, are measured in grams (g) or milligrams (mg). There are 1,000 g in 1 kg. There are 1 million mg in 1 kg. Balances are used to measure mass.

Math Focus

8. Convert How many mg are there in 1 g?

____

Mass tool: balance	SI Unit: kilogram (kg) gram (g) milligram (mg)	1 g = 0.001 kg 1 mg = 0.000001 kg

TEMPERATURE

Temperature is a measure of how hot or cold an object is. The SI unit for temperature is the Kelvin (K). However, most people are more familiar with other units of temperature. For example, most people in the United States measure temperatures using degrees Fahrenheit (°F). Scientists often measure temperatures using degrees Celsius (°C). Thermometers are used to measure temperature. ☑

Temperature tool: thermometer	SI Unit: kelvin (K) degrees Celsius (°C)	0°C = 273 K 100°C = 373 K

READING CHECK

9. Define What is temperature?

____

____

____

It is easy to change measurements in °C to K. To change a temperature measurement from °C to K, you simply add 273 to the measurement. For example, 200 °C = 200 + 273 = 473 K. It is more complicated to change measurements in K or °C into °F. That is why scientists do not measure temperatures in °F. ☑

READING CHECK

10. Explain Why do scientists measure temperature in K or °C instead of °F?

____

____

____

How Can You Stay Safe in Science Class?

Science can be exciting, but it can also be dangerous. In order to stay safe while you are doing a science activity, you should always follow your teacher's directions. Read and follow the lab directions carefully, and do not take "shortcuts." Pay attention to safety symbols, such as the ones in the figure below. If you do not understand something that you see in a science activity, ask your teacher for help.

Safety Symbols

Eye protection

Clothing protection

Hand safety

Heating safety

Electrical safety

Chemical safety

Animal safety

Sharp object

Plant safety

TAKE A LOOK

11. Investigate Look around your classroom for safety symbols like the ones in the figure. Give two examples of places where safety symbols are found in your classroom.

____

____

Name ______________________ Class ______________ Date ______________

Section 4 Review

SECTION VOCABULARY

area a measure of the size of a surface or a region

compound light microscope an instrument that magnifies small objects so that they can be seen easily by using two or more lenses

electron microscope a microscope that focuses a beam of electrons to magnify objects

mass a measure of the amount of matter in an object

technology the application of science for practical purposes; the use of tools, machines, materials, and processes to meet human needs

temperature a measure of how hot (or cold) something is; specifically, a measure of the average kinetic energy of the particles in an object

volume a measure of the size of a body or region in three-dimensional space

1. Describe You can find the volume of a box-shaped object by multiplying its length, width, and height together. How can you measure the volume of an object if it is not shaped like a box?

____

____

____

____

2. Identify Fill in the table to show the tool you would use to carry out each measurement.

Task	Tool
Looking at something that is very small	
Measuring how tall your friend is	
Measuring how much water is in a glass	

3. Identify What are two units that scientists use to measure temperature?

____

____

4. Explain How can you stay safe while doing a science activity? Give three ways.

____

____

____

Name ______________________ Class ______________ Date ______________

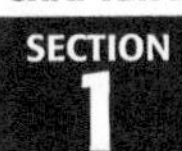

CHAPTER 2 It's Alive!! Or Is It?

SECTION 1 Characteristics of Living Things

BEFORE YOU READ

After you read this section, you should be able to answer these questions:

- What are all living things made of?
- What do all living things have in common?

National Science Education Standards

LS 1b, 1c, 1d, 2a, 2b, 2c, 3a, 3b, 3c

What Are All Living Things Made Of?

If you saw a bright yellow, slimy blob in the grass, would you think it was alive? How could you tell? All living things, or *organisms*, share several characteristics. What does a dog have in common with a bacterium? What do *you* have in common with a bright yellow slime mold?

All living things are made of one or more cells. A **cell** is the smallest unit that can carry out all the activities of life. Every cell is surrounded by a cell membrane. The *cell membrane* separates the cell from the outside environment.

Some organisms are made of trillions of cells. In these organisms, different kinds of cells do different jobs. For example, muscle cells are used for movement. Other organisms are made of only one cell. In these organisms, different parts of the cell have different jobs.

STUDY TIP

Organize As you read this section, make a list of the six characteristics of living things.

STANDARDS CHECK

LS 1b All organisms are composed of cells—the fundamental unit of life. Most organisms are single cells; other organisms, including humans, are multicellular.

1. Identify What are all living things made of?

____

Some organisms, such as the monkeys on the left, are made up of trillions of cells. The protists on the right are made up of one or a few cells. They are so small they can only be seen with a microscope.

How Do Living Things Respond to Change?

All organisms can sense changes in their environments. Each organism reacts differently to these changes. A change that affects how an organism acts is called a **stimulus** (plural, *stimuli*). Stimuli can be chemicals, light, sounds, hunger, or anything that causes an organism to react. ☑

READING CHECK

2. List Give three examples of stimuli.

____

____

____

The touch of an insect is a stimulus for a Venus' flytrap. The stimulus causes the plant to close its leaves quickly.

TAKE A LOOK

3. Complete For a Venus' flytrap, the touch of an insect is a ______________.

Even when things outside the body change, an organism must keep the conditions inside its body the same. The act of keeping a constant environment inside an organism is called **homeostasis**. When an organism maintains homeostasis, all the chemical reactions inside its body can work correctly.

Critical Thinking

4. Predict What would happen if your body couldn't maintain homeostasis?

____

____

____

____

____

____

RESPONDING TO EXTERNAL CHANGES

If it is hot outside, your body starts to sweat to cool down. If it is cold outside, your body starts to shiver to warm up. In each situation, your body reacts to the changes in the environment. It tries to return itself to normal.

Different kinds of organisms react to changes in the environment in different ways. For example, crocodiles lie in the sun to get warm. When they get too warm, they open their mouths wide to release heat.

How Do Organisms Have Offspring?

Every type of organism has *offspring.* The two ways organisms can produce offspring are by sexual or asexual reproduction. Generally, in **sexual reproduction**, two parents make offspring. The offspring get traits from both parents. In **asexual reproduction**, one parent makes offspring. The offspring are identical to the parent. ☑

5. Identify How many parents are generally needed to produce offspring by sexual reproduction?

____

Most plants and animals make offspring by sexual reproduction. However, most single-celled organisms and some multicellular organisms make offspring by asexual reproduction. For example, hydra make offspring by forming buds that break off and grow into new hydra.

Like most animals, bears produce offspring by sexual reproduction. However, some animals, such as hydra, can reproduce asexually.

TAKE A LOOK

6. Identify How do most animals reproduce?

____

____

Why Do Offspring Look Like Their Parents?

All organisms are made of cells. Inside each cell, there is information about all of the organism's traits. This information is found in DNA (**d**eoxyribo**n**ucleic **a**cid). *DNA* carries instructions for the organism's traits. Offspring look like their parents because they get copies of parts of their parent's DNA. Passing traits from parent to offspring is called **heredity**. ☑

READING CHECK

7. Define What is the function of DNA?

____

____

____

____

Why Do Organisms Need Energy?

All organisms need energy to live. Most organisms get their energy from the food they eat. Organisms use this energy to carry out all the activities that happen inside their bodies. For example, organisms need energy to break down food, to move materials in and out of cells, and to build cells.

An organism uses energy to keep up its metabolism. An organism's **metabolism** is all of the chemical reactions that take place in its body. Breaking down food for energy is one of these chemical reactions.

READING CHECK

8. Compare How does growth differ in single-celled organisms and those made of many cells?

____

____

____

____

____

____

How Do Organisms Grow?

All organisms grow during some part of their lives. In a single-celled organism, the cell gets bigger and divides. This makes new organisms. An organism made of many cells gets bigger by making more cells. As these organisms grow, they get new traits. These traits often change how the organism looks. For example, as a tadpole grows into a frog, it develops legs and loses its tail. ☑

Name ______________________ Class ______________ Date ______________

Section 1 Review

NSES LS 1b 1c, 1d, 2a, 2b, 2c, 3a, 3b, 3c

SECTION VOCABULARY

asexual reproduction reproduction that does not involve the union of sex cells and in which one parent produces offspring that are genetically identical to the parent.

cell in biology, the smallest unit that can perform all life processes; cells are covered by a membrane and contain DNA and cytoplasm

heredity the passing of genetic traits from parent to offspring

homeostasis the maintenance of a constant internal state in a changing environment

metabolism the sum of all chemical processes that occur in an organism

sexual reproduction reproduction in which the sex cells from two parents unite to produce offspring that share traits from both parents

stimulus anything that causes a reaction or change in an organism or any part of an organism

1. **Summarize** Complete the Spider Map to show the six characteristics of living things. Add lines to give details on each characteristic.

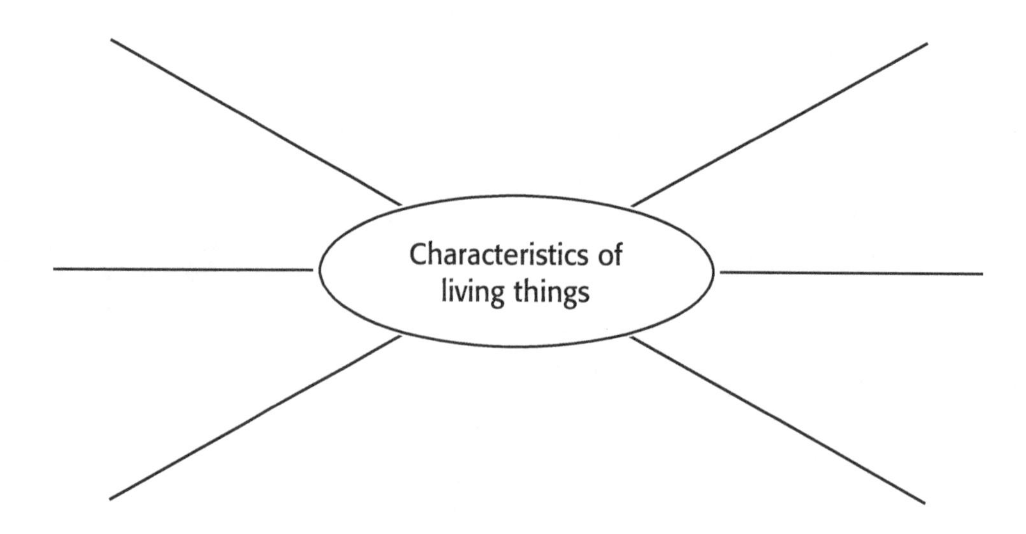

2. **Compare** How does sexual reproduction differ from asexual reproduction?

____

____

____

3. **Explain** How do the buds of an organism such as hydra compare to the parent?

____

____

4. **Identify Relationships** How is a bear's fur related to homeostasis?

____

____

Name ______________________ Class ______________ Date ______________

CHAPTER 2 It's Alive!! Or Is It?

SECTION 2

The Necessities of Life

BEFORE YOU READ

After you read this section, you should be able to answer these questions:

- What things do almost all organisms need?
- Why do living things need food?

National Science Education Standards

LS 1a, 1c, 2c, 3a, 3d, 4b, 4c, 4d

Organize As you read, make a table of the basic needs of most organisms. Fill in examples of how different organisms meet those needs.

What Do Living Things Need?

Would it surprise you to learn that you have the same basic needs as a tree, a frog, and a fly? Almost every organism has the same basic needs: water, air, a place to live, and food.

WATER

Your body is made mostly of water. The cells that make up your body are about 70% to 85% water. Cells need water to keep their inside environments stable. Most of the chemical reactions that happen in cells need water. Your body loses water as you breathe, sweat, or get rid of wastes, such as urine. Because of this, you must replace the water that you lose.

Organisms get water from the fluids they drink and the foods they eat. However, organisms need different amounts of water. You could survive only three days without water. A kangaroo rat never drinks. It lives in the desert and gets all the water it needs from its food.

STANDARDS CHECK

LS 1c Cells carry out the many functions needed to sustain life. They grow and divide, thereby producing more cells. This requires that they take in nutrients, which they use to provide energy for the work that cells do and to make the materials that a cell or organism needs.

Word Help: function
use or purpose

1. Explain Why do cells need water?

____

____

____

____

____

____

AIR

Oxygen, nitrogen, and carbon dioxide are some of the gases in air. Most organisms use oxygen to help them break down food for energy. Other organisms, such as green plants, use carbon dioxide to make food.

TAKE A LOOK

2. Infer Why do you think this diving spider surrounds itself with a bubble in the water?

____

____

____

____

A PLACE TO LIVE

Just as you do, all living things need a place to live. Organisms look for an area that has everything they need to survive. Often, many organisms live in the same area. They all must use the same resources, such as food and water. Many times, an organism will try to keep others out of its area. For example, some birds keep other birds away by singing.

FOOD

All organisms need food. Food gives organisms energy and nutrients to live and grow. However, not all organisms get food in the same way. There are three ways in which organisms can get food. ☑

Some organisms, such as plants, are producers. **Producers** make their own food using energy from their environment. For example, plants, and some bacteria and protists, use the sun's energy to make food from carbon dioxide and water. This process is called *photosynthesis*.

Many organisms are consumers. **Consumers** eat other organisms to get food. For example, a frog is a consumer because it eats insects. All animals are consumers.

A mushroom is a decomposer. Decomposers are a special kind of consumer. **Decomposers** break down dead organisms and animal wastes to get food. Although they are a kind of consumer, decomposers play a different role in an ecosystem than most other consumers. Without decomposers, dead organisms and wastes would pile up all over the Earth!

READING CHECK

3. Explain Why do living things need food?

____

____

____

Critical Thinking

4. Identify Are you a producer, consumer, or decomposer? Explain your answer.

____

____

____

____

____

____

____

TAKE A LOOK

5. Label On the picture, label the producer, consumer, and decomposer.

What Do Organisms Get from Food?

As you just read, organisms can get their food in three different ways. However, all organisms must break down their food to use the nutrients.

Nutrients are molecules. *Molecules* are made of two or more atoms joined together. Most molecules in living things are combinations of carbon, nitrogen, oxygen, phosphorus, and sulfur. Proteins, nucleic acids, lipids, carbohydrates, and ATP are some of the molecules needed by living things.

Discuss With a partner, name 10 organisms and describe what foods they eat. Discuss whether these organisms are producers, consumers, or decomposers.

PROTEINS

Proteins are used in many processes inside a cell. **Proteins** are large molecules made up of smaller molecules called *amino acids*. Living things break down the proteins in food and use the amino acids to make new proteins. ☑

An organism uses proteins in many different ways. Some proteins are used to build or fix parts of an organism's body. Some proteins stay on the outside of a cell, to protect it. Proteins called *enzymes* help to start or speed up reactions inside a cell.

Some proteins help cells do their jobs. For example, a protein called *hemoglobin* is found in our red blood cells. It picks up oxygen and delivers it through the body.

READING CHECK

6. Complete Proteins are made up of ______________________.

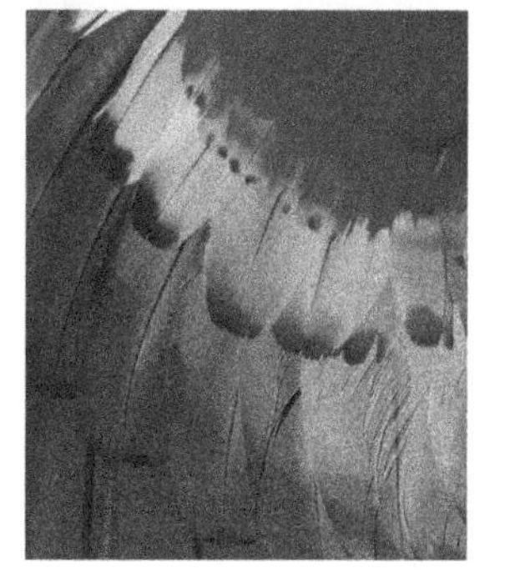

Spider webs, horns, and feathers are made from proteins.

Math Focus

7. Calculate Each red blood cell carries about 250 million molecules of hemoglobin. If every hemoglobin molecule is attached to four oxygen molecules, how many oxygen molecules could one red blood cell carry?

____

NUCLEIC ACIDS

When you bake a cake, you follow instructions to make sure the cake is made correctly. When cells make new molecules, such as proteins, they also follow a set of instructions. The instructions for making any part of an organism are stored in *DNA*.

DNA is a nucleic acid. **Nucleic acids** are molecules made of smaller molecules called *nucleotides*. The instructions carried by DNA tell a cell how to make proteins. The order of nucleotides in DNA tells cells which amino acids to use and which order to put them in.

Critical Thinking

8. Identify Relationships What is the relationship between amino acids and nucleotides?

____

____

____

____

____

____

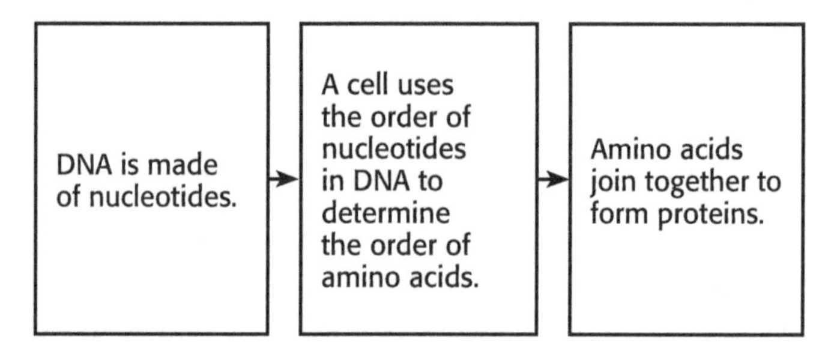

LIPIDS

Lipids are molecules that cannot mix with water. They are a form of stored energy. When lipids are stored in an animal, they are usually solid. These are called *fats*. When lipids are stored in a plant, they are usually liquid. These are called *oils*. When an organism has used up other sources of energy, it can break down fats and oils for more energy.

Lipids also form cell membranes. Cell membranes surround and protect cells. They are made of special lipids called **phospholipids**. When phospholipids are in water, the tail ends of the molecules come together and the head ends face out. This is shown in the figure below.

TAKE A LOOK

9. Describe Describe the structure of a phospholipid, and how it behaves in water.

____

____

____

____

____

Phospholipid Membranes

Head

Tail

The head of a phospholipid molecule is attracted to water, but the tail is not.

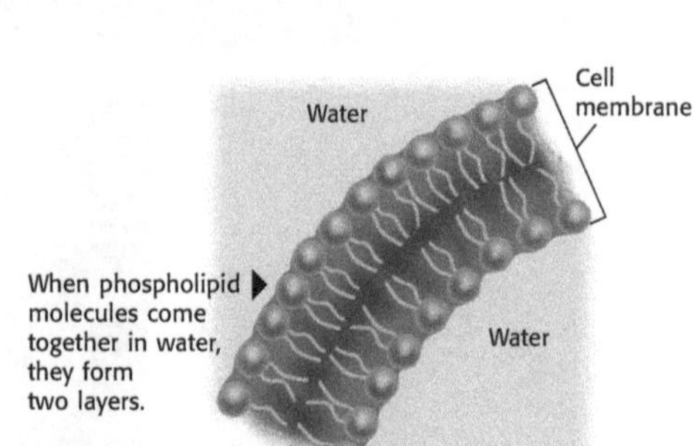

When phospholipid molecules come together in water, they form two layers.

CARBOHYDRATES

Carbohydrates are molecules made of sugars. They provide and store energy for cells. An organism's cells break down carbohydrates to free energy. There are two types of carbohydrates: simple and complex. ☑

READING CHECK

10. Identify What are two types of carbohydrates?

____

____

Simple carbohydrates are made of one or a few sugar molecules. Both table sugar and sugar in fruits are simple carbohydrates. The simple carbohydrate *glucose* is the most common source of energy for cells. The body breaks down simple carbohydrates more quickly than complex carbohydrates.

Complex carbohydrates are made of hundreds of sugar molecules linked together. When organisms such as plants have more sugar than they need, they can store the extra sugar as complex carbohydrates. For example, potatoes store extra sugar as starch. You can also find complex carbohydrates in foods such as whole-wheat bread, pasta, oatmeal, and brown rice.

Type of carbohydrate	Structure	Example
	made of one or a few sugar molecules	
Complex		

TAKE A LOOK

11. Complete Complete the table to explain the two types of carbohydrates.

ATP

After carbohydrates and fats have been broken down, how does their energy get to where it is needed? The cells use **a**denosine **t**riphos**p**hate, or ATP. **ATP** is a molecule that carries energy in cells. The energy released from carbohydrates and fats is passed to ATP molecules. ATP then carries the energy to where it is needed in the cell. ☑

12. Identify What molecule carries energy in cells?

____

Name ______________________ Class ______________ Date ______________

Section 2 Review

NSES LS 1a, 1c, 2c, 3a, 3d, 4b, 4c, 4d

SECTION VOCABULARY

ATP adenosine triphosphate, a molecule that acts as the main energy source for cell processes

carbohydrate a class of energy-giving molecules that includes sugars, starches, and fiber; contains carbon, hydrogen, and oxygen

consumer an organism that eats other organisms or organic matter

decomposer an organism that gets energy by breaking down the remains of dead organisms or animal wastes and consuming or absorbing the nutrients

lipid a type of biochemical that does not dissolve in water; fats and steroids are lipids

nucleic acid a molecule made up of subunits called nucleotides

phospholipid a lipid that contains phosphorus and that is a structural component in cell membranes

producer an organism that can make its own food by using energy from its surroundings

protein a molecule that is made up of amino acids and that is needed to build and repair body structures and to regulate processes in the body

1. List Name four things that organisms need to survive.

____

2. Explain Why are decomposers also consumers?

____

____

3. Identify What two nutrients store energy?

____

4. Describe Describe the structure of a cell membrane.

____

____

____

____

5. Compare Name two ways that simple carbohydrates differ from complex carbohydrates.

____

____

____

____

6. Explain Why is ATP important to cells?

____

____

Name ______________________ Class ______________ Date ____________

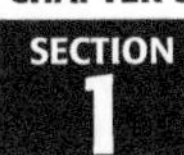

CHAPTER 3 Cells: The Basic Units of Life

SECTION 1

The Diversity of Cells

BEFORE YOU READ

After you read this section, you should be able to answer these questions:

- What is a cell?
- What do all cells have in common?
- What are the two kinds of cells?

National Science Education Standards
LS 1a, 1b, 1c, 2c, 3b, 5a

What Is a Cell?

Most cells are so small that they cannot be seen by the naked eye. So how did scientists find cells? By accident! The first person to see cells wasn't even looking for them.

A **cell** is the smallest unit that can perform all the functions necessary for life. All living things are made of cells. Some living things are made of only one cell. Others are made of millions of cells.

Robert Hooke was the first person to describe cells. In 1665, he built a microscope to look at tiny objects. One day he looked at a piece of cork. Cork is found in the bark of cork trees. Hooke thought the cork looked like it was made of little boxes. He named these boxes *cells*, which means "little rooms" in Latin.

STUDY TIP

Organize As you read this section, make lists of things that are found in prokaryotic cells, things that are found in eukaryotic cells, and things that are found in both kinds of cells.

The first cells that Hooke saw were from cork. These cells were easy to see because plant cells have cell walls. At first, Hooke didn't think animals had cells because he couldn't see them. Today we know that all living things are made of cells.

STANDARDS CHECK

LS 1b All organisms are composed of cells—the fundamental unit of life. Most organisms are single cells; other organisms, including humans, are multicellular.

1. Identify What is the basic unit of all living things?

____

In the late 1600s, a Dutch merchant named Anton van Leeuwenhoek studied many different kinds of cells. He made his own microscopes. With them, he looked at tiny pond organisms called protists. He also looked at blood cells, yeasts, and bacteria.

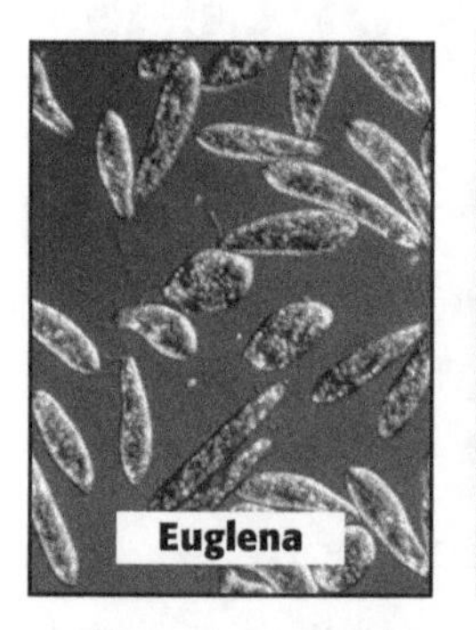

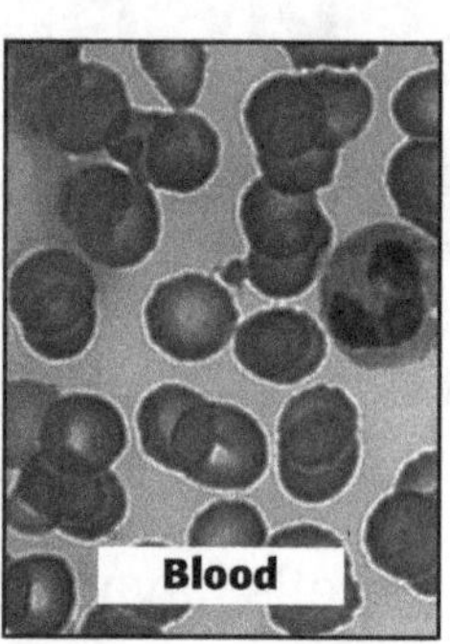

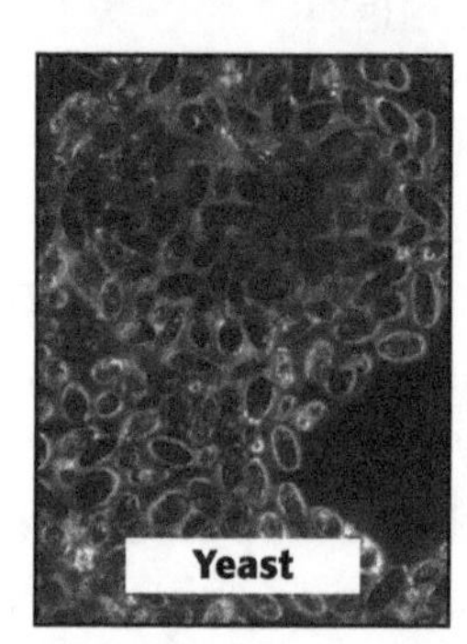

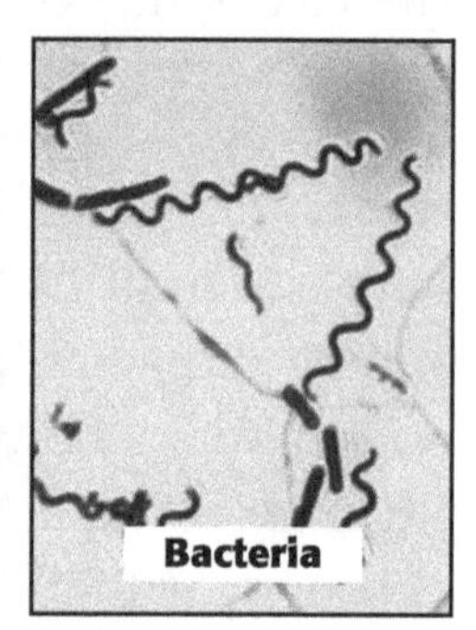

Leeuwenhoek looked at many different kinds of cells with his microscope. He was the first person to see bacteria. Bacterial cells are usually much smaller than most other types of cells.

TAKE A LOOK

2. Identify Which of these cells is probably the smallest? Explain your answer.

____

____

____

What Is the Cell Theory?

Since Hooke first saw cork cells, many discoveries have been made about cells. Cells from different organisms can be very different from one another. Even cells from different parts of the same organism can be very different. However, all cells have several important things in common. These observations are known as the *cell theory*. The cell theory has three parts:

1. All organisms are made of one or more cells.
2. The cell is the basic unit of all living things.
3. All cells come from existing cells.

What Are the Parts of a Cell?

Cells come in many shapes and sizes and can have different functions. However, all cells have three parts in common: a cell membrane, genetic material, and organelles. ☑

3. List What three parts do all cells have in common?

____

____

____

CELL MEMBRANE

All cells are surrounded by a cell membrane. The **cell membrane** is a layer that covers and protects the cell. The membrane separates the cell from its surroundings. The cell membrane also controls all material going in and out of the cell. Inside the cell is a fluid called *cytoplasm*.

GENETIC MATERIAL

All cells contain DNA (deoxyribonucleic acid) at some point in their lives. *DNA* is the genetic material that carries information needed to make proteins, new cells, and new organisms. DNA is passed from parent cells to new cells and it controls the activities of the cell.

The DNA in some cells is found inside a structure called the **nucleus**. Most of your cells have a nucleus.

STANDARDS CHECK

LS 2c Every organism requires a set of instructions for specifying its traits. Heredity is the passage of these instructions from one generation to another.

Word Help: specify to describe or define in detail

4. Explain What is the function of DNA?

____

____

____

____

____

ORGANELLES

Cells have structures called **organelles** that do different jobs for the cell. Most organelles have a membrane covering them. Different types of cells can have different organelles.

Parts of a Cell

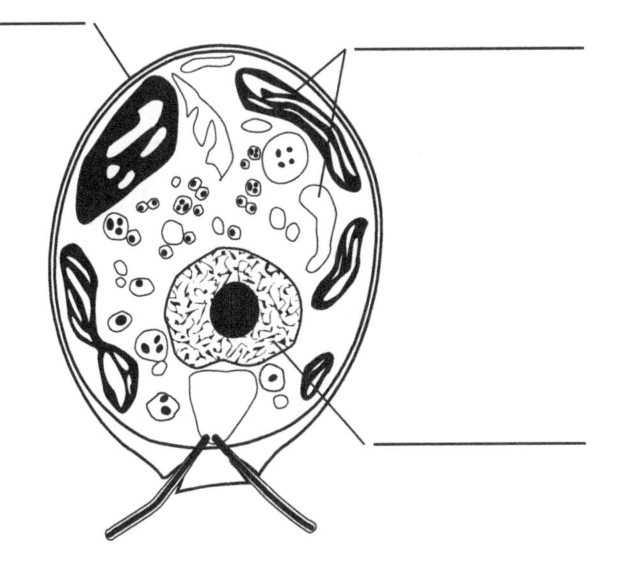

TAKE A LOOK

5. Identify Use the following words to fill in the blank labels on the figure: DNA, cell membrane, organelles.

What Are the Two Kinds of Cells?

There are two basic kinds of cells—cells with a nucleus and cells without a nucleus. Those without a nucleus are called *prokaryotic cells*. Those with a nucleus are called *eukaryotic cells*. ☑

READING CHECK

6. Compare What is one way prokaryotic and eukaryotic cells differ?

____

____

____

What Are Prokaryotes?

A **prokaryote** is an organism made of one cell that does not have a nucleus or other organelles covered by a membrane. Prokaryotes are made of prokaryotic cells. There are two types of prokaryotes: bacteria and archaea.

BACTERIA

The most common prokaryotes are bacteria (singular, *bacterium*). Bacteria are the smallest known cells. These tiny organisms live almost everywhere. Some bacteria live in the soil and water. Others live on or inside other organisms. You have bacteria living on your skin and teeth and in your digestive system. The following are some characteristics of bacteria:

- no nucleus
- circular DNA shaped like a twisted rubber band
- no membrane-covered (or *membrane-bound*) organelles
- a cell wall outside the cell membrane
- a *flagellum* (plural, *flagella*), a tail-like structure that some bacteria use to help them move

Critical Thinking

7. Make Inferences Why do you think bacteria can live in your digestive system without making you sick?

____

____

____

____

A Bacterium

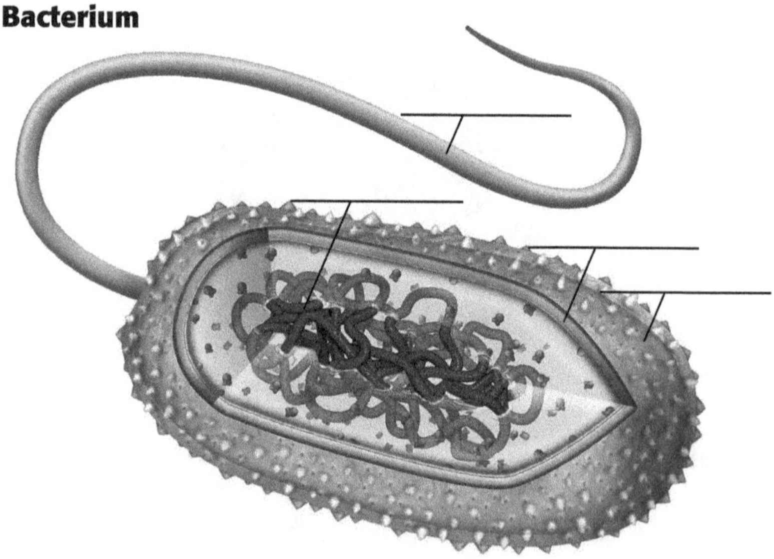

TAKE A LOOK

8. Identify Label the parts of the bacterium using the following terms: DNA, flagellum, cell membrane, cell wall.

ARCHAEA

Archaea (singular, *archaeon*) and bacteria share the following characteristics:

- no nucleus
- no membrane-bound organelles
- circular DNA
- a cell wall

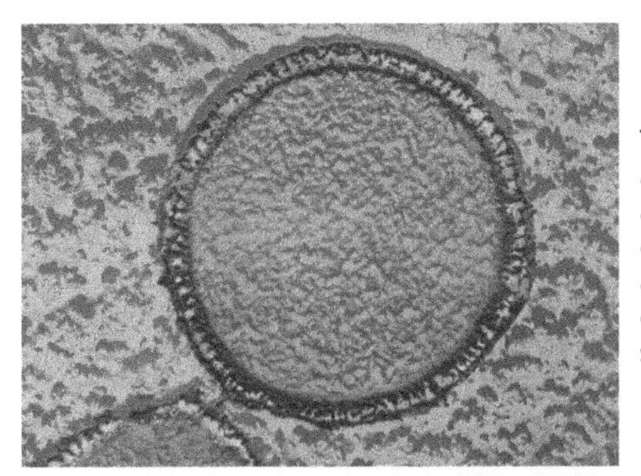

This photograph was taken with an electron microscope. This archaeon lives in volcanic vents deep in the ocean. Temperatures at these vents are very high. Most other living things could not survive there.

Archaea have some other features that no other cells have. For example, the cell wall and cell membrane of archaea are made of different substances from those of bacteria. Some archaea live in places where no other organisms could live. For example, some can live in the boiling water of hot springs. Others can live in toxic places such as volcanic vents filled with sulfur. Still others can live in very salty water in places such as the Dead Sea. ☑

READING CHECK

9. Compare Name two ways that archaea differ from bacteria.

____

____

____

____

____

What Are Eukaryotes?

Eukaryotic cells are the largest cells. They are about 10 times larger than bacteria cells. However, you still need a microscope to see most eukaryotic cells.

Eukaryotes are organisms made of eukaryotic cells. These organisms can have one cell or many cells. Yeast, which makes bread rise, is an example of a eukaryote with one cell. Multicellular organisms, or those made of many cells, include plants and animals.

Unlike prokaryotic cells, eukaryotic cells have a nucleus that holds their DNA. Eukaryotic cells also have membrane-bound organelles. ☑

READING CHECK

10. Identify Name two things eukaryotic cells have that prokaryotic cells do not.

____

____

Eukaryotic Cell

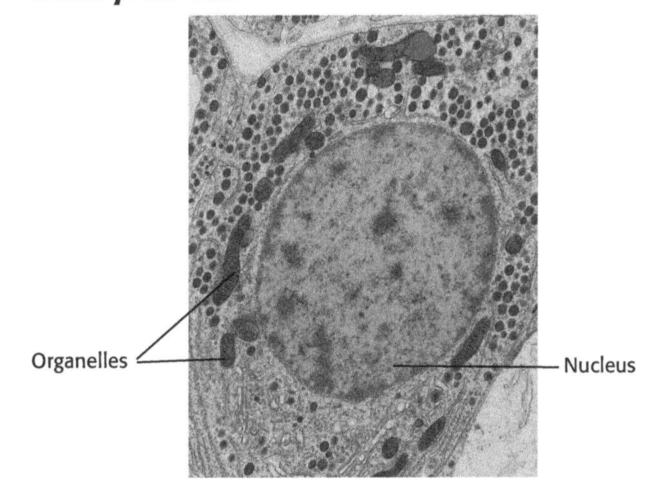

Organelles in a Typical Eukaryotic Cell

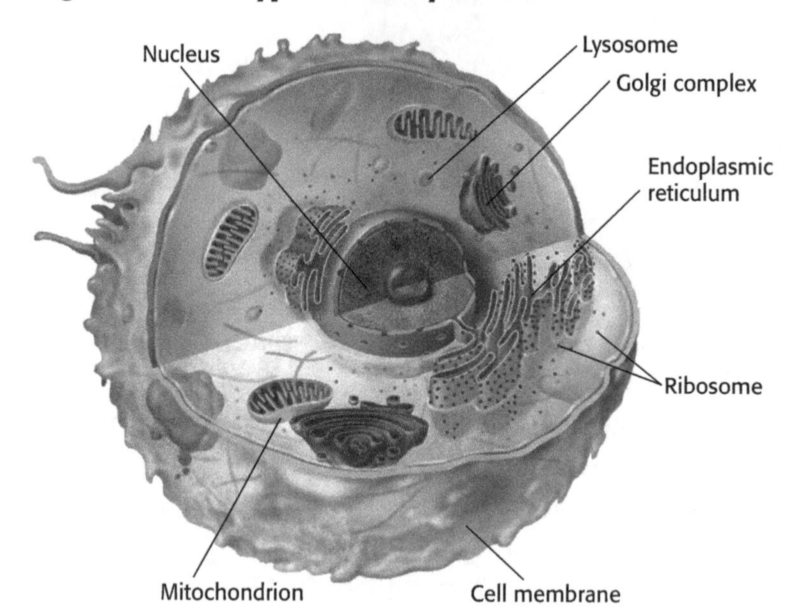

TAKE A LOOK

11. Identify Where is the genetic material found in this cell?

____

Why Are Cells So Small?

Your body is made of trillions of cells. Most cells are so small you need a microscope to see them. More than 50 human cells can fit on the dot of this letter *i*. However, some cells are big. For example, the yolk of a chicken egg is one big cell! Why, then, are most cells small?

Cells take in food and get rid of waste through their outer surfaces. As a cell gets larger, it needs more food to survive. It also produces more waste. This means that more materials have to pass through the surface of a large cell than a small cell.

Critical Thinking

12. Apply Concepts The yolk of a chicken egg is a very large cell. Unlike most cells, egg yolks do not have to take in any nutrients. Why does this allow the cell to be so big?

____

____

____

____

____

____

____

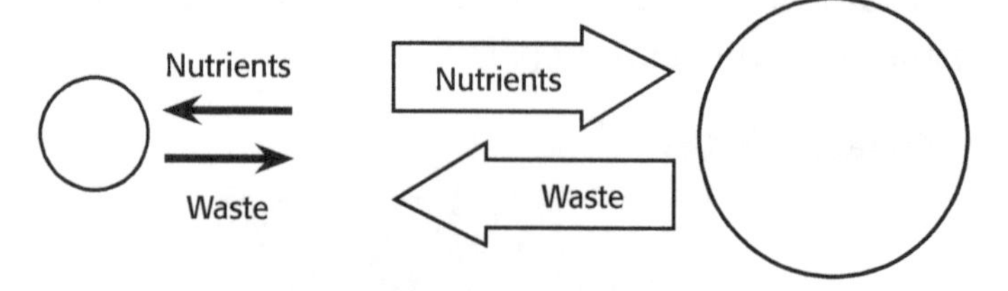

Large cells have to take in more nutrients and get rid of more wastes than small cells.

As a cell's volume increases, its outside surface area grows too. However, volume always grows faster than surface area. If the cell volume gets too big, the surface area will not be large enough for the cell to survive. The cell will not be able to take in enough nutrients or get rid of all its wastes. This means that surface area limits the size of most cells.

SURFACE AREA AND VOLUME OF CELLS

To understand how surface area limits the size of a cell, study the figures below. Imagine that the cubes are cells. You can calculate the surface areas and volumes of the cells using these equations:

$$\textit{volume of cube} = \textit{side} \times \textit{side} \times \textit{side}$$

$$\textit{surface area of cube} = \textit{number of sides} \times \textit{area of side}$$

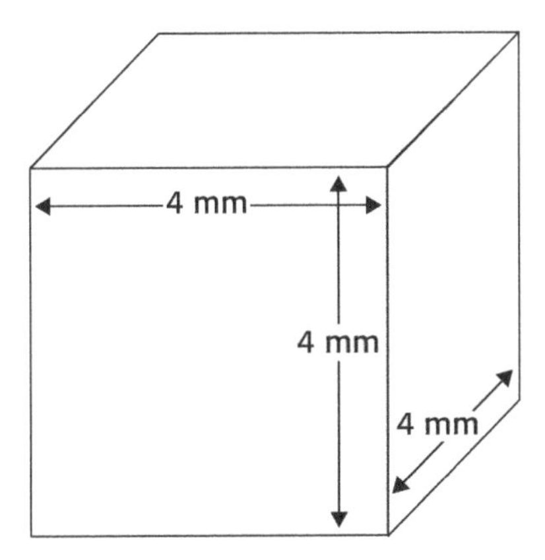

The volume of this cell is 64 mm^3. Its surface area is 96 mm^2.

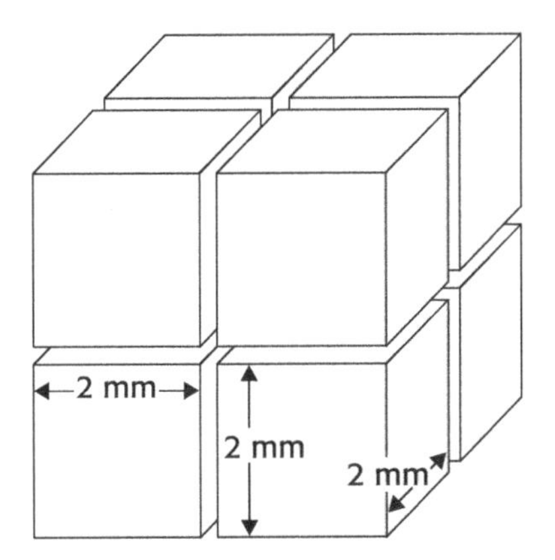

When the large cell is broken up into 8 smaller cells, the total volume stays the same. However, all of the small cells together have more surface area than the large cell. The total surface area of the small cells is 192 mm^2.

The large cell takes in and gets rid of the same amount of material as all of the smaller cells. However, the large cell does not have as much surface area as the smaller cells. Therefore, it cannot take in nutrients or get rid of wastes as easily as each of the smaller cells.

Math Focus

13. Calculate Ratios Scientists say that most cells are small because of the surface area-to-volume ratio. What is this ratio for the large cell?

____

TAKE A LOOK

14. Compare Which cell has a greater surface area compared to its volume—the large cell or one of the smaller cells?

____

Name ______________________ Class ______________ Date ______________

Section 1 Review

NSES LS 1a, 1b, 1c, 2c, 3b, 5a

SECTION VOCABULARY

cell in biology, the smallest unit that can perform all life processes; cells are covered by a membrane and have DNA and cytoplasm

cell membrane a phospholipid layer that covers a cell's surface; acts as a barrier between the inside of a cell and the cell's environment

eukaryote an organism made up of cells that have a nucleus enclosed by a membrane; eukaryotes include animals, plants, and fungi, but not archaea or bacteria

nucleus in a eukaryotic cell, a membrane-bound organelle that contains the cell's DNA and that has a role in processes such as growth, metabolism, and reproduction

organelle one of the small bodies in a cell's cytoplasm that are specialized to perform a specific function

prokaryote an organism that consists of a single cell that does not have a nucleus

1. **Identify** What are the three parts of the cell theory?

____

____

____

2. **Compare** Fill in the Venn Diagram below to compare prokaryotes and eukaryotes. Be sure to label the circles.

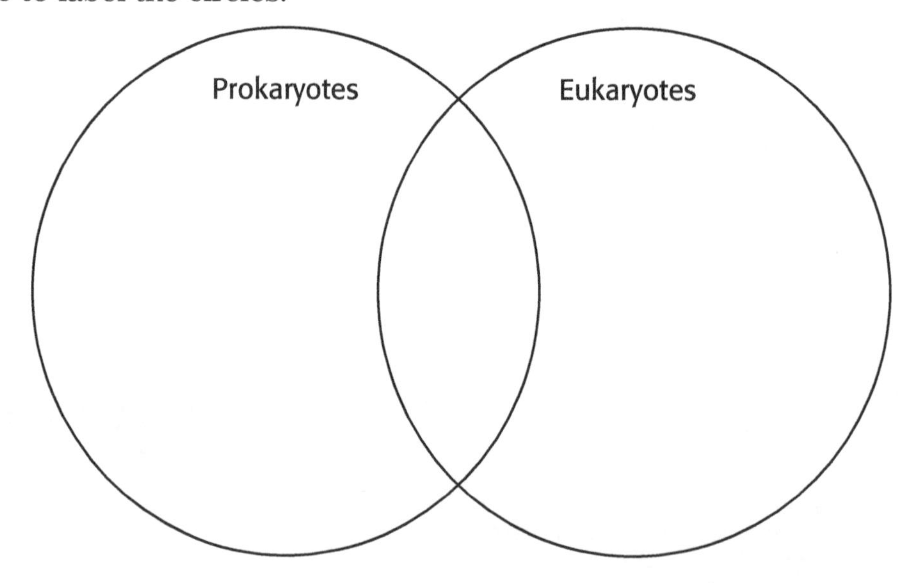

3. **Apply Concepts** You have just discovered a new organism. It has only one cell and was found on the ocean floor, at a vent of boiling hot water. The organism has a cell wall but no nucleus. Explain how you would classify this organism.

____

____

____

Name ______________________ Class ______________ Date ____________

CHAPTER 3 Cells: The Basic Units of Life

SECTION 2

Eukaryotic Cells

BEFORE YOU READ

After you read this section, you should be able to answer these questions:

- What are the parts of a eukaryotic cell?
- What is the function of each part of a eukaryotic cell?

National Science Education Standards
LS 1a, 1b, 1c, 3a, 5a

What Are the Parts of a Eukaryotic Cell?

Plant cells and animal cells are two types of eukaryotic cells. A eukaryotic cell has many parts that help the cell stay alive.

STUDY TIP
Organize As you read this section, make a chart comparing plant cells and animal cells.

CELL WALL

All plant cells have a cell wall. The **cell wall** is a stiff structure that supports the cell and surrounds the cell membrane. The cell wall of a plant cell is made of a type of sugar called cellulose.

Fungi (singular *fungus*), such as yeasts and mushrooms, also have cell walls. The cell walls of fungi are made of a sugar called *chitin*. Prokaryotic cells such as bacteria and archaea also have cell walls. ☑

READING CHECK
1. Identify Name two kinds of eukaryotes that have a cell wall.

____

____

Plant Cell

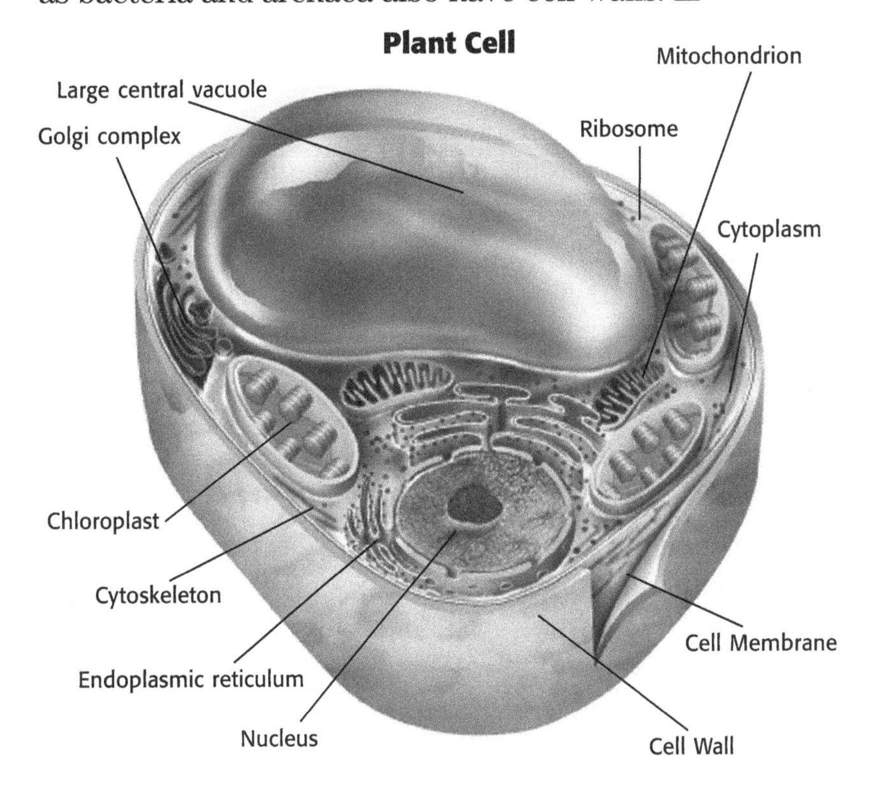

TAKE A LOOK
2. Identify Describe where the cell wall is located.

____

____

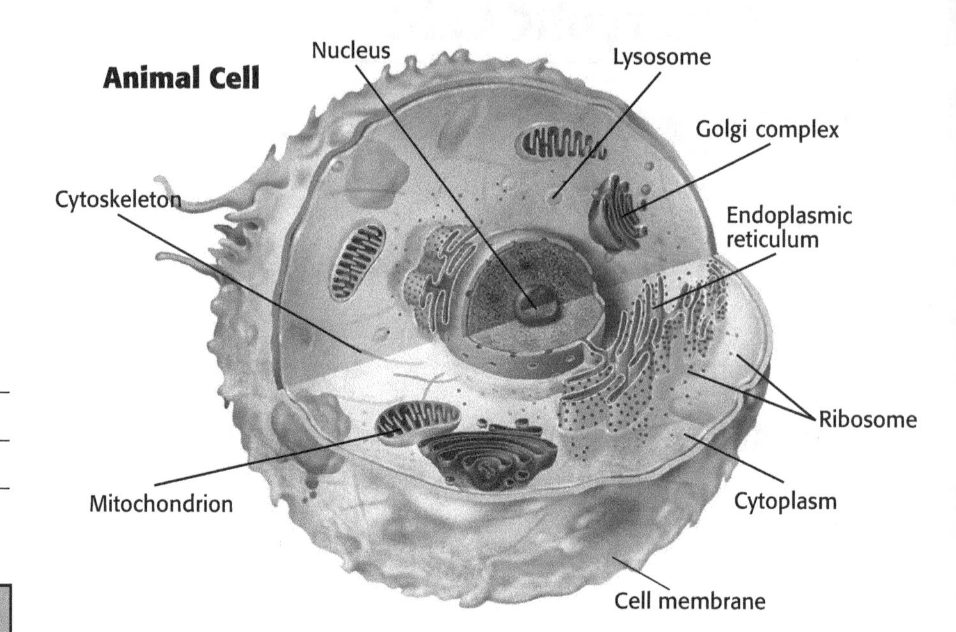

TAKE A LOOK

3. Compare Compare the pictures of an animal cell and a plant cell. Name three structures found in both.

____

____

____

STANDARDS CHECK

LS 1c Cells carry on the many functions needed to sustain life. They grow and divide, thereby producing more cells. This requires that they take in nutrients, which they use to provide energy for the work that cells do and to make the materials that a cell or an organism needs.

Word Help: function
use or purpose

4. Explain What is the main function of the cell membrane?

____

____

____

____

____

____

CELL MEMBRANE

All cells have a cell membrane. The cell membrane is a protective barrier that surrounds the cell. It separates the cell from the outside environment. In cells that have a cell wall, the cell membrane is found just inside the cell wall.

The cell membrane is made of different materials. It contains proteins, lipids, and phospholipids. Proteins are molecules made by the cell for a variety of functions. Lipids are compounds that do not dissolve in water. They include fats and cholesterol. Phospholipids are lipids that contain the element phosphorous.

The proteins and lipids in the cell membrane control the movement of materials into and out of the cell. A cell needs materials such as nutrients and water to survive and grow. Nutrients and wastes go in and out of the cell through the proteins in the cell membrane. Water can pass through the cell membrane without the help of proteins.

RIBOSOMES

Ribosomes are organelles that make proteins. They are the smallest organelles. A cell has many ribosomes. Some float freely in the cytoplasm. Others are attached to membranes or to other organelles. Unlike most organelles, ribosomes are not covered by a membrane. ☑

5. Compare How are ribosomes different from other organelles?

____

____

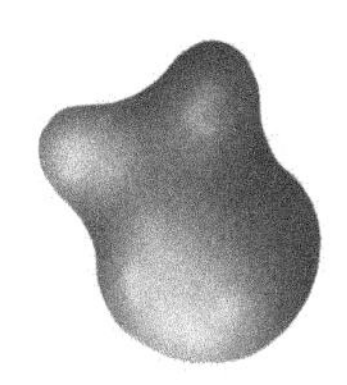

Ribosome
This organelle is where amino acids are hooked together to make proteins.

NUCLEUS

The nucleus is a large organelle in a eukaryotic cell. It contains the cell's genetic material, or DNA. DNA has the instructions that tell a cell how to make proteins.

The nucleus is covered by two membranes. Materials pass through pores in the double membrane. The nucleus of many cells has a dark area called the *nucleolus*.

The Nucleus

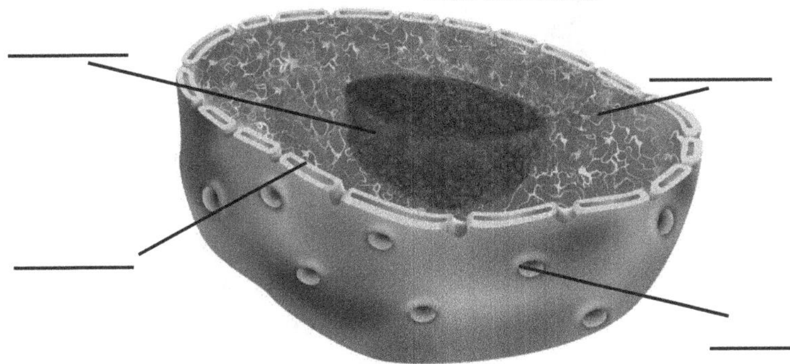

TAKE A LOOK

6. Identify Label the diagram of a nucleus using these terms: pore, DNA, nucleolus, double membrane.

ENDOPLASMIC RETICULUM

Many chemical reactions take place in the cell. Many of these reactions happen on or inside the endoplasmic reticulum. The **endoplasmic reticulum** (ER) is a system of membranes with many folds in which proteins, lipids, and other materials are made.

The ER is also part of the cell's delivery system. Its folds have many tubes and passageways. Materials move through the ER to other parts of the cell.

There are two types of ER: rough and smooth. Smooth ER makes lipids and helps break down materials that could damage the cell. Rough ER has ribosomes attached to it. The ribosomes make proteins. The proteins are then delivered to other parts of the cell by the ER. ☑

READING CHECK

7. Compare What is the difference between smooth ER and rough ER?

____

____

____

____

____

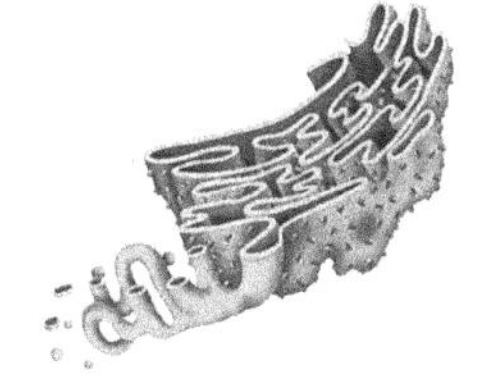

Endoplasmic reticulum
This organelle makes lipids, breaks down drugs and other substances, and packages proteins for the Golgi complex.

MITOCHONDRIA

A **mitochondrion** (plural, *mitochondria*) is the organelle in which sugar is broken down to make energy. It is the main power source for a cell.

A mitochondrion is covered by two membranes. Most of a cell's energy is made in the inside membrane. Energy released by mitochondria is stored in a molecule called ATP. The cell uses ATP to do work.

Mitochondria are about the same size as some bacteria. Like bacteria, mitochondria have their own DNA. The DNA in mitochondria is different from the cell's DNA. ☑

READING CHECK

8. Compare How are mitochondria like bacteria?

____

____

Mitochondrion
This organelle breaks down food molecules to make ATP.

CHLOROPLASTS

Plants and algae have chloroplasts in some of their cells. *Chloroplasts* are organelles in which photosynthesis takes place. *Photosynthesis* is a process by which plants use sunlight, carbon dioxide, and water to make sugar and oxygen. Animal cells do not have chloroplasts.

Chloroplasts are green because they contain a green molecule called *chlorophyll*. Chlorophyll traps the energy of sunlight. Mitochondria then use the sugar made in photosynthesis to make ATP.

Critical Thinking

9. Infer Why don't animal cells need chloroplasts?

____

____

____

____

Chloroplast
This organelle uses the energy of sunlight to make food.

CYTOSKELETON

The cytoskeleton is a web of proteins inside the cell. It acts as both a skeleton and a muscle. The cytoskeleton helps the cell keep its shape. It also helps some cells, such as bacteria, to move.

VESICLES

A **vesicle** is a small sac that surrounds material to be moved. The vesicle moves material to other areas of the cell or into or out of the cell. All eukaryotic cells have vesicles.

GOLGI COMPLEX

The **Golgi complex** is the organelle that packages and distributes proteins. It is the "post office" of the cell. The Golgi complex looks like the smooth ER.

The ER delivers lipids and proteins to the Golgi complex. The Golgi complex can change the lipids and proteins to do different jobs. The final products are then enclosed in a piece of the Golgi complex's membrane. This membrane pinches off to form a vesicle. The vesicle transports the materials to other parts of the cell or out of the cell. ☑

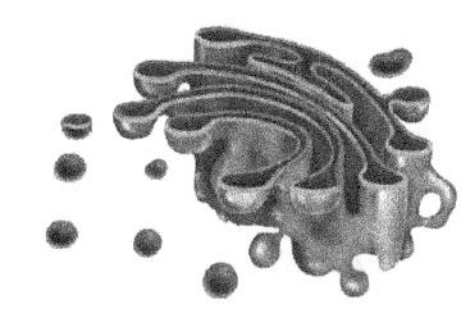

Golgi complex
This organelle processes and transports proteins and other materials out of cell.

READING CHECK

10. Define What is the function of the Golgi complex?

____

____

____

LYSOSOMES

Lysosomes are organelles that contain digestive enzymes. The enzymes destroy worn-out or damaged organelles, wastes, and invading particles.

Lysosomes are found mainly in animal cells. The cell wraps itself around a particle and encloses it in a vesicle. Lysosomes bump into the vesicle and pour enzymes into it. The enzymes break down the particles inside the vesicle. Without lysosomes, old or dangerous materials could build up and damage or kill the cell.

Lysosome
This organelle digests food particles, wastes, cell parts, and foreign invaders.

VACUOLES

A vacuole is a vesicle. In plant and fungal cells, some vacuoles act like lysosomes. They contain enzymes that help a cell digest particles. The large central vacuole in plant cells stores water and other liquids. Large vacuoles full of water help support the cell. Some plants wilt when their vacuoles lose water. ☑

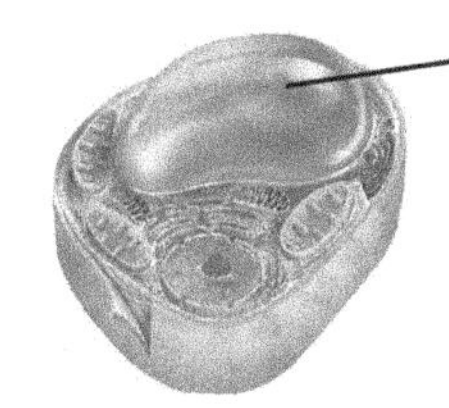

Large central vacuole
This organelle stores water and other materials.

READING CHECK

11. Identify Vacuoles are found in what types of eukaryotic cells?

____

____

Name ______________________ Class ______________ Date ______________

Section 2 Review

NSES **LS 1a, 1b, 1c, 3a, 5a**

SECTION VOCABULARY

cell wall a rigid structure that surrounds the cell membrane and provides support to the cell

endoplasmic reticulum a system of membranes that is found in a cell's cytoplasm and that assists in the production, processing, and transport of proteins and in the production of lipids

Golgi complex cell organelle that helps make and package materials to be transported out of the cell

lysosome a cell organelle that contains digestive enzymes

mitochondrion in eukaryotic cells, the cell organelle that is surrounded by two membranes and that is the site of cellular respiration

ribosome cell organelle composed of RNA and protein; the site of protein synthesis

vesicle a small cavity or sac that contains materials in a eukaryotic cell

1. **Compare** Name three parts of a plant cell that are not found in an animal cell.

____

____

2. **Explain** How does a cell get water and nutrients?

____

____

____

3. **Explain** What would happen to an animal cell if it had no lysosomes?

____

____

____

4. **Apply Concepts** Which kind of cell in the human body do you think would have more mitochondria—a muscle cell or a skin cell? Explain.

____

____

____

____

5. **List** What are two functions of the cytoskeleton?

____

____

SECTION 3
The Organization of Living Things

BEFORE YOU READ

After you read this section, you should be able to answer these questions:

- What are the advantages of being multicellular?
- What are the four levels of organization in living things?
- How are structure and function related in an organism?

National Science Education Standards
LS 1a, 1d, 1e

What Is an Organism?

Anything that can perform life processes by itself is an **organism**. An organism made of a single cell is called a *unicellular organism*. An organism made of many cells is a multicellular organism. The cells in a multicellular organism depend on each other for the organism to survive. ☑

STUDY TIP
Outline As you read, make an outline of this section. Use the heading questions from the section in your outline.

READING CHECK
1. Define What is an organism?

____

____

____

What Are the Benefits of Having Many Cells?

Some organisms exist as one cell. Others can be made of trillions of cells. A *multicellular organism* is an organism made of many cells.

There are three benefits of being multicellular: larger size, longer life, and specialization of cells.

LARGER SIZE

Most multicellular organisms are bigger than one-celled organisms. In general, a larger organism, such as an elephant, has few predators. ☑

READING CHECK
2. Identify Name one way that being large can benefit an organism.

____

____

____

LONGER LIFE

A multicellular organism usually lives longer than a one-celled organism. A one-celled organism is limited to the life span of its one cell. The life span of a multicellular organism, however, is not limited to the life span of any one of its cells.

SPECIALIZATION

In a multicellular organism, each type of cell has a particular job. Each cell does not have to do everything the organism needs. Specialization makes the organism more efficient.

Standards Check

LS 1d Specialized cells perform specialized functions in multicellular organisms. Groups of specialized cells cooperate to form a tissue, such as a muscle. Different tissues are in turn grouped together and form larger functional units, called organs. Each type of cell, tissue, and organ has a distinct structure and set of functions that serve the organism as a whole.

3. List What are the four levels or organization for an organism?

____

____

____

____

What Are the Four Levels of Organization of Living Things?

Multicellular organisms have four levels of organization:

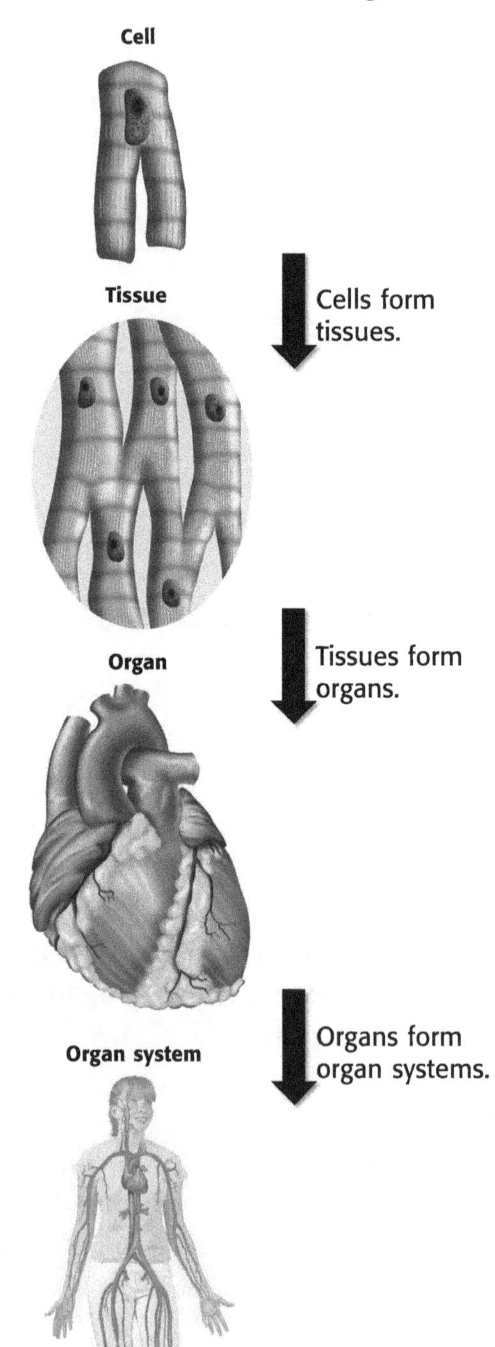

Organ systems form organisms such as you.

TAKE A LOOK

4. Explain Are the cells that make up heart tissue prokaryotic or eukaryotic? How do you know?

____

____

____

____

CELLS WORK TOGETHER AS TISSUES

A **tissue** is a group of cells that work together to perform a specific job. Heart muscle tissue, for example, is made of many heart muscle cells.

TISSUES WORK TOGETHER AS ORGANS

A structure made of two or more tissues that work together to do a certain job is called an **organ**. Your heart, for example, is an organ made of different tissues. The heart has muscle tissues and nerve tissues that work together.

ORGANS WORK TOGETHER AS ORGAN SYSTEMS

A group of organs working together to do a job is called an **organ system**. An example of an organ system is your digestive system. Organ systems depend on each other to help the organism function. For example, the digestive system depends on the cardiovascular and respiratory systems for oxygen.

HOW DOES STRUCTURE RELATE TO FUNCTION?

In an organism, the structure and function of part are related. **Function** is the job the part does. **Structure** is the arrangement of parts in an organism. It includes the shape of a part or the material the part is made of.

The function of the lungs is to bring oxygen to the body and get rid of carbon dioxide. The structure of the lungs helps them to perform their function.

Oxygen-poor blood

Oxygen-rich blood

Blood vessels

The lungs contain tiny, spongy sacs that blood can flow through. Carbon dioxide moves out of the blood and into the sacs. Oxygen flows from the sacs into the blood. If the lungs didn't have this structure, it would be hard for them to perform their function.

Critical Thinking

5. Apply Concepts Do prokaryotes have tissues? Explain.

____

____

____

____

____

Say It

Name With a partner, name as many of the organs in the human body as you can.

Section 3 Review

NSES LS 1a, 1d, 1e

SECTION VOCABULARY

function the special, normal, or proper activity of an organ or part

organ a collection of tissues that carry out a specialized function of the body

organ system a group of organisms that work together to perform body functions

organism a living thing; anything that can carry out life processes independently

structure the arrangement of parts in an organism

tissue a group of similar cells that perform a common function

1. **List** What are three benefits of being multicellular?

____

____

____

2. **Apply Concepts** Could an organism have organs but no tissues? Explain.

____

____

3. **Compare** How are structure and function different?

____

____

____

4. **Explain** What does "specialization of cells" mean?

____

____

5. **Apply Concepts** Why couldn't your heart have only cardiac tissue?

____

____

6. **Explain** Why do multicellular organisms generally live longer than unicellular organisms?

____

____

Name ______________________ Class ______________ Date ______________

CHAPTER 4 The Cell in Action

SECTION 1

Exchange with the Environment

BEFORE YOU READ

After you read this section, you should be able to answer these questions:

- How do cells take in food and get rid of wastes?
- What is diffusion?

National Science Education Standards

LS 1a, 1c

Where Do Cells Get the Materials They Need?

What would happen to a factory if its power were shut off or its supply of materials never arrived? What would happen if the factory couldn't get rid of its garbage? Like a factory, an organism must be able to get energy and raw materials and get rid of wastes. These jobs are done by an organism's cells. Materials move in and out of the cell across the cell membrane. Many materials, such as water and oxygen, can cross the membrane by diffusion.

Compare As you read, make a chart comparing diffusion and osmosis. In your chart, show how they are similar and how they are different.

What Is Diffusion?

The figure below shows what happens when dye is placed on top of a layer of gelatin. Over time, the dye mixes with the gelatin. Why does this happen?

Everything, including the gelatin and the dye, is made of tiny moving particles. Particles tend to move from places where they are crowded to places where they are less crowded. When there are many of one type of particle, this is a high concentration. When there are fewer of one kind of particle, this is a low concentration. The movement from areas of high concentration to areas of low concentration is called **diffusion**. ☑

1. Define What is diffusion?

____

____

____

____

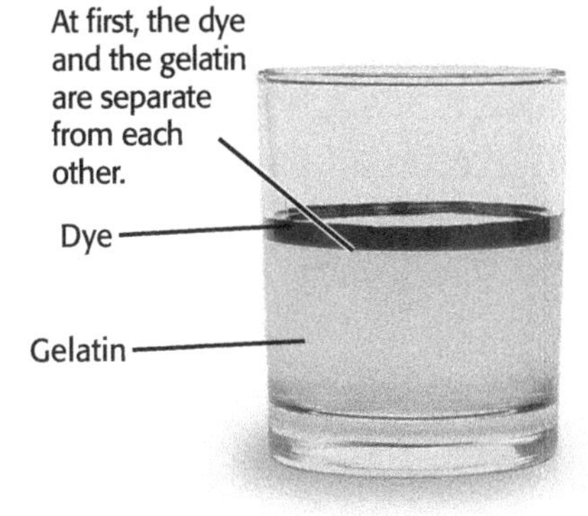

After a while, the particles in the dye move into the gelatin. This process is called diffusion.

TAKE A LOOK

2. Identify How do dye particles move through the water?

____

DIFFUSION OF WATER

Substances, such as water, are made up of particles called *molecules*. Pure water has the highest concentration of water molecules. This means that 100% of the molecules are water molecules. If you mix another substance, such as food coloring, into the water, you lower the concentration of water molecules. This means that water molecules no longer make up 100% of the total molecules.

The figure below shows a container that has been divided by a membrane. The membrane is *semipermeable*—that is, only some substances can pass through it. The membrane lets smaller molecules, such as water, pass through. Larger molecules, such as food coloring, cannot pass through. Water molecules will move across the membrane. The diffusion of water through a membrane is called **osmosis**.

Critical Thinking

3. Apply Concepts Which of the following has a higher concentration of water molecules—200 molecules of water, or a mixture of 300 molecules of water and 100 molecules of food coloring? Explain your answer.

____

____

____

____

____

____

Osmosis

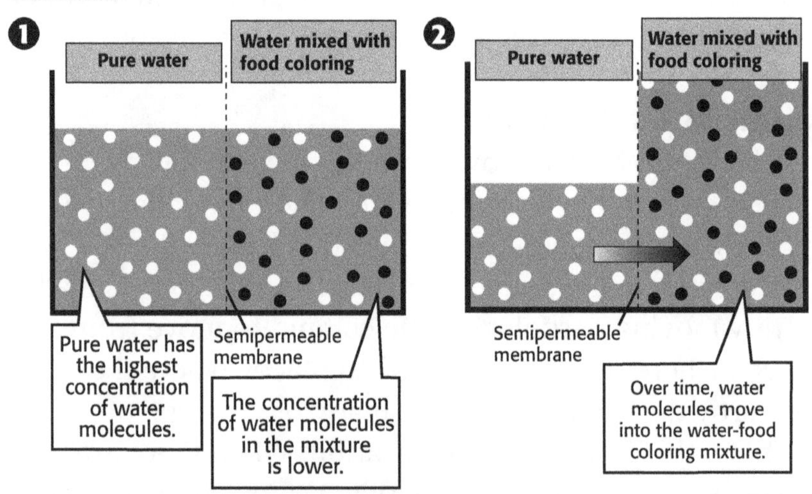

TAKE A LOOK

4. Explain Why does the volume of liquid in the right-hand side of the container increase with time?

____

____

A cell membrane is a type of semipermeable membrane. This means that water can pass through the cell membrane, but most other substances cannot. The cells of organisms are surrounded by and filled with fluids. These fluids are made mostly of water. Water moves in and out of a cell by osmosis. ☑

READING CHECK

5. Identify How does water move into and out of cells?

____

How Do Small Particles Enter and Leave a Cell?

Small particles, such as sugars, can cross the cell membrane through passageways called *channels*. These channels in the cell membrane are made of proteins. Particles can travel through these channels by passive transport or by active transport.

During **passive transport**, particles move through the cell membrane without using energy from the cell. During passive transport, particles move from areas of high concentration to areas of lower concentration. Diffusion and osmosis are examples of passive transport.

During **active transport**, the cell has to use energy to move particles through channels. During active transport, particles usually move from areas of low concentration to areas of high concentration. ☑

READING CHECK

6. Identify What is needed to move particles from areas of low concentration to areas of high concentration?

____

How Do Large Particles Enter and Leave a Cell?

Large particles cannot move across a cell membrane in the same ways as small particles. Larger particles must move in and out of the cell by endocytosis and exocytosis. Both processes require energy from the cell.

Endocytosis happens when a cell surrounds a large particle and encloses it in a vesicle. A *vesicle* is a sac formed from a piece of cell membrane.

Endocytosis

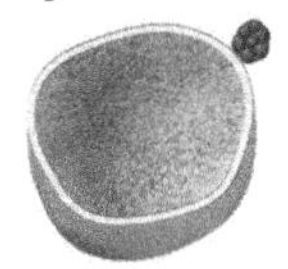

❶ The cell comes into contact with a particle.

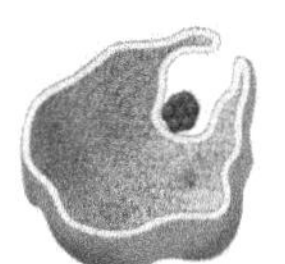

❷ The cell membrane begins to wrap around the particle.

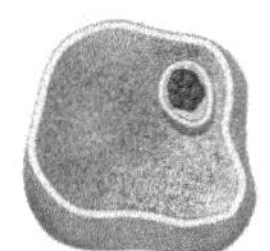

❸ Once the particle is completely surrounded, a vesicle pinches off.

TAKE A LOOK

7. Identify Label the vesicle in the figure.

Exocytosis happens when a cell uses a vesicle to move a particle from within the cell to outside the cell. Exocytosis is how cells get rid of large waste particles.

Exocytosis

❶ Large particles that must leave the cell are packaged in vesicles.

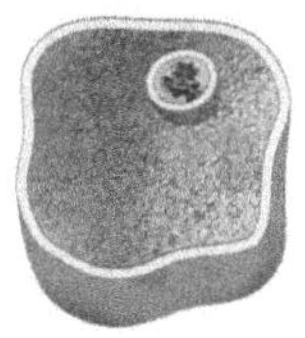

❷ The vesicle travels to the cell membrane and fuses with it.

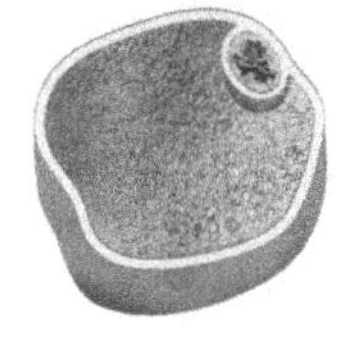

❸ The cell releases the particle to the outside of the cell.

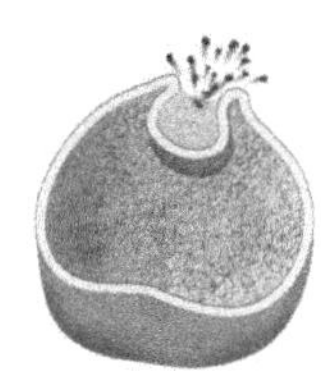

Name ______________________ Class ______________ Date ______________

Section 1 Review

NSES LS 1a, 1c

SECTION VOCABULARY

active transport the movement of substances across the cell membrane that requires the cell to use energy

diffusion the movement of particles from regions of higher density to regions of lower density

endocytosis the process by which a cell membrane surrounds a particle and encloses the particle in a vesicle to bring the particle into the cell

exocytosis the process in which a cell releases a particle by enclosing the particle in a vesicle that then moves to the cell surface and fuses with the cell membrane

osmosis the diffusion of water through a semipermeable membrane

passive transport the movement of substances across a cell membrane without the use of energy by the cell

1. **Compare** How is endocytosis different from exocytosis? How are they similar?

____

____

____

2. **Explain** How is osmosis related to diffusion?

____

____

3. **Compare** What are the differences between active and passive transport?

____

____

____

____

4. **Identify** What structures allow small particles to cross cell membranes?

____

5. **Apply Concepts** Draw an arrow in the figure below to show the direction that water molecules will move in.

semipermeable membrane

water mixed with sugar

pure water

Name ______________________ Class ______________ Date ____________

CHAPTER 4 The Cell in Action

SECTION 2

Cell Energy

BEFORE YOU READ

After you read this section, you should be able to answer these questions:

- How do plant cells make food?
- How do plant and animal cells get energy from food?

National Science Education Standards

LS 1c, 4c

How Does a Plant Make Food?

The sun is the major source of energy for life on Earth. Plants use carbon dioxide, water, and the sun's energy to make food in a process called **photosynthesis**. The food that plants make gives them energy. When animals eat plants, the plants become sources of energy for the animals.

Plant cells have molecules called *pigments* that absorb light energy. Chlorophyll is the main pigment used in photosynthesis. Chlorophyll is found in chloroplasts. The food plants make is a simple sugar called *glucose*. Photosynthesis also produces oxygen. ☑

Compare As you read this section, make a Venn Diagram to compare cellular respiration and fermentation.

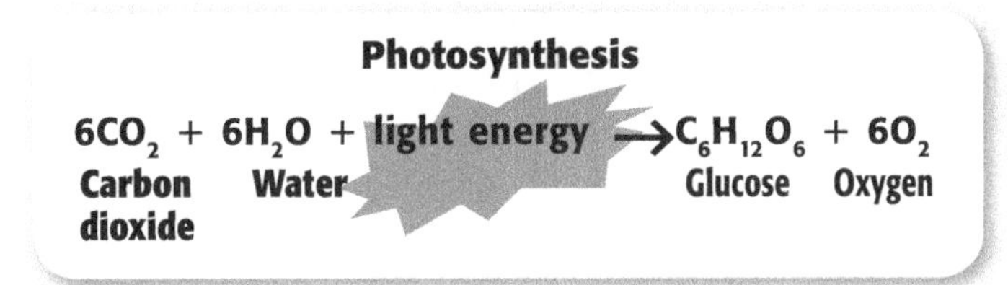

$$6CO_2 + 6H_2O + \text{light energy} \rightarrow C_6H_{12}O_6 + 6O_2$$

Carbon dioxide — Water — Glucose — Oxygen

1. Identify In which cell structures does photosynthesis take place?

____

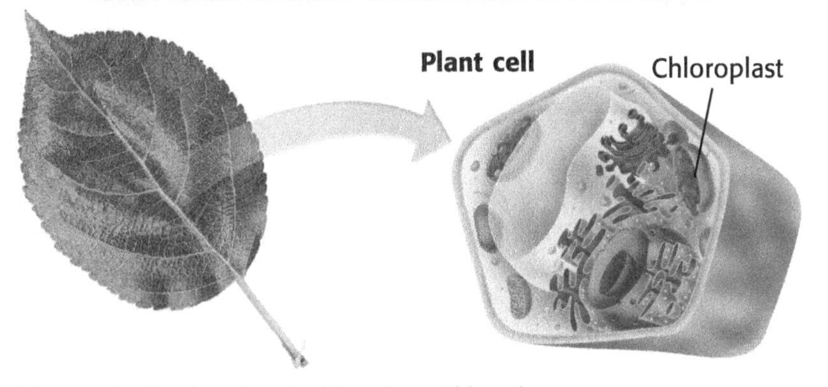

Photosynthesis takes place in chloroplasts. Chloroplasts are found inside plant cells.

TAKE A LOOK

2. Identify What two materials are produced during photosynthesis?

____

____

How Do Organisms Get Energy from Food?

Both plant and animal cells must break down food molecules to get energy from them. There are two ways cells get energy: cellular respiration and fermentation.

During **cellular respiration**, cells use oxygen to break down food. During **fermentation**, food is broken down without oxygen. Cellular respiration releases more energy from food than fermentation. Most eukaryotes, such as plants and animals, use cellular respiration.

STANDARDS CHECK

LS 1c Cells carry on the many functions needed to sustain life. They grow and divide, thereby producing more cells. This requires that they take in nutrients, which they use to provide energy for the work that cells do and to make the materials a cell or an organism needs.

Word Help: function
to work

Word Help: energy
the capacity to do work

3. Identify Name two ways cells can get energy from food.

____

____

What Happens During Cellular Respiration?

When you hear the word *respiration*, you might think of breathing. However, cellular respiration is different from breathing. Cellular respiration is a chemical process that happens in cells. In eukaryotic cells, such as plant and animal cells, cellular respiration takes place in structures called *mitochondria.*

Recall that to get energy, cells must break down glucose. During cellular respiration, glucose is broken down into carbon dioxide (CO_2) and water (H_2O), and energy is released. This energy is stored in a molecule called *ATP* (adenosine triphosphate). The figure below shows how energy is released when a cow eats grass.

TAKE A LOOK

4. Identify What two materials are needed for cellular respiration?

____

____

5. List What three things are produced during cellular respiration?

____

____

____

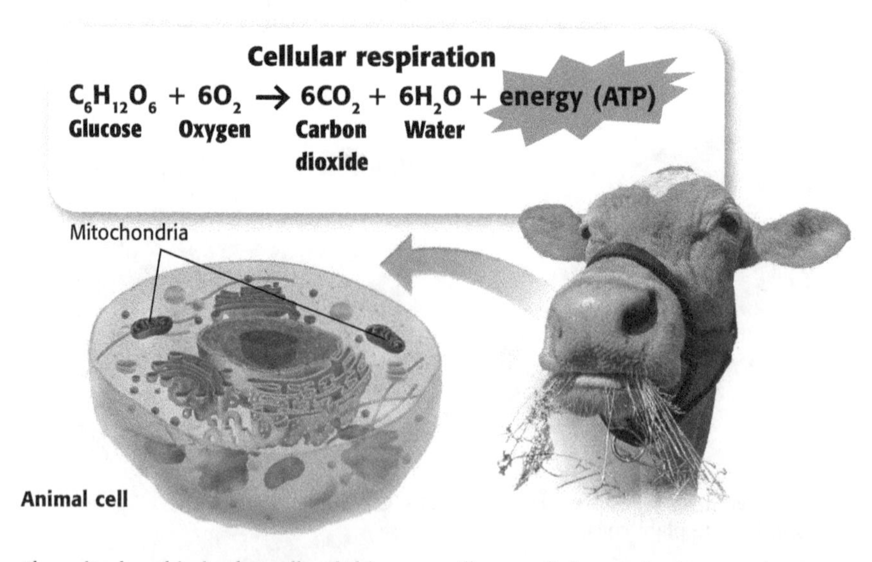

The mitochondria in the cells of this cow will use cellular respiration to release the energy stored in the grass.

The Connection Between Photosynthesis and Cellular Respiration

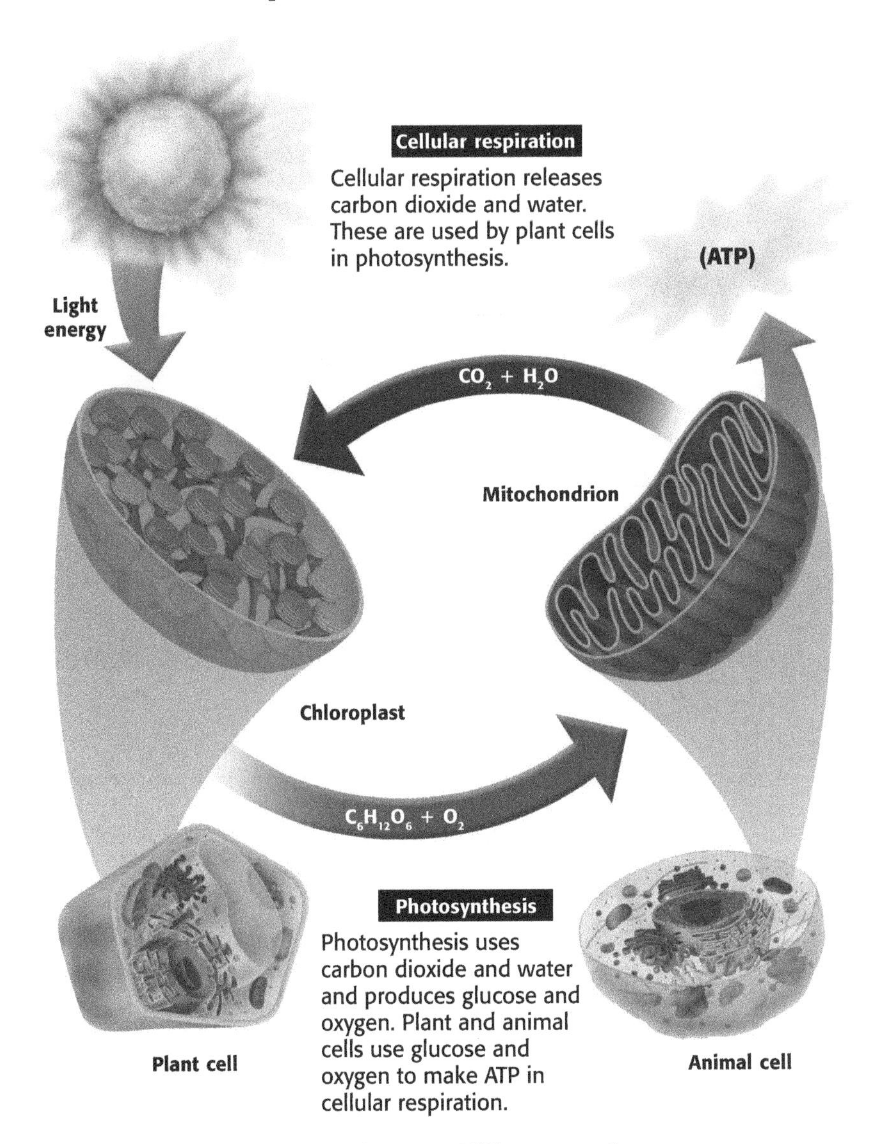

Critical Thinking

6. Apply Concepts What would happen if oxygen were not produced during photosynthesis?

____
____
____
____
____
____
____

TAKE A LOOK

7. Complete Plant and animal cells use glucose and oxygen to make

______________________.

How Is Fermentation Different from Cellular Respiration?

During fermentation, cells break down glucose without oxygen. Some bacteria and fungi rely only on fermentation to release energy from food. However, cells in other organisms may use fermentation when there is not enough oxygen for cellular respiration.

When you exercise, your muscles use up oxygen very quickly. When cells don't have enough oxygen, they must use fermentation to get energy. Fermentation creates a byproduct called *lactic acid*. This is what makes your muscles ache if you exercise too hard or too long.

Research Use the school library or the Internet to research an organism that uses fermentation. What kind of organism is it? Where is it found? Is this organism useful to humans? Present your findings to the class.

Name ______________________ Class ______________ Date ______________

Section 2 Review

NSES LS 1c, 4c

SECTION VOCABULARY

cellular respiration the process by which cells use oxygen to produce energy from food

fermentation the breakdown of food without the use of oxygen

photosynthesis the process by which plants, algae, and some bacteria use sunlight, carbon dioxide, and water to make food.

1. **Identify** What kind of cells have chloroplasts?

   ____

2. **Explain** How do plant cells make food?

   ____

   ____

3. **Explain** Why do plant cells need both chloroplasts and mitochondria?

   ____

   ____

4. **Apply Concepts** How do the processes of photosynthesis and cellular respiration work together?

   ____

   ____

   ____

   ____

5. **Compare** What is one difference between cellular respiration and fermentation?

   ____

   ____

   ____

   ____

6. **Explain** Do your body cells always use cellular respiration to break down glucose? Explain your answer.

   ____

   ____

   ____

SECTION 3

The Cell Cycle

BEFORE YOU READ

After you read this section, you should be able to answer these questions:

- How are new cells made?
- What is mitosis?
- How is cell division different in animals and plants?

National Science Education Standards
LS 1c, 2d

How Are New Cells Made?

As you grow, you pass through different stages in your life. Cells also pass through different stages in their life cycles. These stages are called the **cell cycle**. The cell cycle starts when a cell is made, and ends when the cell divides to make new cells.

Before a cell divides, it makes a copy of its DNA (deoxyribonucleic acid). *DNA* is a molecule that contains all the instructions for making new cells. The DNA is stored in structures called **chromosomes**. Cells make copies of their chromosomes so that new cells have the same chromosomes as the parent cells. Although all cells pass through a cell cycle, the process differs in prokaryotic and eukaryotic cells. ☑

STUDY TIP
Summarize As you read this section, make a diagram showing the stages of the eukaryotic cell cycle.

READING CHECK
1. Explain What must happen before a cell can divide?

____

____

____

How Do Prokaryotic Cells Divide?

Prokaryotes have only one cell. Prokaryotic cells have no nucleus. They also have no organelles that are surrounded by membranes. The DNA for prokaryotic cells, such as bacteria, is found on one circular chromosome. The cell divides by a process called *binary fission.* During binary fission, the cell splits into two parts. Each part has one copy of the cell's DNA.

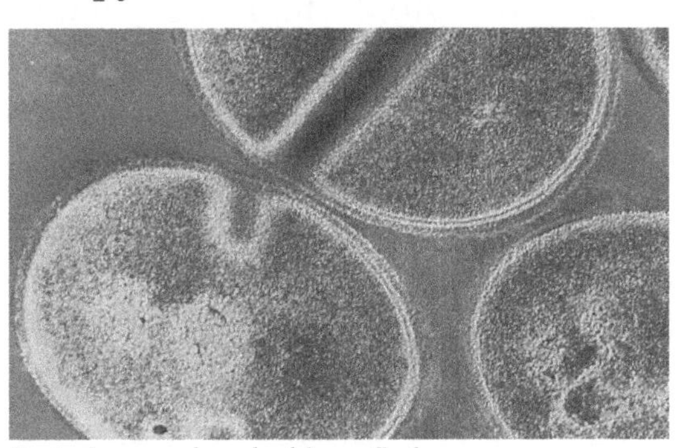

Bacteria reproduce by binary fission.

TAKE A LOOK
2. Complete Prokaryotic cells divide by ______________

______________________.

How Do Eukaryotic Cells Divide?

Different kinds of eukaryotes have different numbers of chromosomes. However, complex eukaryotes do not always have more chromosomes than simpler eukaryotes. For example, potatoes have 48 chromosomes, but humans have 46. Many eukaryotes, including humans, have pairs of similar chromosomes. These pairs are called **homologous chromosomes**. One chromosome in a pair comes from each parent.

Critical Thinking

3. Compare What is the difference between a chromosome and a chromatid?

____

____

____

____

____

Cell division in eukaryotic cells is more complex than in prokaryotic cells. The cell cycle of a eukaryotic cell has three stages: interphase, mitosis, and cytokinesis.

The first stage of the cell cycle is called *interphase*. During interphase, the cell grows and makes copies of its chromosomes and organelles. The two copies of a chromosome are called *chromatids*. The two chromatids are held together at the *centromere*.

This duplicated chromosome consists of two chromatids. The chromatids are joined at the centromere.

Chromatids

Centromere

The second stage of the cell cycle is called **mitosis**. During this stage, the chromatids separate. This allows each new cell to get a copy of each chromosome. Mitosis happens in four phases, as shown in the figure on the next page: prophase, metaphase, anaphase, and telophase.

The third stage of the cell cycle is called **cytokinesis**. During this stage, the cytoplasm of the cell divides to form two cells. These two cells are called *daughter cells*. The new daughter cells are exactly the same as each other. They are also exactly the same as the original cell. ☑

4. Identify What are the three stages of the eukaryotic cell cycle?

____

____

____

THE CELL CYCLE

The figure on the following page shows the cell cycle. In this example, the stages of the cell cycle are shown in a eukaryotic cell that has only four chromosomes.

Interphase Before mitosis begins, chromosomes are copied. Each chromosome is then made of two chromatids.

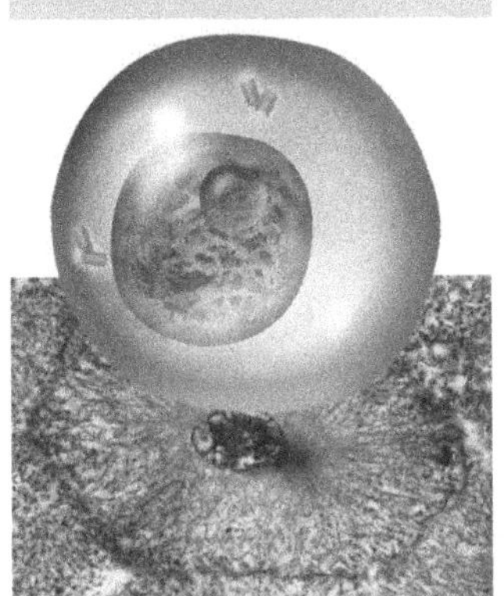

Mitosis Phase 1 (Prophase) Mitosis begins. Chromatids condense from long strands to thick rods.

Mitosis Phase 2 (Metaphase) The nuclear membrane dissolves. Chromosome pairs line up around the equator of the cell.

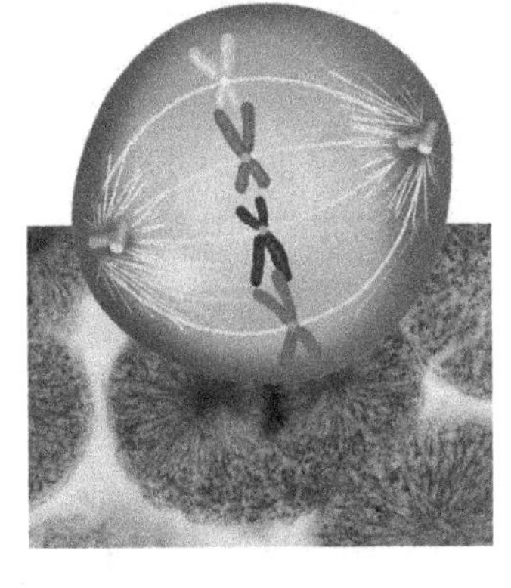

Mitosis Phase 3 (Anaphase) Chromatids separate and move to opposite sides of the cell.

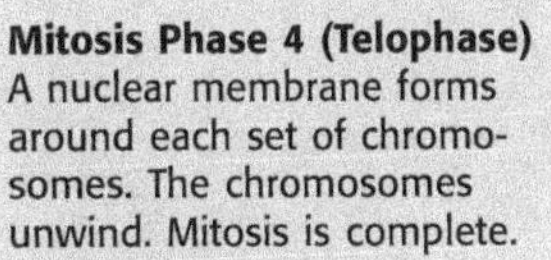

Mitosis Phase 4 (Telophase) A nuclear membrane forms around each set of chromosomes. The chromosomes unwind. Mitosis is complete.

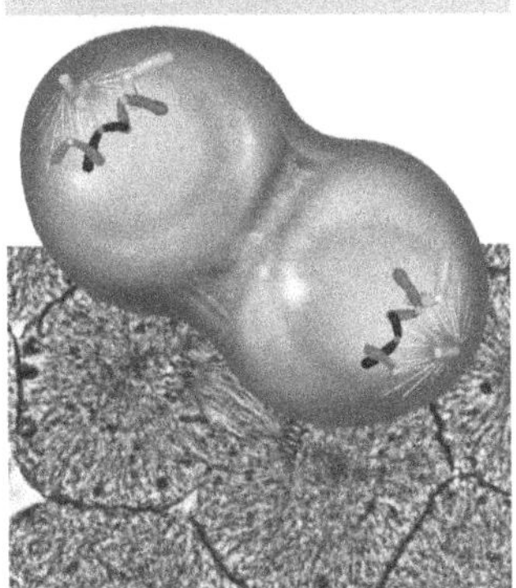

Cytokinesis In cells with no cell wall, the cell pinches in two.

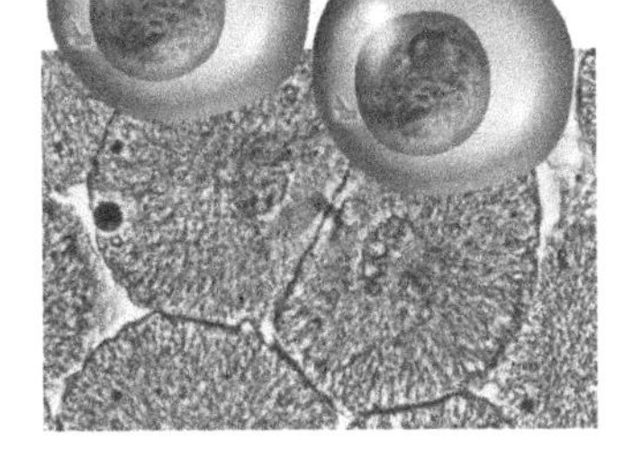

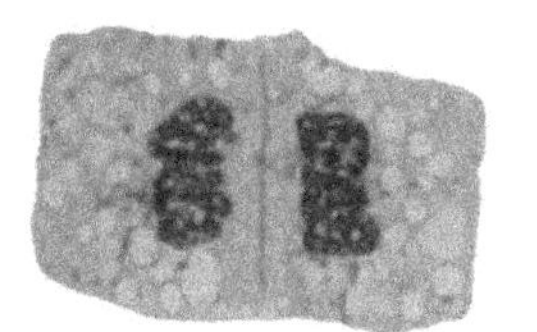

In cells with a cell wall, a cell plate forms and separates the new cells.

Math Focus

5. Calculate Cell A takes 6 h to complete division. Cell B takes 8 h to complete division. After 24 h, how many more copies of cell A than cell B will there be?

____

TAKE A LOOK

6. List What are the four phases of mitosis?

____

____

____

____

7. Identify What structure do plant cells have during cytokinesis that animal cells do not have?

____

Name ______________________ Class ______________ Date ____________

Section 3 Review

NSES LS 1c, 2d

SECTION VOCABULARY

cell cycle the life cycle of a cell **chromosome** in a eukaryotic cell, one of the structures in the nucleus that are made up of DNA and protein; in a prokaryotic cell, the main ring of DNA **cytokinesis** the division of cytoplasm of a cell	**homologous chromosomes** chromosomes that have the same sequence of genes and the same structure **mitosis** in eukaryotic cells, a process of cell division that forms two new nuclei, each of which has the same number of chromosomes

1. Compare How does the DNA of prokaryotic and eukaryotic cells differ?

____

____

2. Summarize Complete the Process Chart to explain the three stages of the cell cycle. Include the four phases of mitosis.

↓
Mitosis begins with prophase. The chromosomes condense.
↓
↓
During telophase the nuclear membrane forms. The chromosomes lengthen and mitosis ends.
↓

3. Explain Why does a cell make a copy of its DNA before it divides?

____

____

____

4. Infer Why is cell division in eukaryotic cells more complex than in prokaryotic cells?

____

____

____

SECTION 1

Mendel and His Peas

BEFORE YOU READ

After you read this section, you should be able to answer these questions:

- What is heredity?
- How did Gregor Mendel study heredity?

National Science Education Standards
LS 2b, 2e

What Is Heredity?

Why don't you look like a rhinoceros? The answer to that question seems simple. Neither of your parents is a rhinoceros. Only a human can pass on its traits to make another human. Your parents passed some of their traits on to you. The passing of traits from parents to offspring is called **heredity**.

About 150 years ago, a monk named Gregor Mendel performed experiments on heredity. His discoveries helped establish the field of genetics. *Genetics* is the study of how traits are passed on, or inherited. ☑

STUDY TIP
Define As you read this section, make a list of all of the underlined and italicized words. Write a definition for each of the words.

READING CHECK
1. Define What is genetics?

____
____
____

Who Was Gregor Mendel?

Gregor Mendel was born in Austria in 1822. He grew up on a farm where he learned a lot about flowers and fruit trees. When he was 21 years old, Mendel entered a monastery. A monastery is a place where monks study and practice religion. The monks at Mendel's monastery also taught science and performed scientific experiments.

Mendel studied pea plants in the monastery garden to learn how traits are passed from parents to offspring. He used garden peas because they grow quickly. They also have many traits, such as height and seed color, that are easy to see. His results changed the way people think about how traits are passed on. ☑

READING CHECK
2. Explain Why did Mendel choose to study pea plants?

____
____
____
____

Gregor Mendel discovered the principles of heredity while studying pea plants.

REPRODUCTION IN PEAS

Like many flowering plants, pea plants have both male and female reproductive parts. Many flowering plants can reproduce by cross-pollination. In most plants, sperm are carried in structures called pollen. In *cross-pollination*, sperm in the pollen of one plant fertilize eggs in the flower of another plant. Pollen can be carried by organisms, such as insects. It may also be carried by the wind from one flower to another.

Some flowering plants must use cross-pollination. They need another plant to reproduce. However, some plants, including pea plants, can also reproduce by self-pollination. In *self-pollination*, sperm from one plant fertilize the eggs of the same plant.

Mendel used self-pollination in pea plants to grow true-breeding plants for his experiments. When a *true-breeding* plant self-pollinates, its offspring all have the same traits as the parent. For example, a true-breeding plant with purple flowers always has offspring with purple flowers.

Critical Thinking

3. Compare What is the difference between cross-pollination and self-pollination?

____

____

____

____

____

____

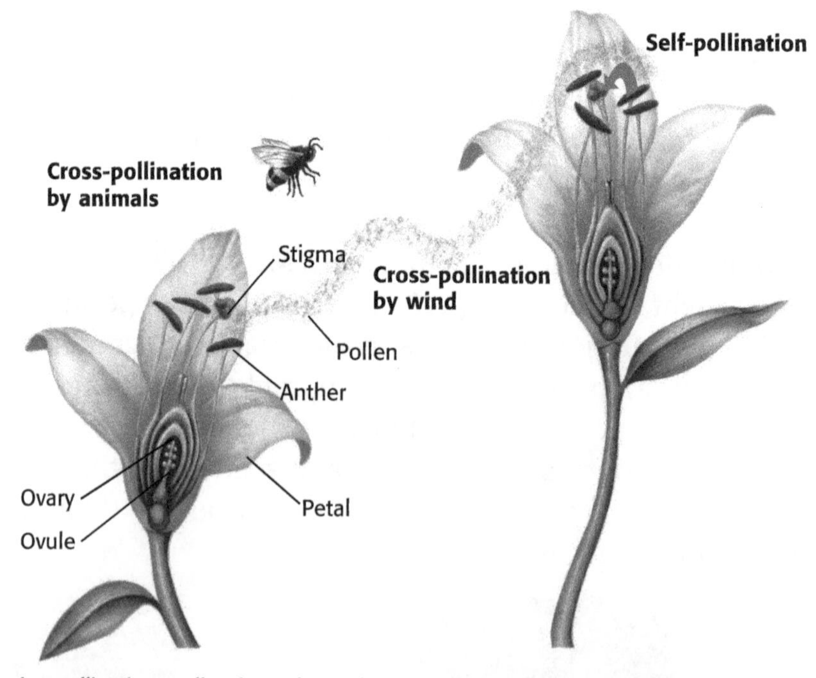

During pollination, pollen from the anther (male) is carried to the stigma (female). Fertilization happens when a sperm from the pollen moves through the stigma and enters an egg in an ovule.

TAKE A LOOK

4. Identify What are two ways pollen can travel from one plant to another during cross-pollination?

____

____

____

____

CHARACTERISTICS

A *characteristic* is a feature that has different forms. For example, hair color is a characteristic of humans. The different forms or colors, such as brown or red hair, are *traits*. ☑

Mendel studied one characteristic of peas at a time. He used plants that had different traits for each characteristic he studied. One characteristic he studied was flower color. He chose plants that had purple flowers and plants that had white flowers. He also studied other characteristics, such as seed shape, pod color, and plant height.

READING CHECK

5. Explain How are characteristics and traits related?

____

____

____

____

CROSSING PEA PLANTS

Mendel was careful to use true-breeding plants in his experiments. By choosing these plants, he would know what to expect if his plants self-pollinated. He decided to find out what would happen if he bred, or crossed, two plants that had different traits.

Say It

Describe How would you describe yourself? Make a list of your physical traits, such as height, hair color, and eye color. List other traits you have that you weren't born with. Share this list with your classmates. Which of these traits did you inherit?

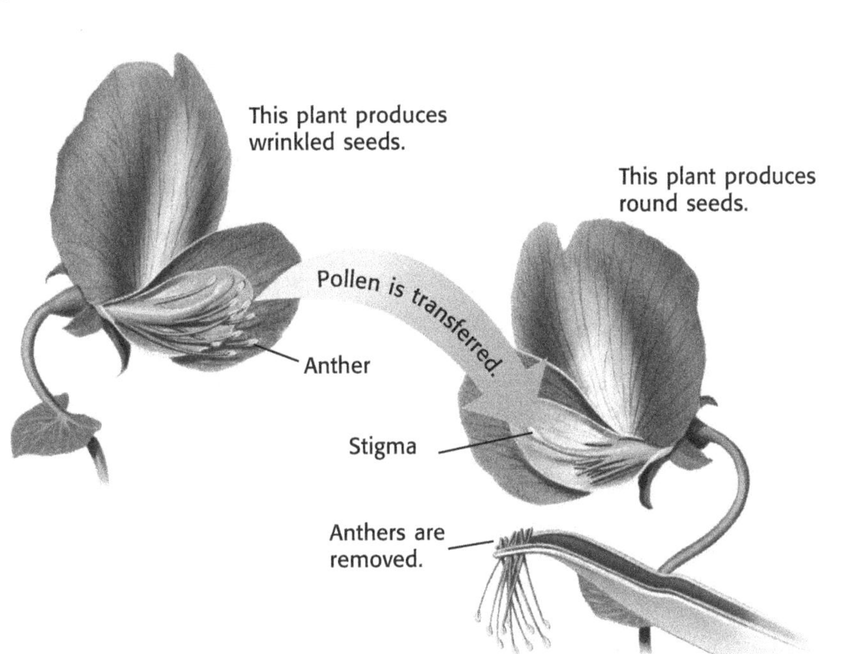

Mendel removed the anthers from a plant that made round seeds. Then, he used pollen from a plant that made wrinkled seeds to fertilize the plant that made round seeds.

TAKE A LOOK

6. Describe How did Mendel make sure that the plant with round seeds did not self-pollinate?

____

____

____

____

____

What Happened in Mendel's First Experiments?

Mendel studied seven different characteristics in his first experiments with peas. He crossed plants that were true-breeding for different traits. For example, he crossed plants that had purple flowers with plants that had white flowers. The offspring from such a cross are called *first-generation plants*. All of the first-generation plants in this cross had purple flowers. What happened to the trait for white flowers?

Mendel got similar results for each cross. One trait was always present in the first generation and the other trait seemed to disappear. Mendel called the trait that appeared the **dominant trait**. He called the other trait the **recessive trait**. To *recede* means "to go away or back off." To find out what happened to the recessive trait, Mendel did another set of experiments. ☑

7. Identify What kind of trait appeared in the first generation?

____

What Happened in Mendel's Second Experiment?

Mendel let the first-generation plants self-pollinate. Some of the offspring were white-flowered, even though the parent was purple-flowered. The recessive trait for white flowers had reappeared in the second generation.

Mendel did the same experiment on plants with seven different characteristics. Each time, some of the second-generation plants had the recessive trait.

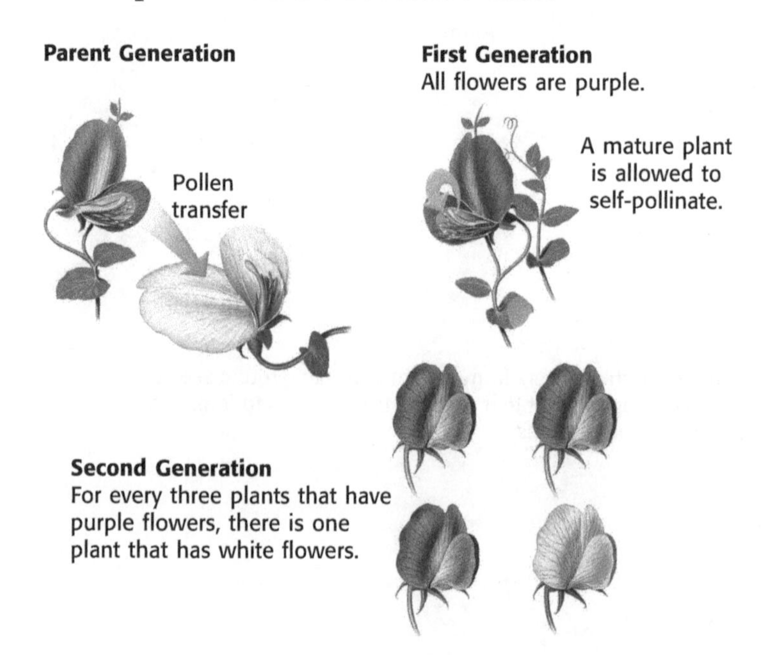

TAKE A LOOK

8. Identify What type of traits appeared in the second generation?

____

____

RATIOS IN MENDEL'S EXPERIMENTS

Mendel counted the number of plants that had each trait in the second generation. He hoped that this might help him explain his results.

As you can see from the table below, the recessive trait did not show up as often as the dominant trait. Mendel decided to figure out the ratio of dominant traits to recessive traits. A *ratio* is a relationship between two numbers. It is often written as a fraction. For example, the second generation produced 705 plants with purple flowers and 224 plants with white flowers. Mendel used this formula to calculate the ratios:

$$\frac{705}{224} = \frac{3.15}{1} \text{ or } 3.15:1$$

Characteristic	Dominant trait	Recessive trait	Ratio
Flower color	705 purple	224 white	3.15:1
Seed color	6,002 yellow	2,001 green	
Seed shape	5,474 round	1,850 wrinkled	
Pod color	428 green	152 yellow	
Pod shape	882 smooth	299 bumpy	
Flower position	651 along stem	207 at tip	
Plant height	787 tall	277 short	

Math Focus

9. Find Ratios Calculate the ratios of the other pea plant characteristics in the table.

Math Focus

10. Round Round off all numbers in the ratios to whole numbers. What ratio do you get?

____

What Did Mendel Conclude?

Mendel knew that his results could be explained only if each plant had two sets of instructions for each characteristic. He concluded that each parent gives one set of instructions to the offspring. The dominant set of instructions determines the offspring's traits.

Name ______________________ Class ______________ Date ______________

Section 1 Review

NSES LS 2b, 2e

SECTION VOCABULARY

dominant trait the trait observed in the first generation when parents that have different traits are bred

heredity the passing of genetic traits from parent to offspring

recessive trait a trait that is apparent only when two recessive alleles for the same characteristic are inherited

1. **Define** What is a true-breeding plant?

____

____

____

2. **Apply Concepts** Cats may have straight or curly ears. A curly-eared cat mated with a straight-eared cat. All the kittens had curly ears. Are curly ears a dominant or recessive trait? Explain your answer.

____

____

3. **Summarize** Complete the cause and effect map to summarize Mendel's experiments on flower color in pea plants.

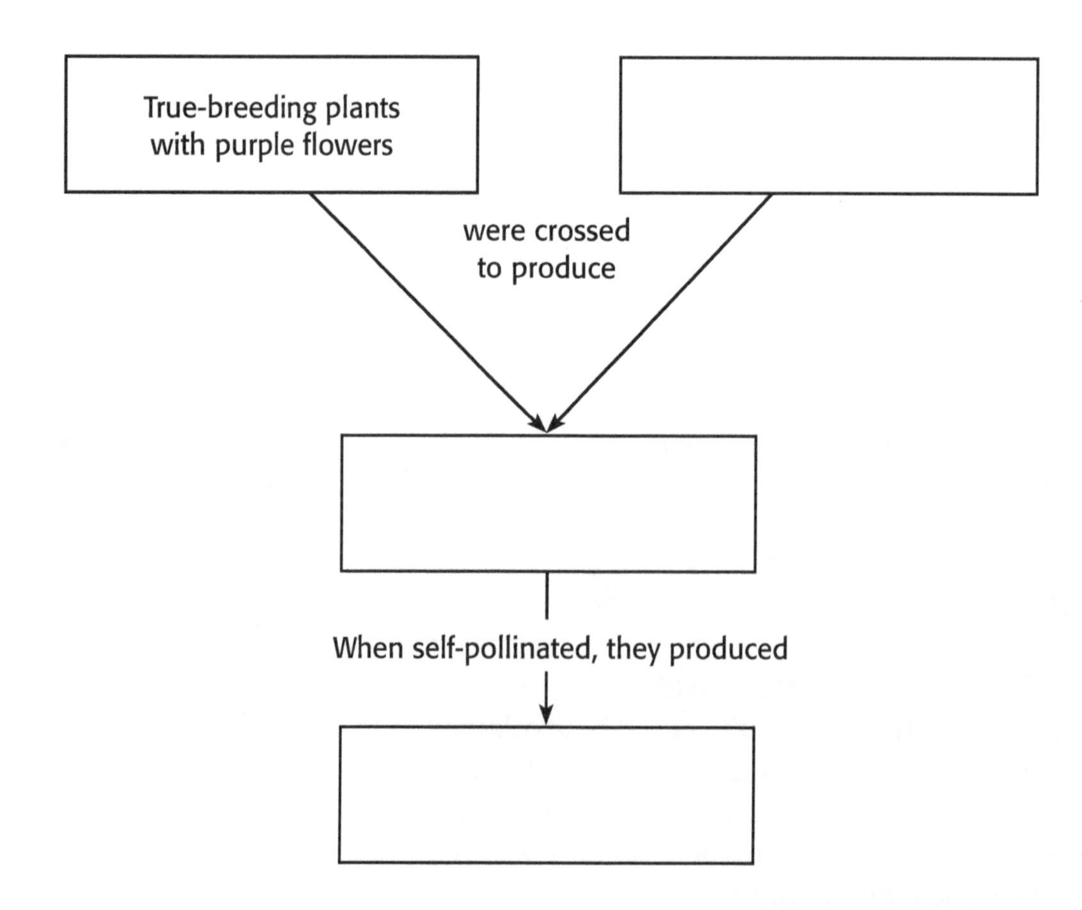

Name ______________________ Class ______________ Date ______________

CHAPTER 5 Heredity

SECTION 2

Traits and Inheritance

BEFORE YOU READ

After you read this section, you should be able to answer these questions:

- What did Mendel's experiments tell him about heredity?
- Are there exceptions to Medel's laws of heredity?

National Science Education Standards
LS 2a, 2b, 2c, 2d, 2e

What Did Mendel Learn About Heredity?

Mendel knew from his pea plant experiments that there must be two sets of instructions for each characteristic. All of the first-generation plants showed the dominant trait. However, they could give the recessive trait to their offspring. Instructions for an inherited trait are called **genes**. Offspring have two sets of genes—one from each parent.

STUDY TIP

Organize As you read, make a Concept Map using the vocabulary words highlighted in the section.

The two sets of genes that parents give to offspring are never exactly the same. The same gene might have more than one version. The different versions of a gene are called **alleles**. ☑

READING CHECK

1. Define What is an allele?

____

____

____

Alleles may be dominant or recessive. A trait for an organism is usually identified with two letters, one for each allele. Dominant alleles are given capital letters (*A*). Recessive alleles are given lowercase letters (*a*). If a dominant allele is present, it will hide a recessive allele. An organism can have a recessive trait only if it gets a recessive allele for that trait from both parents.

PHENOTYPE

An organism's genes affect its traits. The appearance of an organism, or how it looks, is called its **phenotype**. The phenotypes for flower color in Mendel's pea plants were purple and white. The figure below shows one example of a human phenotype. ☑

READING CHECK

2. Define What is a phenotype?

____

____

Albinism is an inherited disorder that affects a person's phenotype in many ways.

GENOTYPE

A **genotype** is the combination of alleles that an organism gets from its parents. A plant with two dominant or two recessive alleles (*PP*, *pp*) is *homozygous. Homo* means "the same." A plant with one dominant allele and one recessive allele (*Pp*) is *heterozygous. Hetero* means "different." The allele for purple flowers (*P*) in pea plants is dominant. The plant will have purple flowers even if it has only one *P* allele. ☑

READING CHECK

3. Identify What kind of alleles does a heterozygous individual have?

____

____

PUNNETT SQUARES

A Punnett square is used to predict the possible genotypes of offspring from certain parents. It can be used to show the alleles for any trait. In a Punnett square, the alleles for one parent are written along the top of the square. The alleles for the other parent are written along the side of the square. The possible genotypes of offspring are found by combining the letters at the top and side of each square.

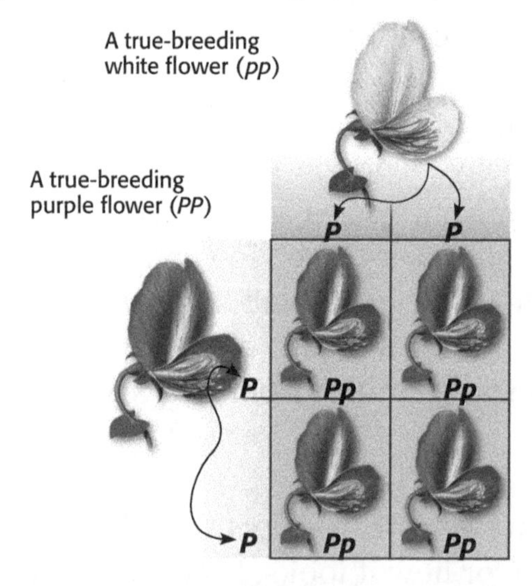

All of the offspring for this cross have the same genotype—*Pp*.

TAKE A LOOK

4. Identify Is the plant with white flowers homozygous or heterozygous? How can you tell?

____

____

____

The figure shows a Punnett square for a cross of two true-breeding plants. One has purple flowers and the other has white flowers. The alleles for a true-breeding purple-flowered plant are written as *PP*. The alleles for a true-breeding white flowered plant are written as *pp*. Offspring get one of their two alleles from each parent. All of the offspring from this cross will have the same genotype: *Pp*. Because they have a dominant allele, all of the offspring will have purple flowers.

MORE EVIDENCE FOR INHERITANCE

In his second experiments, Mendel let the first-generation plants self-pollinate. He did this by covering the flowers of the plant. This way, no pollen from another plant could fertilize its eggs. The Punnett square below shows a cross of a plant that has the genotype *Pp*.

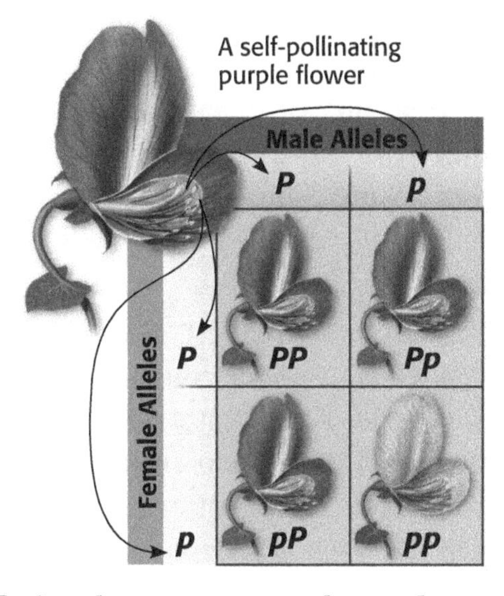

This Punnett square shows the possible results from the cross *Pp* × *Pp*.

TAKE A LOOK

5. List What are the possible genotypes of the offspring in this cross?

____

____

Notice that one square shows the genotype *Pp* and another shows *pP*. These are exactly the same genotype. They both have one *p* allele and one *P* allele. The combinations *PP*, *Pp*, and *pP* have the same phenotype—purple flowers. This is because they all have at least one dominant allele, *P*. ☑

Only one combination, *pp*, produces plants that have white flowers. The ratio of dominant phenotypes to recessive phenotypes is 3:1. This means that three out of four offspring from that cross will have purple flowers. This is the same ratio Mendel found.

READING CHECK

6. Explain Why do the genotypes *PP*, *Pp*, and *pP* all have the same phenotype?

____

____

____

What Is the Chance That Offspring Will Receive a Certain Allele?

Each parent has two alleles for each gene. When an individual reproduces, it passes one of its two alleles to its offspring. When a parent has two different alleles for a gene, such as *Pp*, offspring may receive either of the alleles. Both alleles have an equal chance to be passed from the parent to the offspring.

Think of a coin toss. When you toss the coin, there is a 50% chance you will get heads, and a 50% chance you will get tails. The chance of the offspring receiving one allele or another from a parent is as random as a coin toss.

PROBABILITY

The mathematical chance that something will happen is known as **probability**. Probability is usually written as a fraction or percentage. If you toss a coin, the probability of tossing tails is 1/2, or 50%. In other words, you will get tails half of the time.

What is the probability that you will toss two heads in a row? To find out, multiply the probability of tossing the first head (1/2) by the probability of tossing the second head (1/2). The probability of tossing two heads in a row is 1/4.

GENOTYPE PROBABILITY

Finding the probability of certain genotypes for offspring is like predicting the results of a coin toss. To have white flowers, a pea plant must receive a *p* allele from each parent. Each offspring of a *Pp* × *Pp* cross has a 50% chance of receiving either allele from either parent. So, the probability of inheriting two *p* alleles is 1/2 × 1/2. This equals 1/4, or 25%.

	P	p
P		
p		

Math Focus

7. Complete Complete the Punnett square to show the cross between two heterozygous parents. What percentage of the offspring are homozygous?

____

Are There Exceptions to Mendel's Principles?

Mendel's experiments helped show the basic principles of how genes are passed from one generation to the next. Mendel studied sets of traits such as flower color and seed shape. The traits he studied in pea plants are easy to predict because there are only two choices for each trait. ☑

Traits in other organisms are often harder to predict. Some traits are affected by more than one gene. A single gene may affect more than one trait. As scientists learned more about heredity, they found exceptions to Mendel's principles.

READING CHECK

8. Explain Why were color and seed shape in pea plants good traits for Mendel to study?

____

____

____

INCOMPLETE DOMINANCE

Sometimes, one trait isn't completely dominant over another. These traits do not blend together, but each allele has an influence on the traits of offspring. This is called *incomplete dominance.* For example, the offspring of a true-breeding red snapdragon and a true-breeding white snapdragon are all pink. This is because both alleles for the gene influence color.

Critical Thinking

9. Infer If snapdragons showed complete dominance like pea plants, what would the offspring look like?

____

____

____

____

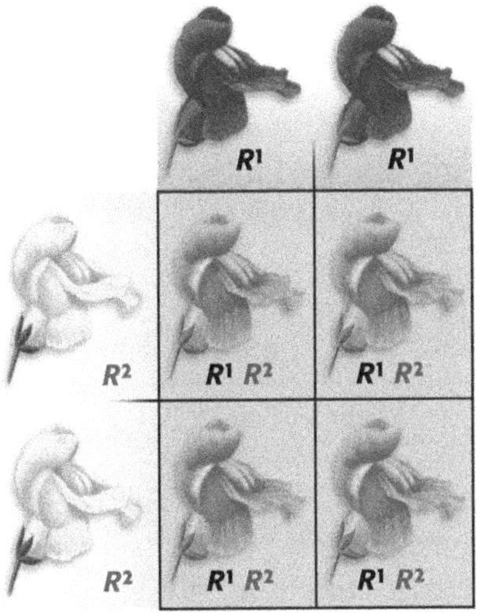

	R^1	R^1
R^2	R^1 R^2	R^1 R^2
R^2	R^1 R^2	R^1 R^2

The offspring of two true-breeding show incomplete dominance.

ONE GENE, MANY TRAITS

In Mendel's studies, one gene controlled one trait. However, some genes affect more than one trait. For example, some tigers have white fur instead of orange. These white tigers also have blue eyes. This is because the gene that controls fur color also affects eye color.

Critical Thinking

10. Compare How is the allele for fur color in tigers different from the allele for flower color in pea plants?

____

____

____

____

____

____

MANY GENES, ONE TRAIT

Some traits, such as the color of your skin, hair, and eyes, are the result of several genes acting together. In humans, different combinations of many alleles can result in a variety of heights. ☑

READING CHECK

11. Identify Give an example of a single trait that is affected by more than one gene.

____

THE IMPORTANCE OF ENVIRONMENT

Genes are not the only things that can affect an organism's traits. Traits are also affected by factors in the environment. For example, human height is affected not only by genes. Height is also influenced by nutrition. An individual who has plenty of food to eat may be taller than one who does not.

Name ______________________ Class ______________ Date ______________

Section 2 Review

NSES LS 2a, 2b, 2c, 2d, 2e

SECTION VOCABULARY

allele one of the alternative forms of a gene that governs a characteristic, such as hair color **gene** one set of instructions for an inherited trait **genotype** the entire genetic makeup of an organism; also the combination of genes for one or more specific traits	**phenotype** an organism's appearance or other detectable characteristic **probability** the likelihood that a possible future event will occur in any given instance of the event

1. **Identify Relationships** How are genes and alleles related?

____

____

2. **Explain** How is it possible for two individuals to have the same phenotype but different genotype for a trait?

____

____

____

____

3. **Punnett Square** Mendel allowed a pea plant that was heterozygous for yellow seeds (*Y*) to self-pollinate. Fill in the Punnett square below for this cross. What percentage of the offspring will have green (*y*) seeds?

____

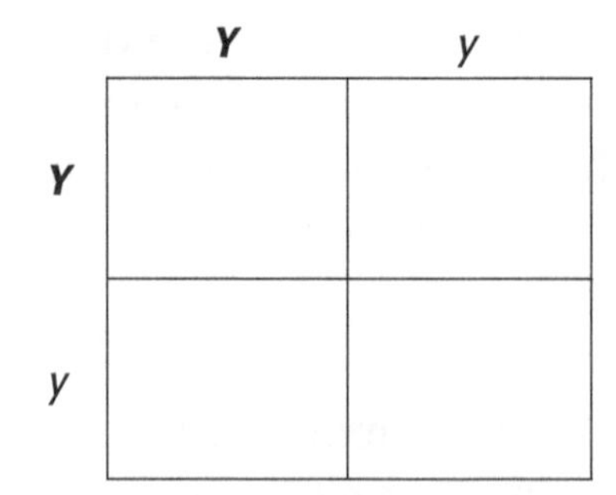

4. **Discuss** How is human height an exception to Mendel's principles of heredity?

____

____

____

Name ______________________ Class ______________ Date ______________

Meiosis

CHAPTER 5 Heredity
SECTION 3

BEFORE YOU READ

After you read this section, you should be able to answer these questions:

- What are sex cells?
- How does meiosis help explain Mendel's results?

National Science Education Standards
LS 1c, 1d, 2a, 2b, 2c, 2d

How Do Organisms Reproduce?

When organisms reproduce, their genetic information is passed on to their offspring. There are two kinds of reproduction: asexual and sexual.

STUDY TIP

Summarize Make flashcards that show the steps of meiosis. On the front of the cards, write the steps of meiosis. On the back of the cards, write what happens at each step. Practice arranging the steps in the correct order.

ASEXUAL REPRODUCTION

In asexual reproduction, only one parent is needed to produce offspring. Asexual reproduction produces offspring with exact copies of the parent's genotype.

SEXUAL REPRODUCTION

In sexual reproduction, cells from two parents join to form offspring. Sexual reproduction produces offspring that share traits with both parents. However, the offspring are not exactly like either parent.

What Are Homologous Chromosomes?

Recall that genes are the instructions for inherited traits. Genes are located on chromosomes. Each human body cell has a total of 46 chromosomes, or 23 pairs. A pair of chromosomes that carry the same sets of genes are called **homologous chromosomes**. One chromosome from a pair comes from each parent. ☑

READING CHECK

1. Define What are homologous chromosomes?

____

____

____

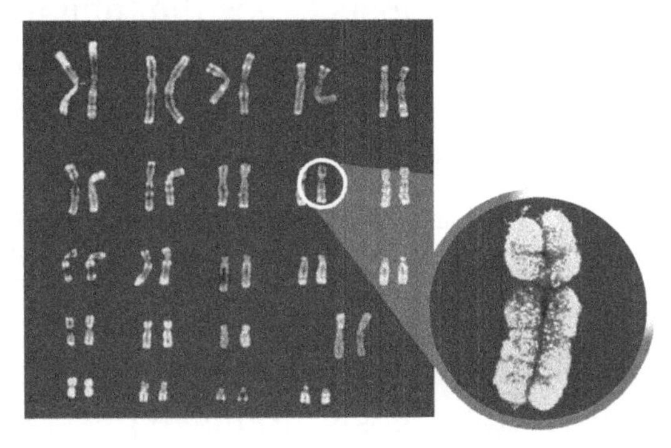

Human body cells have 23 pairs of chromosomes. One member of a pair of homologous chomosomes has been magnified.

TAKE A LOOK

2. Identify How many total chromosomes are in each human body cell?

____

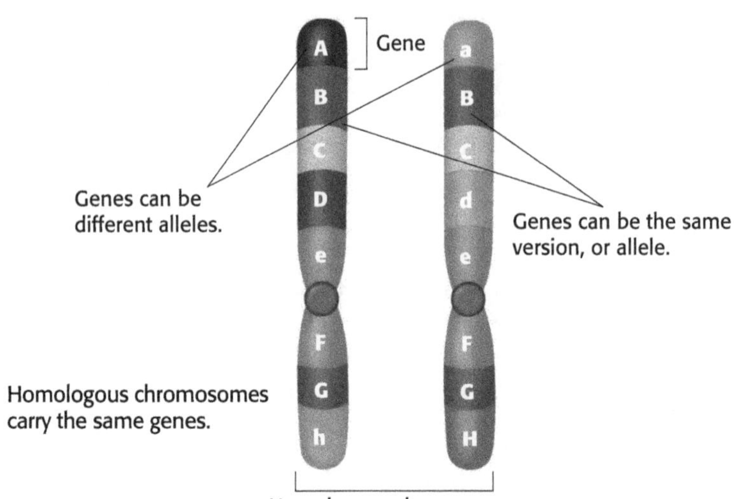

What Are Sex Cells?

In sexual reproduction, cells from two parents join to make offspring. However, only certain cells can join. Cells that can join to make offspring are called *sex cells*. An egg is a female sex cell. A sperm is a male sex cell. Unlike ordinary body cells, sex cells do not have homologous chromosomes. ☑

3. Explain How are sex cells different from ordinary body cells?

____

____

____

Imagine a pair of shoes. Each shoe is like a chromosome and the pair represents a homologous pair of chromosomes. Recall that your body cells have a total of 23 pairs of "shoes," or homologous chromosomes. Each sex cell, however, has only one of the chromosomes from each homologous pair. Sex cells have only one "shoe" from each pair. How do sex cells end up with only one chromosome from each pair?

STANDARDS CHECK

LS 2b In many species, including humans, females produce eggs and males produce sperm...An egg and sperm unite to begin development of a new individual. The individual receives genetic information from its mother (via the egg) and its father (via the sperm). Sexually produced offspring never are identical to either of their parents.

4. Define What is the function of meiosis?

____

____

____

____

How Are Sex Cells Made?

Sex cells are made during meiosis. **Meiosis** is a copying process that produces cells with half the usual number of chromosomes. Meiosis keeps the total number of chromosomes the same from one generation to the next.

In meiosis, each sex cell that is made gets only one chromosome from each homologous pair. For example, a human egg cell has 23 chromosomes and a sperm cell has 23 chromosomes. When these sex cells later join together during reproduction, they form pairs. The new cell has 46 chromosomes, or 23 pairs. The figure on the next page describes the steps of meiosis. To make the steps easy to see, only four chromosomes are shown.

Steps of Meiosis

First cell division

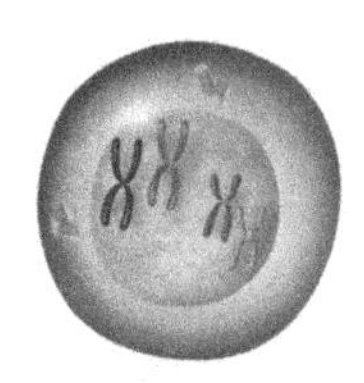

1 The chromosomes are copied before meiosis begins. The identical copies, or chromatids, are joined together.

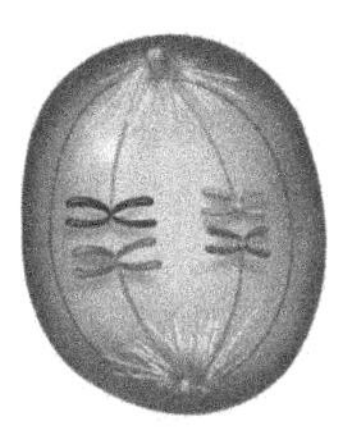

2 The nuclear membrane disappears. Pairs of homologous chromosomes line up at the equator of the cell.

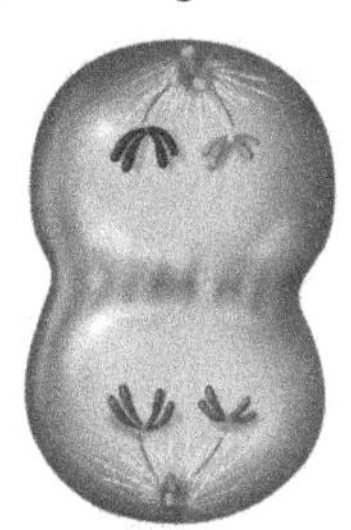

3 The chromosomes separate from their homologous partners. Then they move to the opposite ends of the cell.

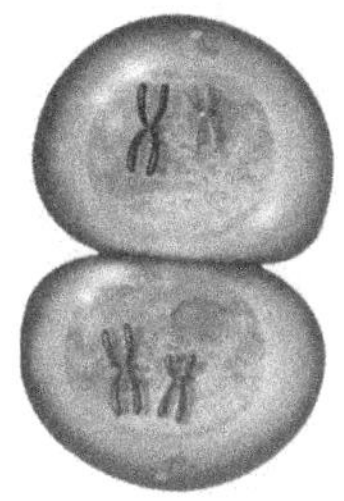

4 The nuclear membrane re-forms, and the cell divides. The paired chromatids are still joined.

Second cell division

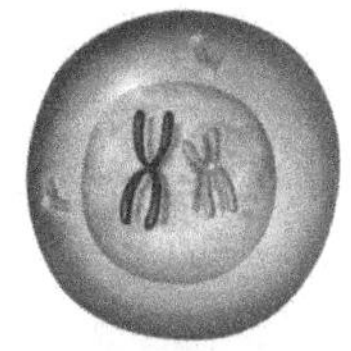

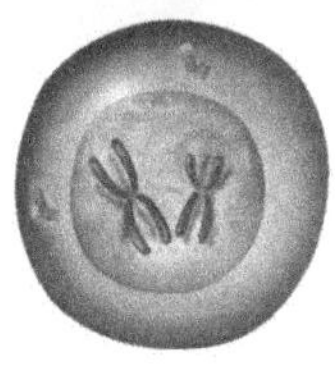

5 Each cell contains one member of the homologous chromosome pair. The chromosomes are not copied again between the two cell divisions.

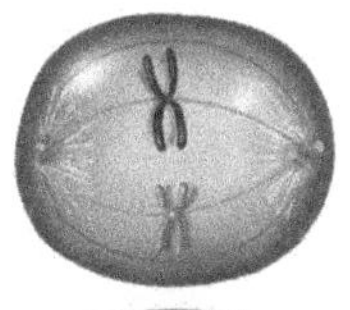

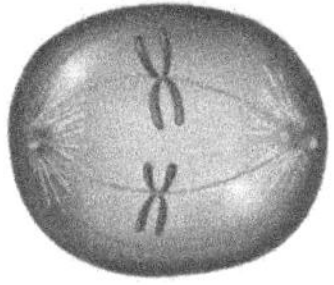

6 The nuclear membrane disappears. The chromosomes line up along the equator of each cell.

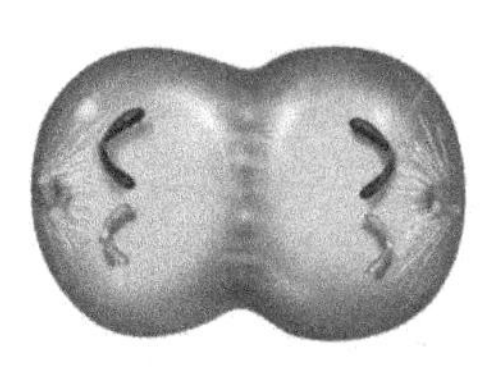

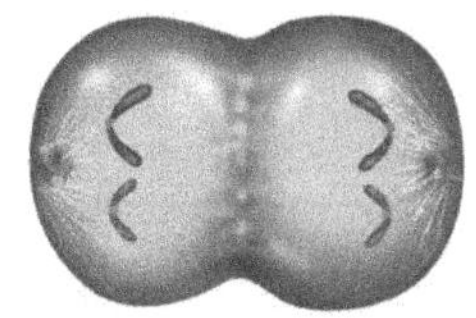

7 The chromatids pull apart and move to opposite ends of the cell. The nuclear membranes re-form, and the cells divide.

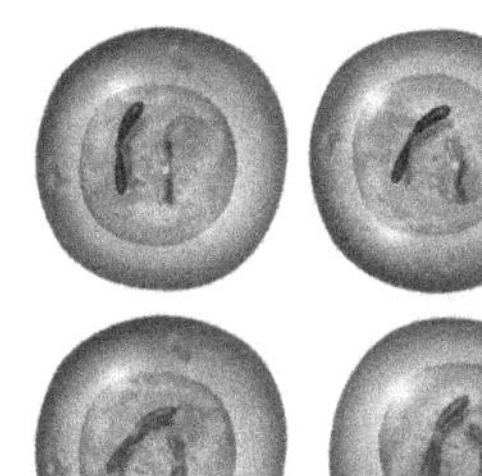

8 Four new cells have formed from the original cell. Each new cell has half the number of chromosomes as the original cell.

Critical Thinking

5. Predict What would happen if meiosis did not occur?

____

____

____

____

____

TAKE A LOOK

6. Identify How many times does the cell nucleus divide during meiosis?

____

7. Identify At the end of meiosis, how many sex cells have been produced from one cell?

____

How Does Meiosis Explain Mendel's Results?

Mendel knew that eggs and sperm give the same amount of information to offspring. However, he did not know how traits were actually carried in the cell. Many years later, a scientist named Walter Sutton was studying grasshopper sperm cells. He knew about Mendel's work. When he saw chromosomes separating during meiosis, he made an important conclusion: genes are located on chromosomes.

The figure below shows what happens to chromosomes during meiosis and fertilization in pea plants. The cross shown is between two true-breeding plants. One produces round seeds and the other produces wrinkled seeds.

Critical Thinking

8. Identify Relationships How did Sutton's work build on Mendel's work?

____
____
____
____
____
____
____

Meiosis and Dominance

Male Parent In the plant cell nucleus below, each homologous chromosome has an allele for seed shape. Each allele carries the same instructions: to make wrinkled seeds.

Female Parent In the plant cell nucleus below, each homologous chromosome has an allele for seed shape. Each allele carries the same instructions: to make round seeds.

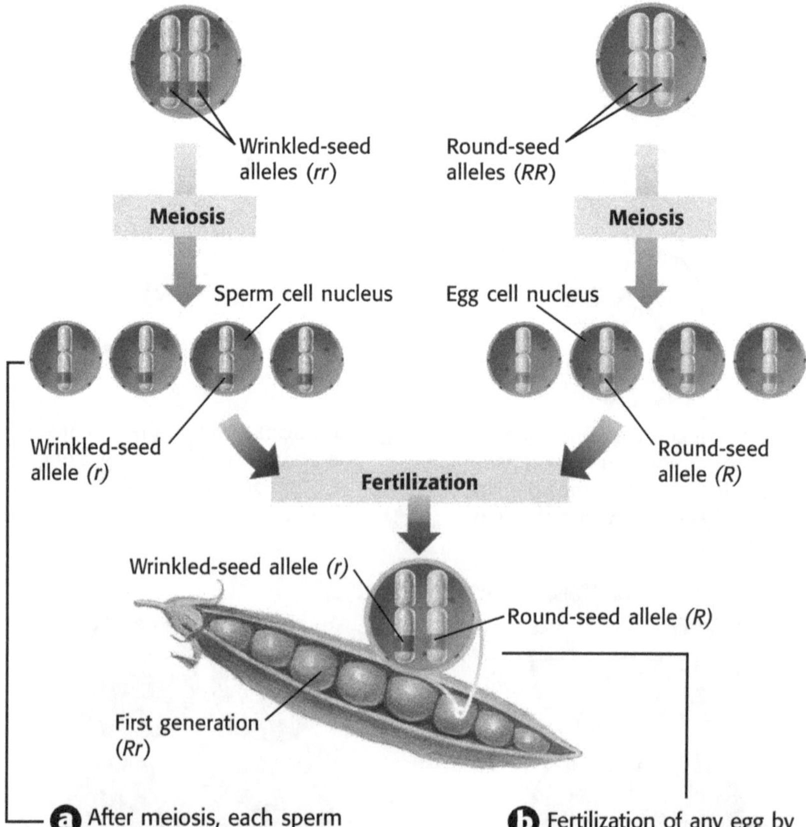

a After meiosis, each sperm cell has a recessive allele for wrinkled seeds. Each egg cell has a dominant allele for round seeds.

b Fertilization of any egg by any sperm gives the same genotype (*Rr*) and the same phenotype (round). This result is exactly what Mendel found in his studies.

TAKE A LOOK

9. Explain In this figure, how many genotypes are possible for the offspring? Explain your answer.

____
____
____
____
____

What Are Sex Chromosomes?

Information contained on chromosomes determines many of our traits. **Sex chromosomes** carry genes that determine sex. In humans, females have two X chromosomes. Human males have one X chromosome and one Y chromosome. ☑

During meiosis, one of each of the chromosome pairs ends up in a sex cell. Females have two X chromosomes in each body cell. When meiosis produces egg cells, each egg gets one X chromosome. Males have both an X chromosome and a Y chromosome in each body cell. Meiosis produces sperm with either an X or a Y chromosome.

An egg fertilized by a sperm with an X chromosome will produce a female. If the sperm contains a Y chromosome, the offspring will be male.

READING CHECK

10. Identify What combination of sex chromosomes makes a human male?

____

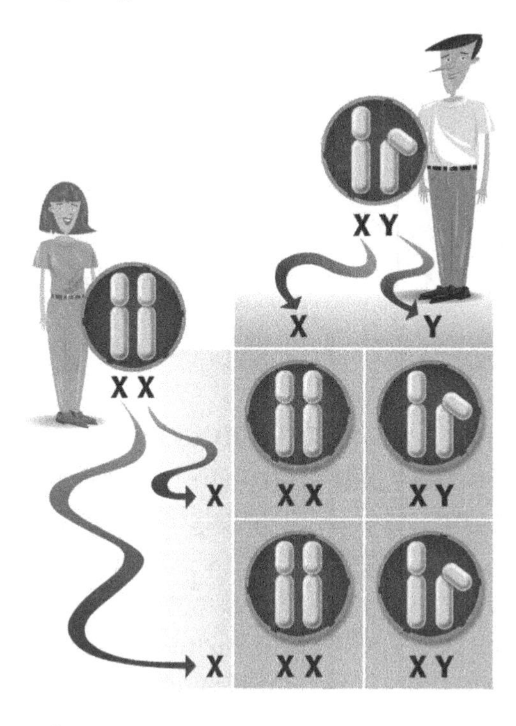

Egg and sperm join to form either the XX or XY combination.

TAKE A LOOK

11. Identify Circle the offspring in the figure that will be female.

SEX-LINKED DISORDERS

Hemophilia is a disorder that prevents blood from clotting. People with hemophilia bleed for a long time after small cuts. This disorder can be fatal. Hemophilia is an example of a sex-linked disorder. The genes for *sex-linked disorders* are carried on the X chromosome. Colorblindness is another example of a sex-linked disorder. Men are more likely than women to have sex-linked disorders. Why is this? ☑

READING CHECK

12. Define What is a sex-linked disorder?

____

____

____

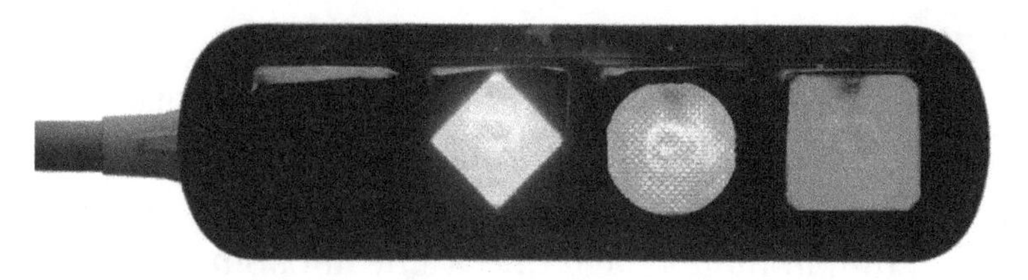

This stoplight in Canada was made to help the colorblind see signals easily.

The Y chromosome does not carry all of the genes that an X chromosome does. Females have two X chromosomes, so they carry two copies of each gene found on the X chromosome. This makes a backup gene available if one becomes damaged. Males have only one copy of each gene on their one X chromosome. If a male gets an allele for a sex-linked disorder, he will have the disorder, even if the allele is recessive.

TAKE A LOOK

13. Complete A particular sex-linked disorder is recessive. Fill in the Punnett Square to show how the disorder is passed from a carrier to its offspring. The chromosome carrying the trait for the disorder is underlined.

14. Identify Which individual will have the disorder?

____

	X	*Y*
X̲		
X		

GENETIC COUNSELING AND PEDIGREES

Genetic disorders can be traced through a family tree. If people are worried that they might pass a disease to their children, they may consult a genetic counselor.

These counselors often use a diagram called a pedigree. A **pedigree** is a tool for tracing a trait through generations of a family. By making a pedigree, a counselor can often predict whether a person is a carrier of a hereditary disease.

The pedigree on the next page traces a disease called *cystic fibrosis*. Cystic fibrosis causes serious lung problems. People with this disease have inherited two recessive alleles. Both parents need to be carriers of the gene for the disease to show up in their children.

Pedigree for a Recessive Disease

Males Females

Vertical lines connect children to their parents.

or A solid square or circle shows that the person has a certain trait.

or A half-filled square or circle shows that the person is a carrier for the trait.

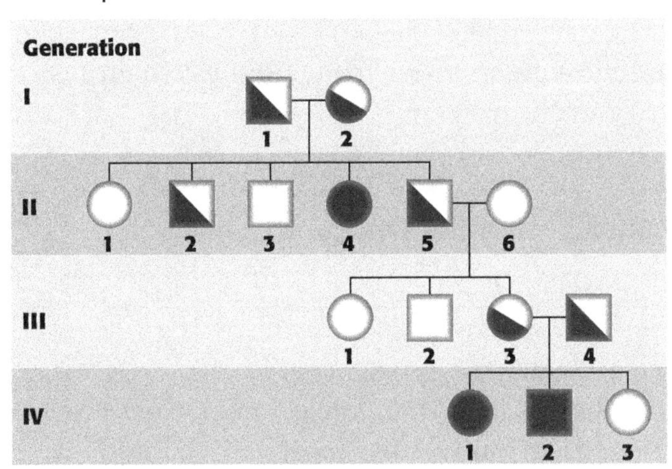

TAKE A LOOK

15. Identify Circle all of the individuals in the pedigree who have the disorder. Draw a line under the individuals that carry the trait, but do not have the disorder.

You could draw a pedigree to trace almost any trait through a group of people who are biologically related. For example, a pedigree can show how you inherited your hair color. Many different pedigrees could be drawn for related individuals.

What Is Selective Breeding?

For thousands of years, humans have bred plants and animals to produce individuals with traits that they liked. This is known as *selective breeding*. Breeders may choose a plant or animal with traits they would like to see in the offspring. They breed that individual with another that also has those traits. For example, farmers might breed fruit trees that bear larger fruits.

You may see example of selective breeding every day. Different breeds of dogs, such as chihuahuas and German sheperds, were produced by selective breeding. Many flowers, such as roses, have been bred to produce large flowers. Wild roses are usually much smaller than roses you would buy at a flower store or plant nursery.

Discuss In a small group, come up with other examples of organisms that humans have changed through selective breeding. What traits do you think people wanted the organism to have? How is this trait helpful to humans?

Name ________________________ Class ______________ Date ____________

Section 3 Review

NSES LS 1c, 1d, 2a, 2b, 2c, 2d

SECTION VOCABULARY

homologous chromosomes chromosomes that have the same sequence of genes and the same structure

meiosis a process in cell division during which the number of chromosomes decreases to half the original number by two divisions of the nucleus, which results in the production of sex cells (gametes or spores)

pedigree a diagram that shows the occurrence of a genetic trait in several generations of a family

sex chromosomes one of the pair of chromosomes that determine the sex of an individual

1. **Identify Relationships** Put the following in order from smallest to largest: chromosome, gene, cell.

____

2. **Explain** Does meiosis happen in all cells? Explain your answer.

____

The pedigree below shows a recessive trait that causes a disorder. Use the pedigree to answer the questions that follow.

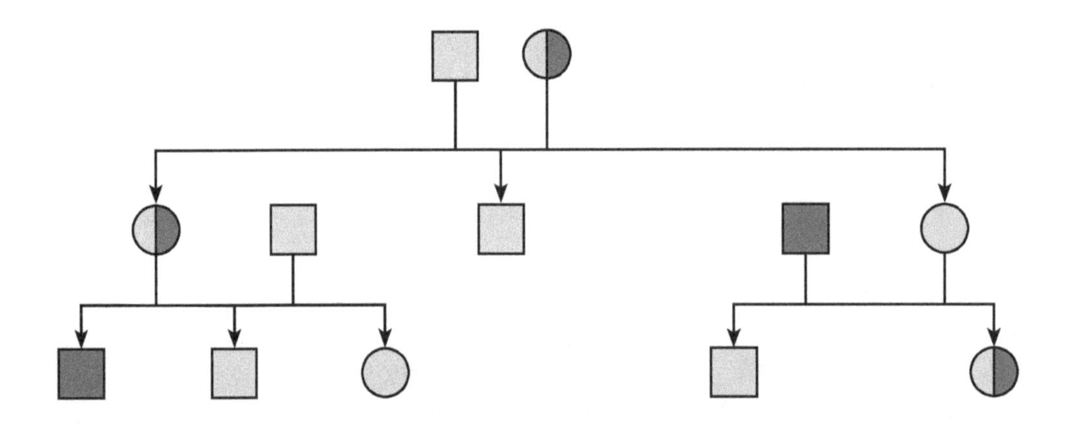

3. **Identify** Circle all individuals on the pedigree that are heterozygous for the trait. Are these individuals male or female?

____

4. **Identify** Put a square around all individuals that have the disorder. Are these individuals male or female?

____

5. **Interpret** Is the trait sex-linked? Explain your answer.

____

____

Name _______________ Class _______________ Date _______________

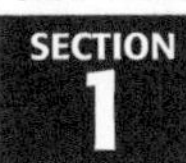

Genes and DNA

SECTION 1

What Does DNA Look Like?

BEFORE YOU READ

After you read this section, you should be able to answer these questions:

- What units make up DNA?
- What does DNA look like?
- How does DNA copy itself?

National Science Education Standards
LS 1a, 2d, 5a

What Is DNA?

Remember that *inherited traits* are traits that are passed from generation to generation. To understand how inherited traits are passed on, you must understand the structure of DNA. **DNA** (***d**eoxyribo**n**ucleic **a**cid*) is the molecule that carries the instructions for inherited traits. In cells, DNA is wrapped around proteins to form *chromosomes*. Stretches of DNA that carry the information for inherited traits are called *genes*.

STUDY TIP

Clarify Concepts As you read the text, make a list of ideas that are confusing. Discuss these with a small group. Ask your teacher to explain things that your group is unsure about.

What Is DNA Made Of?

DNA is made up of smaller units called nucleotides. A *nucleotide* is made of three parts: a sugar, a phosphate, and a base. The sugar and the phosphate are the same for each nucleotide. However, different nucleotides may have different bases.

There are four different bases found in DNA nucleotides. They are *adenine*, *thymine*, *guanine*, and *cytosine*. Scientists often refer to a base by its first letter: *A* for adenine, *T* for thymine, *G* for guanine, and *C* for cytosine. Each base has a different shape.

The Four Nucleotides of DNA

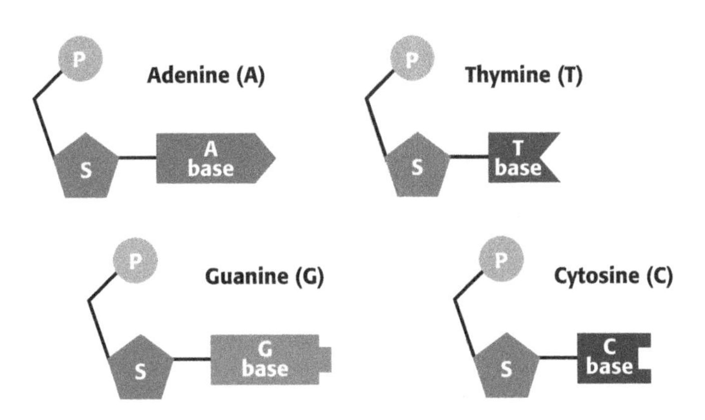

TAKE A LOOK

1. Identify What are two things that are the same in all nucleotides?

____

____

What Does DNA Look Like?

As you can see in the figure below, a strand of DNA looks like a twisted ladder. This spiral shape is called a *double helix*. The two sides of the ladder are made of the sugar and phosphate parts of nucleotides. The sugars and phosphates alternate along each side of the ladder. The rungs of the DNA ladder are made of pairs of bases. ☑

The bases in DNA can only fit together in certain ways, like puzzle pieces. Adenine on one side of a DNA strand always pairs with thymine on the other side. Guanine always pairs with cytosine. This means that adenine is *complementary* to thymine, and guanine is complementary to cytosine. Because the pairs of bases in DNA are complementary, the two sides of a strand of DNA are also complementary.

READING CHECK

2. Identify What are the sides of the DNA "ladder" made of?

____

____

Critical Thinking

3. Apply Concepts Imagine that you are a scientist studying DNA. You measure the number of cytosines and thymines in a small strand of DNA. There are 45 cytosines and 55 thymines. How many guanines are there in the strand? How many adenines are there?

____

____

TAKE A LOOK

4. Identify Give the ways that DNA bases can pair up.

____

____

____

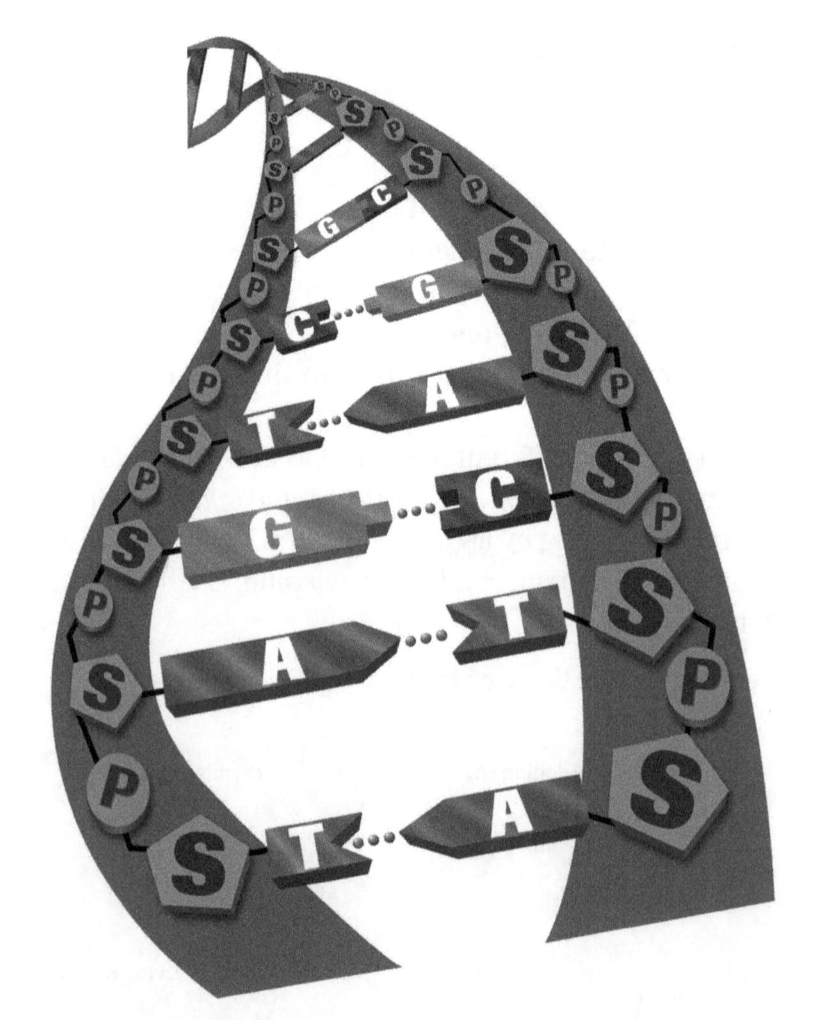

Each side of a DNA molecule is complementary to the other side.

How Does DNA Copy Itself?

Before a cell divides, it makes a copy of its genetic information for the new cell. The pairing of bases allows the cell to *replicate*, or make copies of, DNA. Remember that bases are complementary and can only fit together in certain ways. Therefore, the order of bases on one side of the DNA strand controls the order of bases on the other side of the strand. For example, the base order CGAC can only fit with the order GCTG. ☑

When DNA replicates, the pairs of bases separate and the DNA splits into two strands. The bases on each side of the original strand are used as a pattern to build a new strand. As the bases on the original strands are exposed, the cell adds nucleotides to form a new strand.

Finally, two DNA strands are formed. Half of each of the two DNA strands comes from the original strand. The other half is built from new nucleotides.

5. Explain What happens to DNA before a cell divides?

____

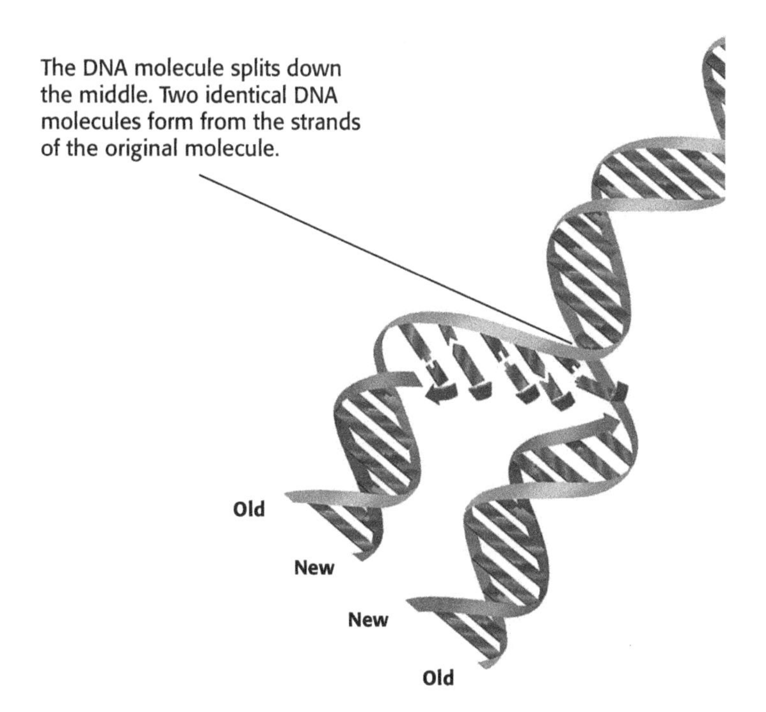

TAKE A LOOK

6. Compare What is the difference between an "old" and a "new" strand of DNA?

____

____

____

DNA is copied every time a cell divides. Each new cell gets a complete copy of the entire DNA strand. Proteins in the cell unwind, copy, and rewind the DNA.

Name ______________________ Class ______________ Date ______________

Section 1 Review

NSES LS 1a, 2d, 5a

SECTION VOCABULARY

DNA deoxyribonucleic acid, a molecule that is present in all living cells and that contains the information that determines the traits that a living thing inherits and needs to live	**nucleotide** in a nucleic-acid chain, a subunit that consists of a sugar, a phosphate, and a nitrogenous base

1. **Identify** Where are genes located? What do they do?

____

____

2. **Compare** How are the four kinds of DNA nucleotides different from each other?

____

3. **Apply Concepts** The diagram shows part of a strand of DNA. Using the order of bases given in the top of the strand, write the letters of the bases that belong on the bottom strand.

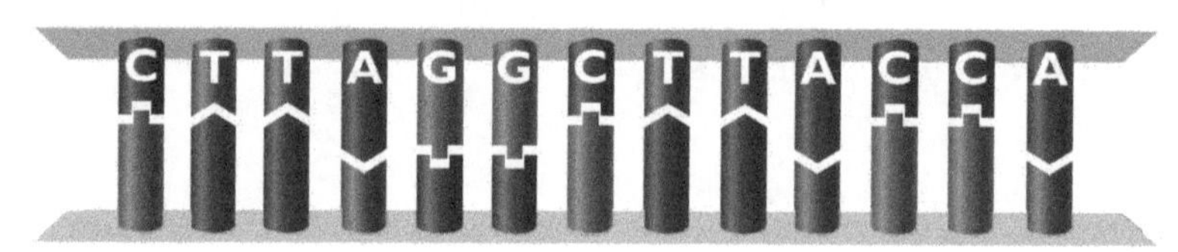

____

4. **Describe** How is DNA related to chromosomes?

____

____

5. **Identify Relationships** How are proteins involved in DNA replication?

____

____

6. **List** What are three parts of a nucleotide?

____

____

____

Name _______________ Class _______________ Date _______________

CHAPTER 6 Genes and DNA

SECTION 2

How DNA Works

BEFORE YOU READ

After you read this section, you should be able to answer these questions:

- What does DNA look like in different cells?
- How does DNA help make proteins?
- What happens if a gene changes?

National Science Education Standards
LS 1c, 1e, 1f, 2b, 2c, 2d, 2e, 5c

What Does DNA in Cells Look Like?

The human body contains trillions of cells, which carry out many different functions. Most cells are very small and can only be seen with a microscope. A typical skin cell, for example, has a diameter of about 0.0025 cm. However, almost every cell contains about 2 m of DNA. How can so much DNA fit into the nucleus of such a small cell? The DNA is bundled.

Compare After you read this section, make a table comparing chromatin, chromatids, and chromosomes.

Math Focus

1. Convert About how long is the DNA in a cell in inches?
1 in. = 2.54 cm

____

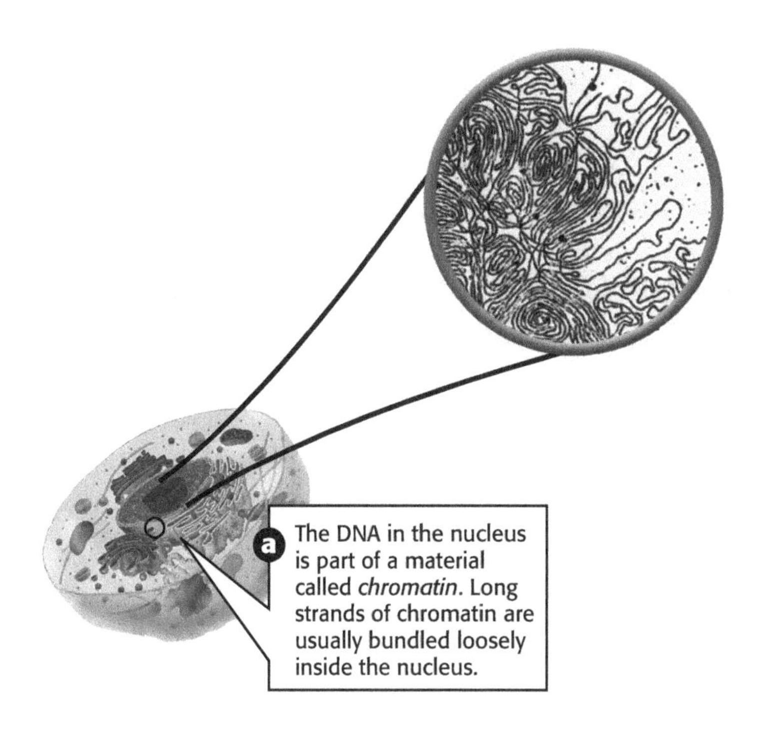

The DNA in the nucleus is part of a material called *chromatin*. Long strands of chromatin are usually bundled loosely inside the nucleus.

TAKE A LOOK

2. Identify In what form is the DNA in the nucleus?

____

FITTING DNA INTO THE CELL

Large amounts of DNA can fit inside a cell because the DNA is tightly bundled by proteins. The proteins found with DNA help support the structure and function of DNA. Together, the DNA and the proteins it winds around make up a chromosome. ☑

READING CHECK

3. Identify What are two things that are found in a chromosome?

____

____

DNA's structure allows it to hold a lot of information. Remember that a gene is made of a string of nucleotides. That is, it is part of the 2 m of DNA in a cell. Because there is an enormous amount of DNA, there can be a large variety of genes.

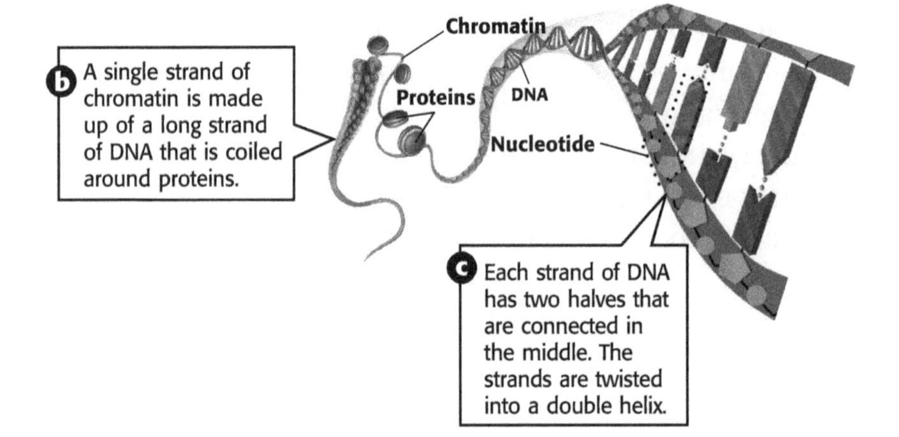

b A single strand of chromatin is made up of a long strand of DNA that is coiled around proteins.

c Each strand of DNA has two halves that are connected in the middle. The strands are twisted into a double helix.

TAKE A LOOK

4. Describe What is chromatin made of?

____

____

DNA IN DIVIDING CELLS

When a cell divides, its genetic material is spread equally into each of the two new cells. How can each of the new cells receive a full set of genetic material? It is possible because DNA replicates before a cell divides.

Remember that when DNA replicates, the strand of DNA splits down the middle. New strands are made when free nucleotide bases bind to the exposed strands. Each of the new strands is identical to the original DNA strand. This is because the DNA bases can join only in certain ways. *A* always pairs with *T*, and *C* always pairs with *G*.

Critical Thinking

5. Predict Consequences Imagine that DNA did not replicate before cell division. What would happen to the amount of DNA in each of the new cells formed during cell division?

____

____

____

____

____

When a cell is ready to divide, it has already copied its DNA. The copies stay attached as two chromatids. The two identical chromatids form a chromosome.

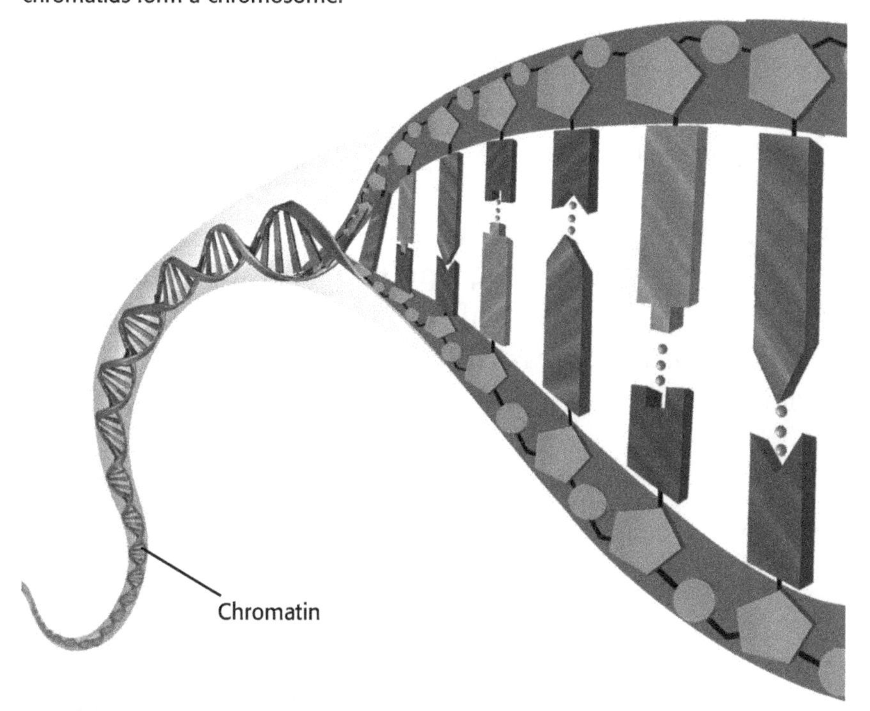

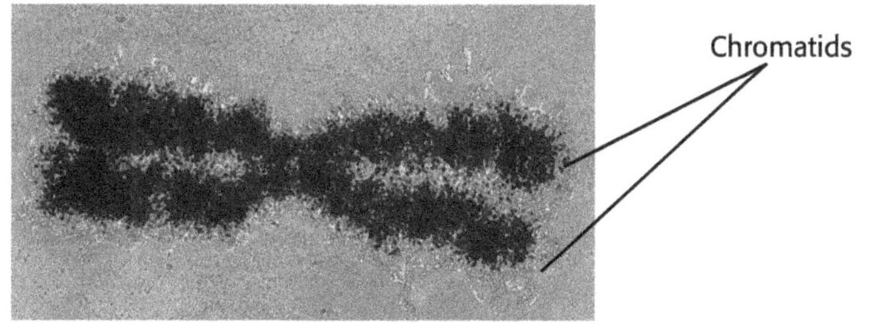

STANDARDS CHECK

LS 2d Hereditary information is contained in the genes, located in the chromosomes of each cell. Each gene carries a single unit of information. An inherited trait of an individual can be determined by one or by many genes, and a single gene can influence more than one trait. A human cell contains many thousands of different genes.

6. Identify Where is DNA found in a eukaryotic cell?

____

How Does DNA Help Make Proteins?

Proteins are found throughout cells. They cause most of the differences that you can see among organisms. A single organism can have thousands of different proteins.

Proteins act as chemical messengers for many of the activities in cells, helping the cells to work together. They also affect traits, such as the color of your eyes and how tall you will grow.

Proteins are made from many subunits called *amino acids*. A long string of amino acids forms a protein.

The order of bases in DNA is a code. The code tells how to make proteins. A group of three DNA bases acts as a code for one amino acid. For example, the group of DNA bases CAA *codes for*, or stands for, the amino acid valine. A gene usually contains instructions for making one specific protein.

Math Focus

7. Calculate How many DNA bases are needed to code for five amino acids?

____

HELP FROM RNA

RNA, or ***ribonucleic acid***, is a chemical that helps DNA make proteins. RNA is similar to DNA. It can act as a temporary copy of part of a DNA strand. One difference between DNA and RNA is that RNA contains the base *uracil* instead of thymine. Uracil is often represented by *U*. ☑

READING CHECK

8. Identify What is one difference between RNA and DNA?

____

____

____

How Are Proteins Made in Cells?

The first step in making a protein is to copy one side of part of the DNA. This mirrorlike copy is made of RNA. It is called *messenger RNA* (mRNA). It moves out of the nucleus and into the cytoplasm of the cell.

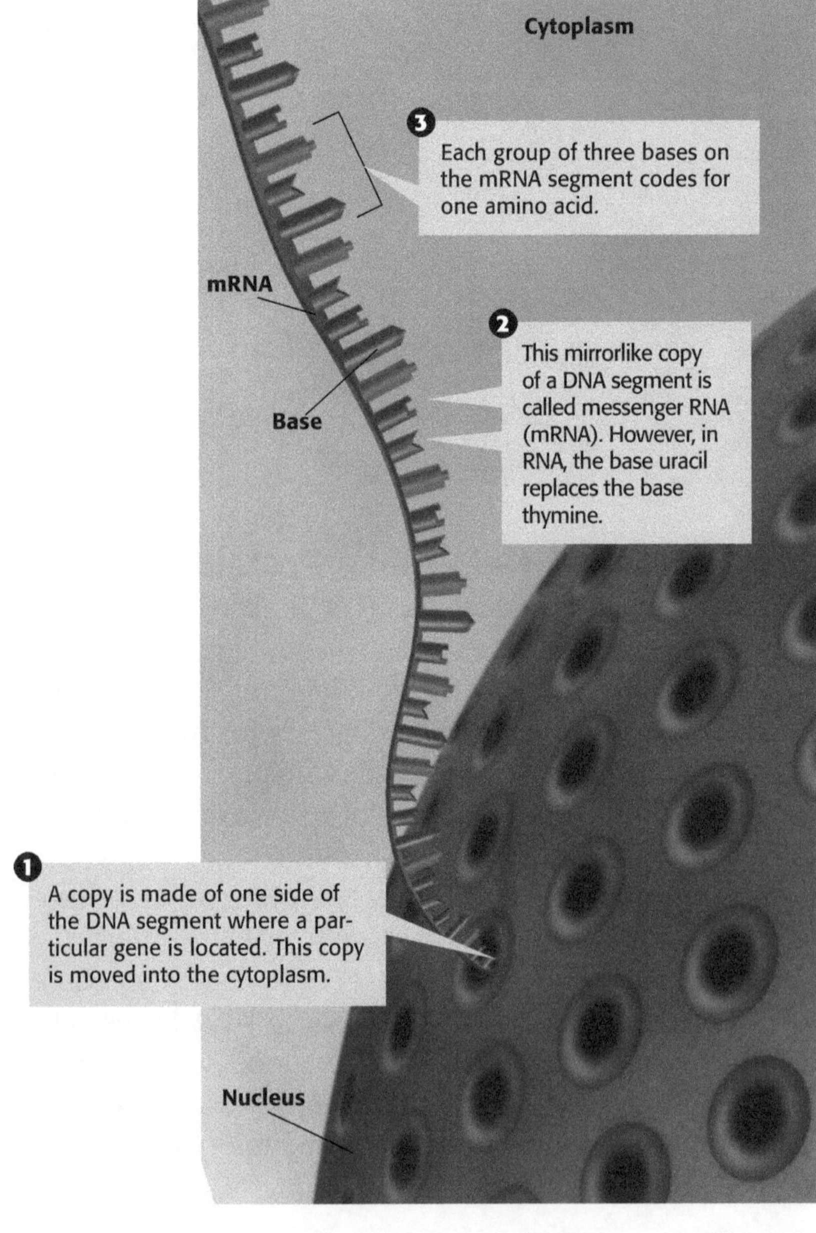

TAKE A LOOK

9. Compare How does the shape of RNA differ from the shape of DNA?

____

____

____

RIBOSOMES

In the cytoplasm, the messenger RNA enters a protein assembly line. The "factory" that runs this assembly line is a ribosome. A **ribosome** is a cell organelle composed of RNA and protein. The mRNA moves through a ribosome as a protein is made.

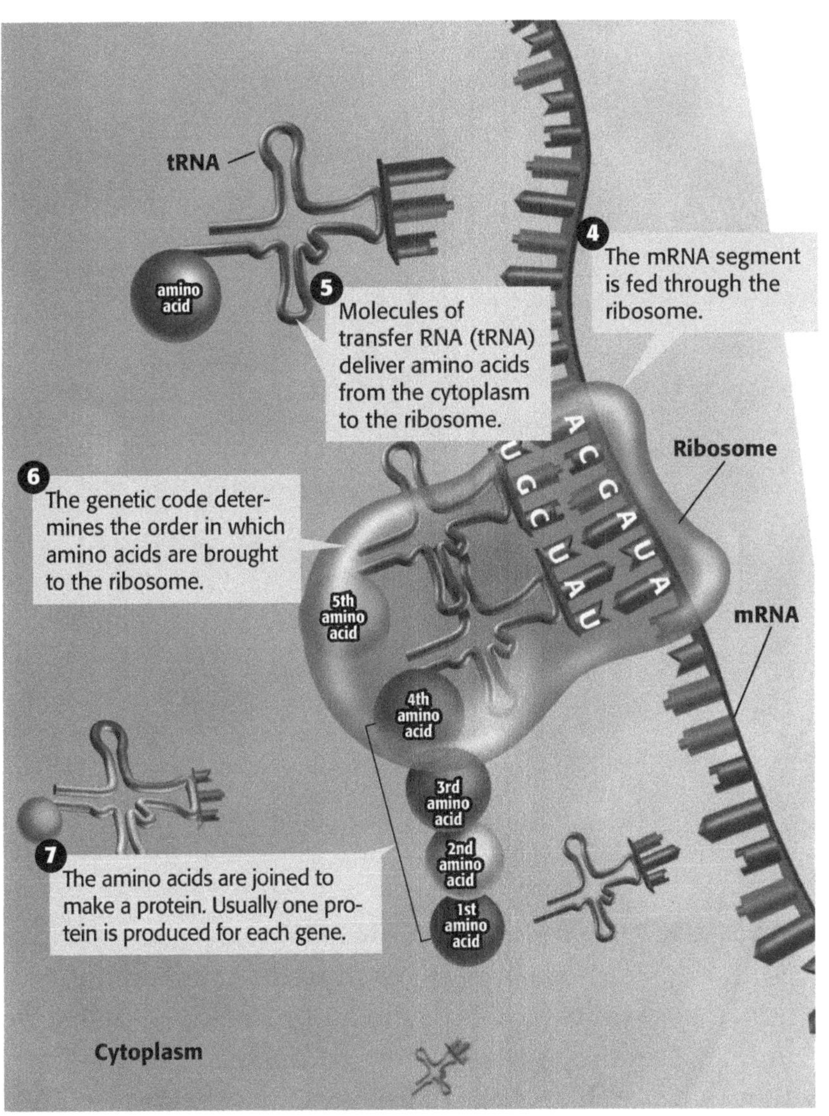

Critical Thinking

10. Explain Proteins are made in the cytoplasm, but DNA never leaves the nucleus of a cell. How does DNA control how proteins are made?

____

____

____

____

____

____

____

TAKE A LOOK

11. Identify What does tRNA do?

____

____

____

What Happens If Genes Change?

Read this sentence: "Put the book on the desk." Does it make sense? What about this sentence: "Rut the zook in the tesk."? Changing only a few letters in a sentence can change what the sentence means. It can even keep the sentence from making any sense at all! In a similar way, even small changes in a DNA sequence can affect the protein that the DNA codes for. A change in the nucleotide-base sequence of DNA is called a **mutation**. ☑

READING CHECK

12. Define What is a mutation?

____

____

____

HOW MUTATIONS HAPPEN

Some mutations happen because of mistakes when DNA is copied. Other mutations happen when DNA is damaged. Things that can cause mutations are called *mutagens*. Examples of mutagens include X rays and ultraviolet radiation. Ultraviolet radiation is one type of energy in sunlight. It can cause suntans and sunburns.

Original sequence

Base pair replaced

Base pair added

Base pair removed

Mutations can happen in different ways. A nucleotide may be replaced, added, or removed.

TAKE A LOOK

13. Compare What happens to one strand of DNA when there is a change in a base on the other strand?

____

____

Brainstorm Whether a mutation is helpful or harmful to an organism often depends on the organism's environment. In a group, discuss how the same mutation could be helpful in one environment but harmful in another.

HOW MUTATIONS AFFECT ORGANISMS

Mutations can cause changes in traits. Some mutations produce new traits that can help an organism survive. For example, a mutation might allow an organism to survive with less water. If there is a drought, the organism will be more likely to survive.

Many mutations produce traits that make an organism less likely to survive. For example, a mutation might make an animal a brighter color. This might make the animal easier for predators to find.

Some mutations are neither helpful nor harmful. If a mutation does not cause a change in a protein, then the mutation will not help or hurt the organism.

PASSING ON MUTATIONS

Cells make proteins that can find and fix many mutations. However, not all mutations can be fixed.

If a mutation happens in egg or sperm cells, the changed gene can be passed from one generation to the next. For example, sickle cell disease is caused by a genetic mutation that can be passed to future generations.

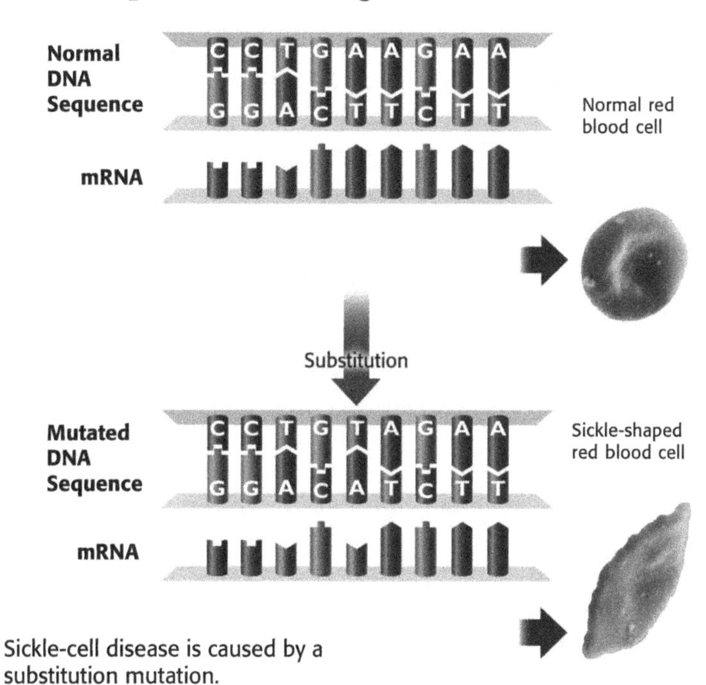

Sickle-cell disease is caused by a substitution mutation.

TAKE A LOOK

14. Identify What kind of mutation causes sickle cell disease: deletion, insertion, or substitution?

____

How Can We Use Genetic Knowledge?

Scientists use their knowledge of genetics in many ways. Most of these ways are helpful to people. However, other ways can cause ethical and scientific concerns.

GENETIC ENGINEERING

Scientists have learned how to change individual genes within organisms. This is called *genetic engineering*. In some cases, scientists transfer genes from one organism to another. For example, scientists can transfer genes from people into bacteria. The bacteria can then make proteins for people who are sick. ☑

READING CHECK

15. Define What is genetic engineering?

____

____

GENETIC IDENTIFICATION

Your DNA is unique, so it can be used like a fingerprint to identify you. *DNA fingerprinting* identifies the unique patterns in a person's DNA. Scientists can use these genetic fingerprints as evidence in criminal cases. They can also use genetic information to determine whether people are related.

Name ______________________ Class ______________ Date ______________

Section 2 Review

NSES LS 1c, 1e, 1f, 2b, 2c, 2d, 2e, 5c

SECTION VOCABULARY

mutation a change in the nucleotide-base sequence of a gene or DNA molecule

ribosome a cell organelle composed of RNA and protein; the site of protein synthesis

RNA ribonucleic acid, a molecule that is present in all living cells and that plays a role in protein production

1. Identify What structures in cells contain DNA and proteins?

____

2. Calculate How many amino acids can a sequence of 24 DNA bases code for?

____

3. Explain Fill in the flow chart below to show how the information in the DNA code becomes a protein.

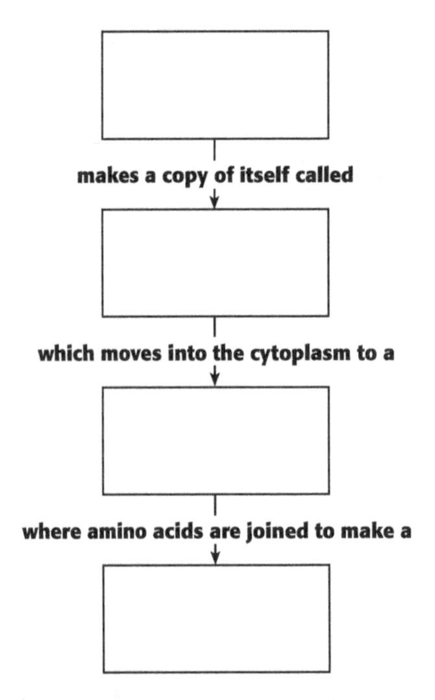

4. Draw Conclusions How can a mutation in a DNA base sequence cause a change in a gene and a trait? What determines whether the mutation is passed on to offspring?

____

____

____

5. Identify Give two ways that genetic fingerprinting can be used.

____

____

Name ______________________ Class ______________ Date ______________

CHAPTER 7 The Evolution of Living Things

SECTION 1

Change over Time

BEFORE YOU READ

After you read this section, you should be able to answer these questions:

- How are organisms different from one another?
- How do scientists know that species change over time?
- How can you tell if species are related?

National Science Education Standards

LS 2e, 3a, 3d, 4a, 5a, 5b, 5c

What Are Adaptations?

The pictures below show three different kinds of frogs. These frogs all have some things in common. For example, they all have long back legs to help them move. However, as you can see in the pictures, there are also many differences between the frogs. Each frog has some physical features that can help it survive.

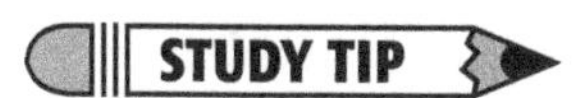

STUDY TIP

Learn New Words As you read this section, circle the words you don't understand. When you figure out what they mean, write the words and their definitions in your notebook.

The red-eyed tree frog has green skin. It hides in the leaves of trees during the day. It comes out at night.

The skin of the smokey jungle frog looks like leaves. It can hide on the forest floor.

The strawberry poison frog has brightly colored skin. Its bright coloring warns predators that it is poisonous.

TAKE A LOOK

1. Explain How does the coloring of the strawberry poison frog help it survive?

____

____

____

____

A feature that helps an organism survive and reproduce in its environment is called an **adaptation**. Some adaptations, such as striped skin or a long neck, are physical features. Other adaptations are behaviors that help an organism find food, protect itself, or reproduce.

Living things with the same features may be members of the same species. A **species** is a group of organisms that can mate with one another to produce fertile offspring. *Fertile* offspring are offspring that can reproduce. For example, all smokey jungle frogs are members of one species. Therefore, smokey jungle frogs can mate with each other to produce offspring. The offspring can also reproduce.

A group of individuals of the same species living in the same place is a *population*. For example, all of the smokey jungle frogs living in a certain jungle are a population.

Critical Thinking

2. Infer Give an example of a behavior that could help an organism survive.

____

____

____

CHANGE OVER TIME

Scientists estimate that there has been life on Earth for over 3 billion years. Earth has changed a great deal during its history. Living things have changed in this time too. Since life first appeared on Earth, many species have died out and many new species have appeared.

Scientists observe that species change over time. They also observe that the inherited characteristics in populations change over time. Scientists think that as populations change, new species form. New species descend from older species. The process in which populations change over time is called **evolution**. ☑

READING CHECK

3. Define What is evolution?

____

____

____

What Is the Evidence that Organisms Have Changed?

Much of the evidence that organisms have changed over time is buried in sedimentary rock. *Sedimentary rock* forms when pieces of sand, dust, or soil are laid down in flat layers. Sedimentary rocks may contain fossils. These fossils provide evidence that organisms have changed over Earth's history.

FOSSILS

Fossils are the remains or imprints of once-living organisms. Some fossils formed from whole organisms or from parts of organisms. Some fossils are signs, such as footprints, that an organism once existed. ☑

4. Identify What are two kinds of fossils?

____

____

Fossils can form when layers of sediment cover a dead organism. Minerals in the sediment may seep into the organism and replace its body with stone. Fossils can also form when an organism dies and leaves an imprint of itself in sediment. Over time, the sediment can become rock and the imprint can be preserved as a fossil.

Some fossils, like this trilobite, form from the bodies of organisms. The trilobite was an ancient marine animal.

Some fossils, like these ferns, form when an organism leaves an imprint in sediment. Over time, the sediment becomes rock and the imprint is preserved as a fossil.

TAKE A LOOK

5. Compare Which of the fossils looks most like an organism that lives today? Give the modern organism that the fossil looks like.

____

____

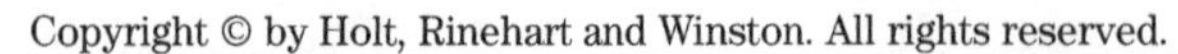

THE FOSSIL RECORD

Comparing fossils provides evidence that organisms have changed over time. Rocks from different times in Earth's history contain fossils of different organisms. Fossils in newer layers of rock tend to be similar to present-day organisms. Fossils from older layers are less similar to present-day organisms. By studying fossils, scientists have made a timeline of life known as the **fossil record**. ☑

READING CHECK

6. Explain How do fossils give evidence that organisms have changed over time?

____

____

____

____

How Do Scientists Compare Organisms?

Scientists observe that all living things have some features in common. They also observe that all living things inherit features in a similar way. Therefore, scientists think that all living species are descended from a common ancestor. Scientists compare features of fossils and of living organisms to determine whether ancient species are related to modern species.

COMPARING STRUCTURES

When scientists study the anatomy, or structures, of different organisms, they find that some organisms share traits. These organisms may share a common ancestor. For example, the figure below shows that humans, cats, dolphins, and bats have similar structures in their front limbs. These similarities suggest that humans, cats, dolphins, and bats have a common ancestor.

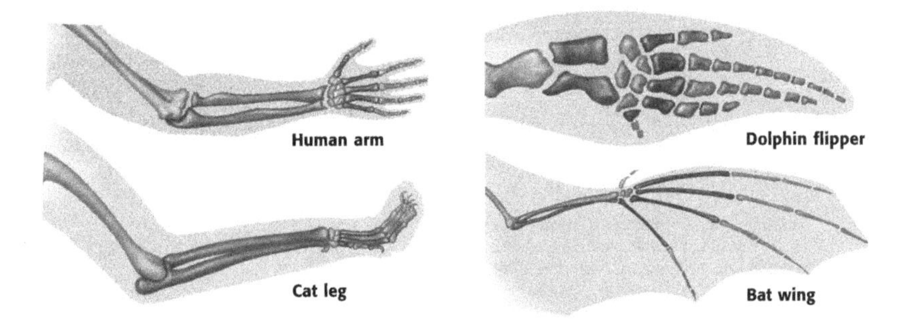

Comparing Structures The front limb bones of humans, cats, dolphins, and bats show some similarities. This suggests that all of these species share a common ancestor.

TAKE A LOOK

7. Identify Color the bones in each front limb to show which bones are similar. Use a different color for each bone.

Scientists can study the structures in modern organisms and compare them to structures in fossils. In this way, scientists can gather evidence that living organisms are related to organisms that lived long ago.

COMPARING CHEMICAL DATA

Remember that the genetic information in an organism's DNA determines the organism's traits. RNA and proteins also affect the traits of an organism. Scientists can compare these chemicals in organisms. The more alike these chemicals are between any two species, the more recently the two species shared a common ancestor.

Comparing DNA, RNA, and proteins can be very useful in determining whether species are related. However, this method can only be used on organisms that are alive today.

Critical Thinking

8. Infer Why can DNA, RNA, and proteins be used to compare only living organisms?

____

____

____

What Is the Evidence that Organisms Are Related?

Examining an organism carefully can give scientists clues about its ancestors. For example, whales look like some fish. However, unlike fish, whales breathe air, give birth to live young, and produce milk. These traits show that whales are mammals. Therefore, scientists think that whales evolved from ancient mammals. By examining fossils from ancient mammals, scientists have been able to determine how modern whales may have evolved. ☑

READING CHECK

9. Describe Why do scientists think that whales evolved from ancient mammals and not ancient fish?

____

____

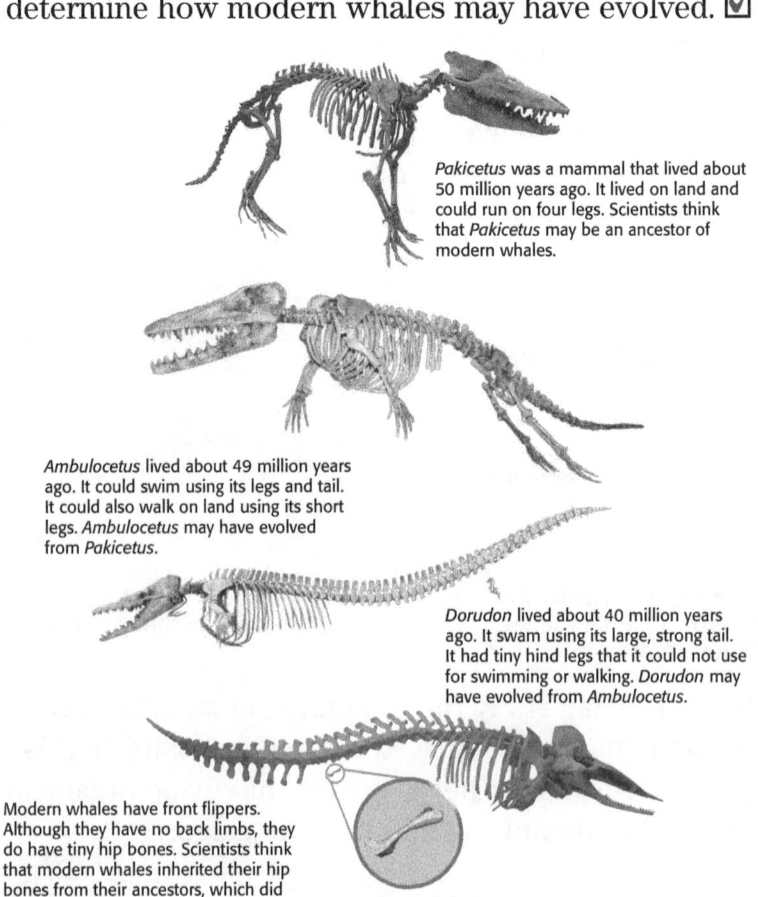

Pakicetus was a mammal that lived about 50 million years ago. It lived on land and could run on four legs. Scientists think that *Pakicetus* may be an ancestor of modern whales.

Ambulocetus lived about 49 million years ago. It could swim using its legs and tail. It could also walk on land using its short legs. *Ambulocetus* may have evolved from *Pakicetus*.

Dorudon lived about 40 million years ago. It swam using its large, strong tail. It had tiny hind legs that it could not use for swimming or walking. *Dorudon* may have evolved from *Ambulocetus*.

Modern whales have front flippers. Although they have no back limbs, they do have tiny hip bones. Scientists think that modern whales inherited their hip bones from their ancestors, which did have back limbs.

TAKE A LOOK

10. Compare Give two ways that modern whales are different from *Pakicetus*, and two ways that they are similar.

____

____

____

____

EVIDENCE FROM FOSSILS

Fossils provide several pieces of evidence that some species of ancient mammals are related to each other and to modern whales. First, each species shares some traits with an earlier species. Second, some species show new traits that are also found in later species. Third, each species had traits that allowed it to survive in a particular time and place in Earth's history.

EVIDENCE FROM MODERN WHALES

Some features of modern whales also suggest that they are related to ancient mammals. For example, modern whales do not have hind limbs. However, they do have tiny hip bones. Scientists think that modern whales inherited these hip bones from the whales' four-legged ancestors. Scientists often use this kind of evidence to determine the relationships between organisms.

How Do Scientists Show the Relationships Between Organisms?

As scientists analyze fossils and living organisms, they develop hypotheses about how species are related. They use *branching diagrams* to show the relationships between species.

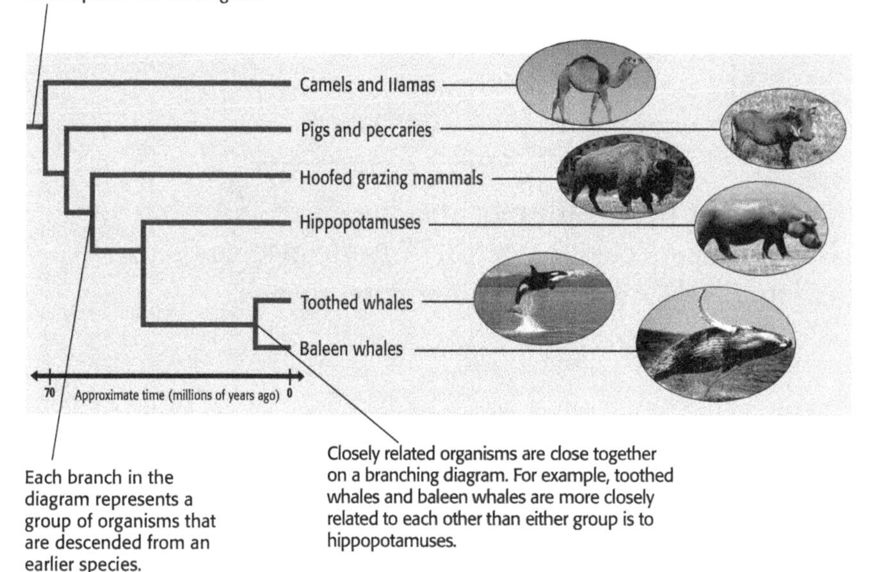

Scientists use branching diagrams to show how organisms are related.

STANDARDS CHECK

LS 5a Millions of species of animals, plants, and microorganisms are alive today. Although different species might look dissimilar, the unity among organisms becomes apparent from an analysis of internal structures, the similarity of their chemical processes, and the evidence of common ancestry.

Word Help: structure
the arrangement of the parts of a whole

Word Help: process
a set of steps, events, or changes

Word Help: evidence
information showing whether an idea or belief is true or valid

11. Identify Give one piece of evidence that indicates ancient mammals are related to modern whales.

____

____

TAKE A LOOK

12. Use a Model According to the diagram, which animals are most closely related to whales?

____

13. Use a Model According to the diagram, which animals are least closely related to whales?

____

Name ______________________ Class ______________ Date ______________

Section 1 Review

NSES **LS 2e, 3a, 3d, 4a, 5a, 5b, 5c**

SECTION VOCABULARY

adaptation a characteristic that improves an individual's ability to survive and reproduce in a particular environment

evolution the process in which inherited characteristics within a population change over generations such that new species sometimes arise

fossil the trace or remains of an organism that lived long ago, most commonly preserved in sedimentary rock

fossil record a historical sequence of life indicated by fossils found in layers of Earth's crust

species a group of organisms that are closely related and can mate to produce fertile offspring

1. Identify What evidence suggests that humans and bats have a common ancestor?

____

____

____

2. Make a Model Humans are most closely related to chimpanzees. Humans are least closely related to orangutans. Fill in the blank spaces on the branching diagram to show how humans, orangutans, chimpanzees, and gorillas are related.

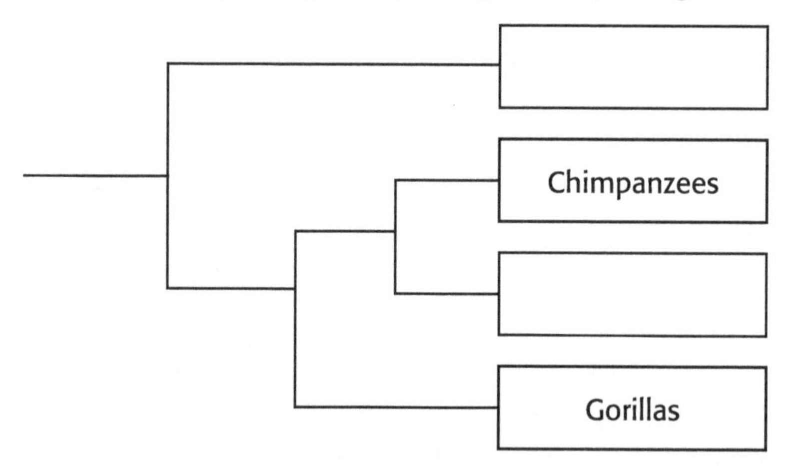

3. Infer A scientist studies the DNA of three different species. The DNA from species A is more similar to DNA from species B than DNA from species C. Which species are probably the most closely related? Explain your answer.

____

____

____

____

Name ______________________ Class ______________ Date ______________

CHAPTER 7 The Evolution of Living Things

SECTION 2

How Does Evolution Happen?

National Science Education Standards
LS 2a, 2b, 2e, 3d, 5a, 5b

BEFORE YOU READ

After you read this section, you should be able to answer these questions:

- Who was Charles Darwin?
- What ideas affected Darwin's thinking?
- What is natural selection?

Who Was Charles Darwin?

In 1831, Charles Darwin graduated from college. Although he eventually earned a degree in religion, Darwin was most interested in the study of plants and animals.

Darwin's interest in nature led him to sign on for a five-year voyage around the world. He was a naturalist on the HMS *Beagle*, a British ship. A *naturalist* is someone who studies nature. During the trip, Darwin made observations that helped him form a theory about how evolution happens. These ideas caused scientists to change the way they thought about the living world.

STUDY TIP

Summarize After you read this section, make a chart showing the four steps of natural selection. In the chart, explain what happens at each step.

DARWIN'S JOURNEY

On the trip, Darwin observed plants and animals from many parts of the world. One place Darwin found interesting was the Galápagos Islands. These islands are located about 1,000 km west of Ecuador, a country in South America. Many unusual organisms live on the Galápagos Islands.

Math Focus

1. Convert About how far are the Galápagos Islands from Ecuador in miles?
1 km = 0.62 mi

____

This line shows the course of the HMS *Beagle*.

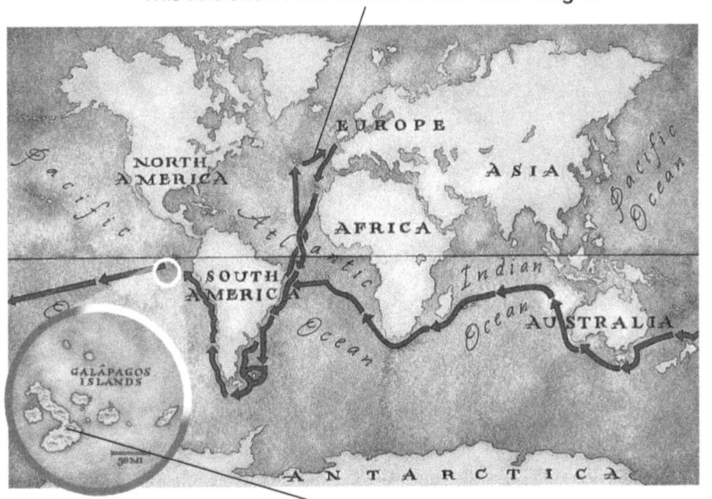

Darwin studied plants and animals on the Galápagos Islands.

TAKE A LOOK

2. Describe Which continent are the Galápagos Islands closest to?

____

DARWIN'S FINCHES

Darwin observed that the animals and plants on the Galápagos Islands were similar to those in Ecuador. However, they were not identical. For example, Darwin closely observed birds called finches. The finches on the Galápagos Islands were slightly different from the finches in Ecuador. In addition, the finches on each island in the Galápagos differed from the finches on the other islands. ☑

READING CHECK

3. Describe What did Darwin observe about the finches on the Galápagos Islands?

____

____

____

____

Darwin hypothesized that the island finches were descendents of South American finches. He thought the first finches on the islands were blown there from South America by a storm. He suggested that over many generations, the finch populations evolved adaptations that helped them survive in the different island environments. For example, the beaks of different finch species are adapted to the kind of food the species eat.

The large ground finch has a wide, strong beak. It can easily crack open large, hard seeds. Its beak works like a nutcracker.

The cactus finch has a tough beak. It uses its beak to eat cactus parts and insects. Its beak works like a pair of needle-nose pliers.

The warbler finch has a small, narrow beak. It can catch small insects with its beak. Its beak works like a pair of tweezers.

Critical Thinking

4. Infer What can you guess about the environment in which the cactus finch lives based on the information in the figure? Explain your answer.

____

____

____

____

____

How Did Darwin Develop the Theory of Evolution by Natural Selection?

After Darwin returned to England, he spent many years thinking about his experiences on the trip. In 1859, Darwin published a famous book called *On the Origin of Species by Means of Natural Selection*. In his book, Darwin proposed the theory that evolution happens by natural selection.

Natural selection happens when organisms that are well adapted to their environment survive, but less well-adapted organisms do not. When the better-adapted organisms reproduce, they pass their useful traits on to their offspring. Over time, more members of the population have these traits. Darwin combined ideas about breeding, population, and Earth's history to come up with a theory to explain his observations. ☑

READING CHECK

5. Define What is natural selection?

____

____

____

____

____

IDEAS ABOUT BREEDING

In Darwin's time, farmers and breeders had produced many kinds of farm animals and plants. They learned that if they bred plants or animals that had a desirable trait, some of the offspring might have the trait. A **trait** is a form of an inherited characteristic. The practice in which humans select plants or animals for breeding based on desired traits is called **selective breeding**.

Selective breeding showed Darwin that the traits of organisms can change and that certain traits can spread through populations. For example, most pets, such as the dogs below, have been bred for a variety of desired traits. Over the past 12,000 years, people have selectively bred dogs to produce more than 150 breeds. ☑

READING CHECK

6. Explain How did ideas about selective breeding affect Darwin's thinking about evolution?

____

____

____

____

____

People have selectively bred dogs for different traits. Today, there are over 150 dog breeds.

IDEAS ABOUT POPULATION

During Darwin's time, a scientist named Thomas Malthus was studying human populations. He observed that there were more babies being born than there were people dying. He thought that the human population could grow more rapidly than food supplies could grow. This would result in a worldwide food shortage. Malthus also pointed out that the size of human populations is limited by problems such as starvation and disease. ☑

READING CHECK

7. Identify According to Thomas Malthus, what are two things that can limit the size of human populations?

____

____

Darwin realized that Malthus's ideas can apply to all species, not just humans. He knew that any species can produce many offspring. He also knew starvation, disease, competition, and predation limited the populations of all species. Only a limited number of individuals live long enough to reproduce.

Darwin reasoned that the survivors had traits that helped them survive in their environment. He also thought that the survivors would pass on some of their traits to their offspring.

IDEAS ABOUT EARTH'S HISTORY

New information about Earth's history also affected Darwin's ideas about evolution. During Darwin's time, most geologists thought that Earth was very young. But important books, such as *Principles of Geology* by Charles Lyell, were changing ideas about the Earth. Lyell's book gave evidence that Earth is much older than anyone once thought. ☑

READING CHECK

8. Explain How did Charles Lyell's book change how scientists thought about Earth's history?

____

____

____

Darwin thought that evolution happens slowly. Darwin reasoned that if Earth was very old, there would be enough time for organisms to change slowly.

Idea	How it contributed to Darwin's theory
Selective breeding	
	helped Darwin realize that not all of an organism's offspring will survive to reproduce
	helped Darwin realize that slow changes can produce large differences over a long period of time

TAKE A LOOK

9. Describe Fill in the blank spaces in the table.

HOW NATURAL SELECTION WORKS

Natural selection has four steps: *overproduction, inherited variation, struggle to survive,* and *successful reproduction.*

❶ **Overproduction** A tarantula's egg sac can hold 500 to 1,000 eggs. Some of the eggs will survive and develop into adult spiders. Some will not.

❷ **Inherited Variation** Every individual has its own combination of traits. Each tarantula is similar, but not identical, to its parents.

❸ **Struggle to Survive** Some tarantulas may have traits that make it more likely that they will survive. For example, a tarantula may be better able to fight off predators, such as this wasp.

❹ **Successful Reproduction** The tarantulas that are best adapted to their environment are likely to survive and reproduce. Their offspring may inherit the traits that help them to survive.

Say It

Give Examples The figure shows one example of how the four steps of natural selection can work. In a group, talk about three or more other examples of how natural selection can affect populations.

TAKE A LOOK

10. Identify Why are some tarantulas more likely to survive than others?

____

____

GENETICS AND EVOLUTION

Darwin knew that organisms inherit traits, but not how they inherit traits. He also knew that there is great variation among organisms, but not how that variation happens. Today, scientists know that genes determine the traits of an organism. These genes are exchanged and passed on from parent to offspring.

Name ______________________ Class ______________ Date ______________

Section 2 Review

NSES LS 2a, 2b, 2e, 3d, 5a, 5b

SECTION VOCABULARY

natural selection the process by which individuals that are better adapted to their environment survive and reproduce more successfully than less well adapted individuals do; a theory to explain the mechanism of evolution

selective breeding the human practice of breeding animals or plants that have certain desired traits

trait a genetically determined characteristic

1. **Explain** How did the ideas in Charles Lyell's book affect Darwin's thinking about evolution?

____

____

____

2. **Identify** In what way are the different finch species of the Galápagos Islands adapted to the different environments on the islands?

____

____

3. **Compare** How is natural selection different from selective breeding?

____

____

____

____

4. **Describe** How did Darwin apply Malthus's ideas about human populations to the theory of evolution by natural selection?

____

____

____

____

5. **List** What are the four steps of natural selection?

____

____

____

____

Name ______________________ Class ______________ Date ______________

CHAPTER 7 The Evolution of Living Things

SECTION 3

Natural Selection in Action

BEFORE YOU READ

After you read this section, you should be able to answer these questions:

- Why do populations change?
- How do new species form?

National Science Education Standards

LS 2a, 2e, 3d, 4d, 5b

Why Do Populations Change?

The theory of evolution by natural selection explains how changes in the environment can cause populations to change. Organisms that are well-adapted to their environment survive to reproduce. Organisms that are less well-adapted do not.

Some environmental changes are caused by people. Others happen naturally. No matter how they happen, though, environmental changes can cause populations to change.

STUDY TIP

Summarize As you read, underline the important ideas in each paragraph. When you finish reading, write a short summary of the section using the ideas that you underlined.

ADAPTATION TO HUNTING

Hunting is one of the factors that can affect the survival of animals. In Africa, people hunt male elephants for their tusks, which are made of ivory. Because of natural genetic variations, some male elephants do not grow tusks. People do not hunt these tuskless elephants, so tuskless elephants tend to live longer than elephants with tusks. Therefore, tuskless elephants are more likely to reproduce and pass the tuskless trait to their offspring. ☑

Over time, the tuskless trait has become more common. For example, in 1930, about 99% of male elephants in one area had tusks. Today, only about 85% of male elephants in that area have tusks.

1. Explain Why are tuskless elephants more likely to reproduce in Africa?

____

____

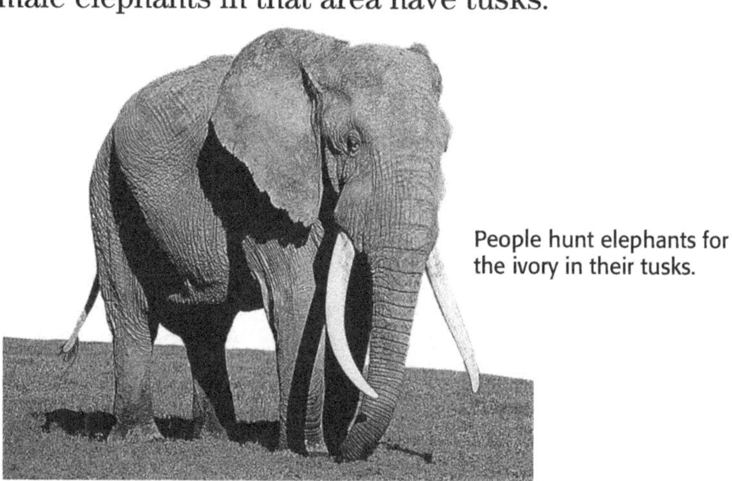

People hunt elephants for the ivory in their tusks.

INSECTICIDE RESISTANCE

Many people use chemicals to control insect pests. These chemicals, called *insecticides*, kill insects. Sometimes, an insecticide that used to work well no longer affects an insect population. The insect population has evolved a resistance to the insecticide. This happens by natural selection, as shown in the figure below. ☑

READING CHECK

2. Define What is an insecticide?

____

____

1 When it is first used, the insecticide kills most of the insects. However, a few insects have genes that make them resistant to the insecticide. These insects survive.

TAKE A LOOK

3. Infer What would happen to the population of insects if none were resistant to the insecticide?

____

2 The insects that are resistant to the insecticide pass on their genes to their offspring. Over time, almost all of the insects in the population have the insecticide-resistance gene.

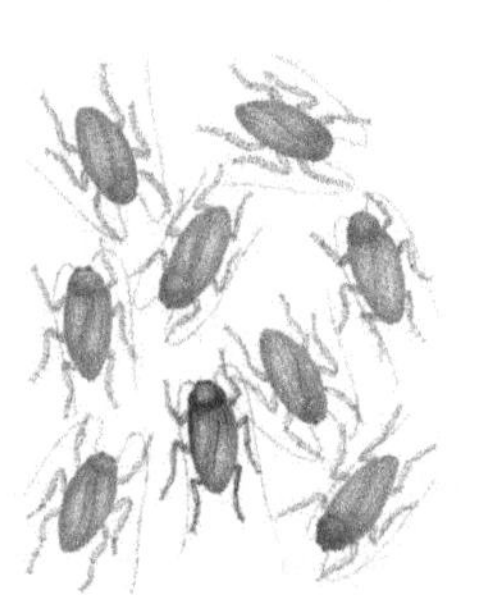

3 When the same insecticide is used on the insects, only a few of the insects are killed. This is because most of the insects are resistant to the insecticide.

Insect populations can evolve quickly. This happens for two reasons: insects have many offspring and they have a short generation time. **Generation time** is the average time between one generation and the next. In general, the longer the generation time for a population, the slower the population can evolve.

Critical Thinking

4. Apply Concepts The females of a certain species of mammal prefer to mate with less-colorful males. What will probably happen to the proportion of colorful males in the population with time?

____

____

COMPETITION FOR MATES

Organisms that reproduce sexually have to compete with one another for mates. For example, many female birds prefer to mate with colorful males. This means that colorful males have more offspring than less-colorful males. In most organisms, color is a genetic trait that is passed on to offspring. Therefore, colorful male birds are likely to produce colorful offspring. Over time, the proportion of colorful birds in the population will increase.

How Does Natural Selection Make New Species?

The formation of a new species as a result of evolution is called **speciation**. Three events often lead to speciation: separation, adaptation, and division. ☑

READING CHECK

5. Define Write the definition of speciation in your own words.

____

____

____

SEPARATION

Speciation may begin when a part of a population becomes separated from the rest. This can happen in many ways. For example, a newly formed canyon, mountain range, or lake can separate the members of a population.

ADAPTATION

After two groups have been separated, each group continues to be affected by natural selection. Different environmental factors may affect each population. Therefore, different traits can be favored in each population. Over many generations, different traits may spread through each population. ☑

READING CHECK

6. Explain Why may separated populations develop different traits?

____

____

____

____

____

DIVISION

Natural selection can cause two separated populations to become very different from each other. With time, the members of the two populations may be unable to mate successfully. The two populations may then be considered different species. The figure below shows how species of Galápagos finches may have evolved through separation, adaptation, and division.

1 Separation Some finches left the South American mainland and reached one of the Galápagos Islands.

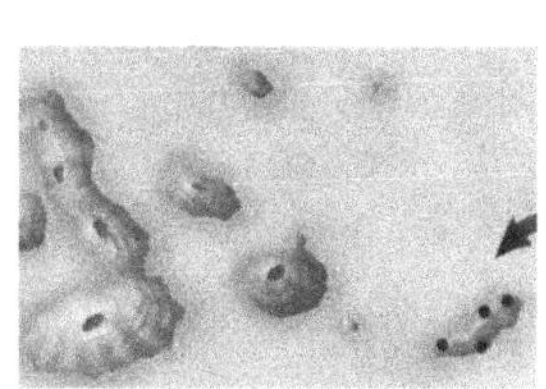

2 Adaptation The finches on the island reproduced. Over time, they adapted to the environment on the island.

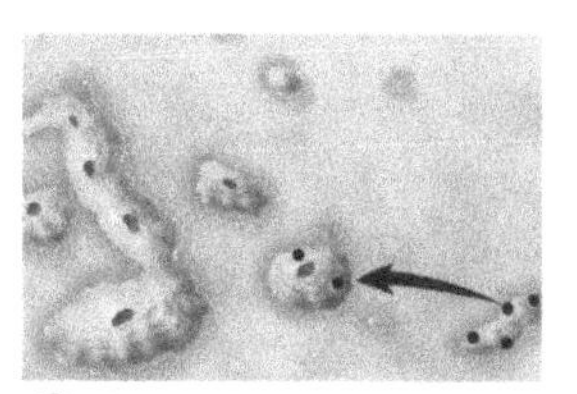

3 Separation Some finches flew to a second island.

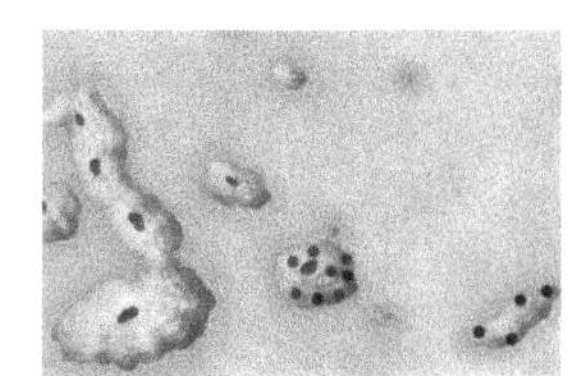

4 Adaptation These finches reproduced on the second island. Over time, they adapted to the second island's environment.

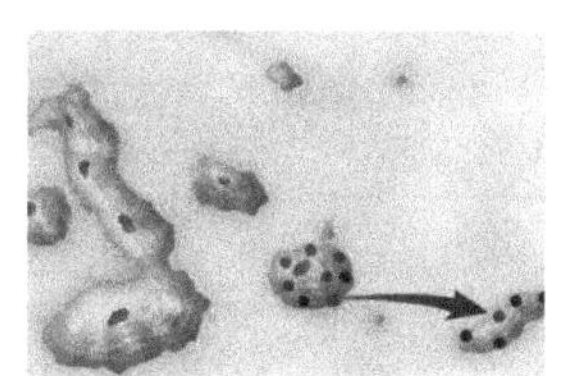

5 Division After many generations, the finches on the second island were unable to successfully mate with the finches on the first island. The populations of finches on the two islands had become different species.

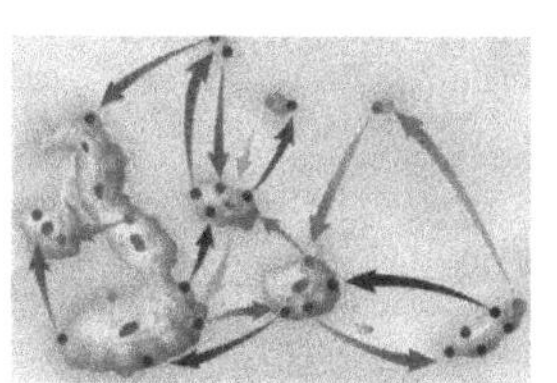

6 Speciation This process may have happened many times as finches flew to the different islands in the Galápagos.

TAKE A LOOK

7. Identify Where did all of the finches on the Galápagos Islands originally come from?

____

Name ______________________ Class ______________ Date ______________

Section 3 Review

NSES **LS 2a, 2e, 3d, 4d, 5b**

SECTION VOCABULARY

generation time the period between the birth of one generation and the birth of the next generation	**speciation** the formation of new species as a result of evolution

1. List What are three events that can lead to speciation?

____

____

____

2. Infer What kinds of environmental factors may affect organisms that live on a rocky beach? Give three examples.

____

____

____

3. Identify Give three examples of things that can cause groups of individuals to become separated.

____

____

____

4. Explain Why can insects adapt to pesticides quickly? Give two reasons.

____

____

5. Apply Concepts Which of the organisms described below can probably evolve more quickly in response to environmental changes? Explain your answer.

Organism	Generation time	Average number of offspring per generation
A	6 years	50
B	2 years	100
C	10 years	5

____

____

____

Name ______________________ Class ______________ Date ______________

CHAPTER 8 The History of Life on Earth

SECTION 1

Evidence of the Past

BEFORE YOU READ

After you read this section, you should be able to answer these questions:

- What are fossils?
- What is the geologic time scale?
- How have conditions on Earth changed with time?

National Science Education Standards
LS 1a, 3d, 5b, 5c

What Are Fossils?

In 1995, geologist Paul Sereno found a dinosaur skull in a desert in Africa. The skull was 1.5 m long. The dinosaur it came from may have been the largest land predator that ever existed!

Scientists like Paul Sereno look for clues to help them learn what happened on Earth in the past. These scientists are called paleontologists. *Paleo* means "old," and *onto* means "life." Therefore, *paleontologists* study "old life," or life that existed in the past. One of the most important ways paleontologists learn about the past is by studying fossils.☑

A **fossil** is the traces or remains of an organism that have been preserved over thousands of years. Some fossils form when an organism's body parts, such as shells or bones, are preserved in rock. Other fossils are signs, such as imprints, that an organism once existed. The figure below shows one way that this kind of fossil can form.

STUDY TIP
Organize After you read this section, make a Concept Map using the vocabulary terms from the section.

1. Define What is a paleontologist?

____
____
____

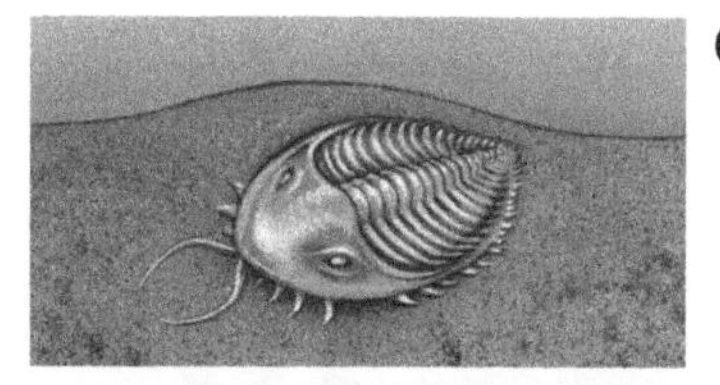

❶ An organism dies and is buried by sediment, such as mud or clay.

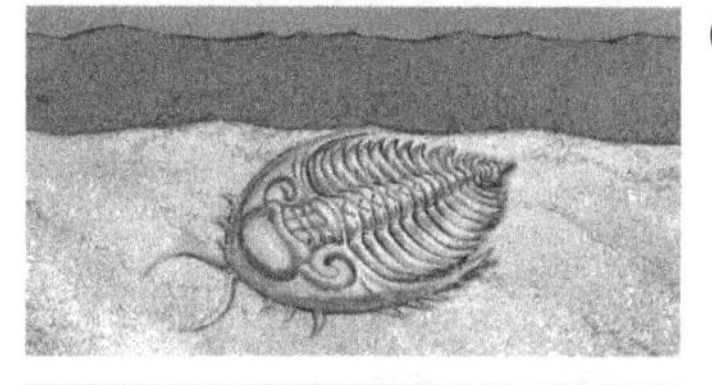

❷ After the organism dies, its body rots away. An imprint, or *mold*, is left in the sediment. The mold has the same shape as the organism's body.

❸ Over time, the mold fills with a different kind of sediment. The sediment forms a *cast* of the organism. The sediment can turn into rock, preserving the cast for millions of years.

TAKE A LOOK

2. Infer If a paleontologist found the cast of this fossil, would the paleontologist be able to tell what the inside of the organism looked like? Explain your answer.

____
____
____
____

How Do Scientists Know the Ages of Fossils?

There are two main methods that scientists use to learn how old a fossil is: relative dating and absolute dating. During **relative dating**, a scientist estimates how old a rock or fossil is compared to other rocks or fossils. ☑

READING CHECK

3. Define What is relative dating?

Almost all fossils are found in a kind of rock called *sedimentary rock*. Sedimentary rocks form in layers. In most bodies of sedimentary rock, the oldest layers are found at the bottom. The newest layers are found at the top. Scientists can use this information to estimate the relative ages of fossils in the layers. Fossils in the bottom layers are usually older than fossils in the top layers.

The second way that scientists learn the age of fossils is by absolute dating. During **absolute dating**, scientists learn the age of a fossil in years. Absolute dating is more precise than relative dating.

Critical Thinking

4. Compare How is absolute dating different from relative dating? Give two ways.

One kind of absolute dating uses atoms. *Atoms* are tiny particles that make up all matter. Some atoms are unstable and can *decay*, or break down. When an atom decays, it becomes a different kind of atom. It may also give off smaller particles, energy, or both. The new atom that forms is stable—it does not decay.

Each kind of unstable atom decays at a certain rate. This rate is called the atom's *half-life*. Scientists know the half-lives of many different kinds of unstable atoms. They can measure the ratio of these unstable atoms to stable atoms in a rock. They can use this ratio to calculate the age of the rock and any fossils in it.

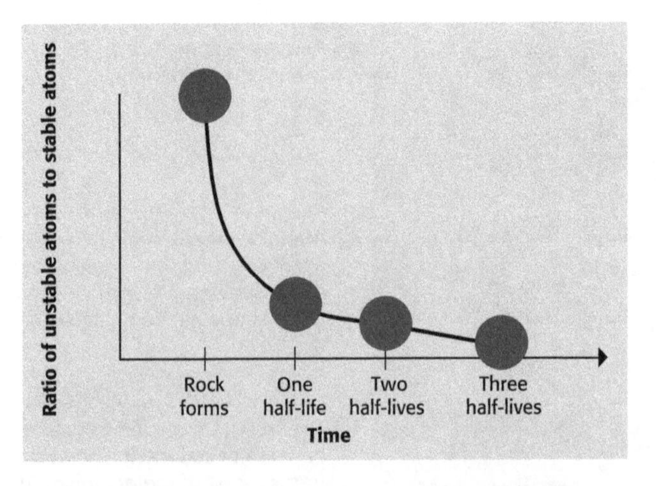

This figure shows how scientists can use half-lives to date rocks. After each half-life, the ratio of unstable to stable atoms in the rock decreases. Scientists can learn the age of the rock by measuring the ratio of unstable to stable atoms in it.

Math Focus

5. Read a Graph How does the ratio of unstable to stable atoms change over time?

What Is the Geologic Time Scale?

Think about some important things that have happened to you. You probably know the day, month, and year in which many of these events happened. The divisions of the calendar make it easy for you to remember these things. They also make it easy for you to talk to someone else about things that happened in the past.

In a similar way, geologists use a type of calendar to divide the Earth's history. This calendar is called the **geologic time scale**. The Earth is very old. Therefore, the divisions of time in the geologic time scale are very long.

Geologists use the geologic time scale to help make sense of how life on Earth has changed. When a new fossil is discovered, geologists figure out how old it is. Then, they place it in the geologic time scale at the correct time. In this way, they can construct a history of life on Earth. The figure below shows part of the geologic time scale.

Era	Period	Millions of years ago
Cenozoic era	Quaternary	1.8
	Tertiary	65.5
Mesozoic era	Cretaceous	146
	Jurassic	200
	Triassic	251
Paleozoic era	Permian	299
	Carboniferous	359
	Devonian	416
	Silurian	444
	Ordovician	488
	Cambrian	542
Precambrian time		4,600

Critical Thinking

6. Predict Consequences All geologists use the same basic geologic time scale. What might happen if every geologist used a different geologic time scale to study the Earth's history? Explain your answer.

____
____
____
____
____
____
____
____

Investigate Many geologic time-scale divisions are named after certain places on Earth. Research one of the periods in the Paleozoic era to find out where its name comes from. Share your findings with a small group.

TAKE A LOOK

7. List Give the three periods in the Mesozoic era in order from oldest to most recent.

____
____
____

STANDARDS CHECK

LS 5c Extinction of a species occurs when the environment changes and the adaptive characteristics of a species are insufficient to allow its survival. Fossils indicate that many organisms that lived long ago are extinct. Extinction of species is common; most of the species that have lived on Earth no longer exist.

Word Help: occur
to happen

Word Help: environment
the surrounding natural conditions that affect an organism

Word Help: insufficient
not enough

Word Help: survival
the continuing to live or exist; the act of continuing to live

Word Help: indicate
to be or give a sign of; to show

8. Infer How could global cooling cause a species to go extinct?

____

____

____

____

How Do Geologists Divide the Geologic Time Scale?

Many of the divisions of the geologic time scale are based on information from fossils. As scientists discover new fossils, they may add information to the geologic time scale. For example, most fossils that have been found are from organisms that have lived since Precambrian time. Therefore, little is known about life on Earth before this time. As scientists find more fossils from Precambrian time, more information may be added to the time scale.

Most of the divisions in the geologic time scale mark times when life on Earth changed. During many of these times, species died out completely, or became **extinct**. A *mass extinction* happens when many species go extinct at the same time.

Scientists do not know what causes all mass extinctions. The table below gives some events that may cause mass extinctions.

Event	How it causes mass extinction
Comet or meteorite impact	can throw dust and ash into the atmosphere, causing global cooling
Many volcanic eruptions	can throw dust and ash into the atmosphere, causing global cooling
Changing ocean currents	can cause climate change by changing how heat is distributed on Earth's surface

How Has the Earth's Surface Changed?

Geologists have found fossils of tropical plants in Antarctica. These kinds of plants cannot live in Antarctica today because the climate is too cold. The fossils show that Antarctica's climate must once have been warmer.

In the early 1900s, German scientist Alfred Wegener studied the shapes of the continents and the fossils found on them. From his observations, he proposed a hypothesis that all the continents were once joined together. They formed a large land mass called *Pangaea*. Since then, the continents have moved over the Earth's surface to their present locations.

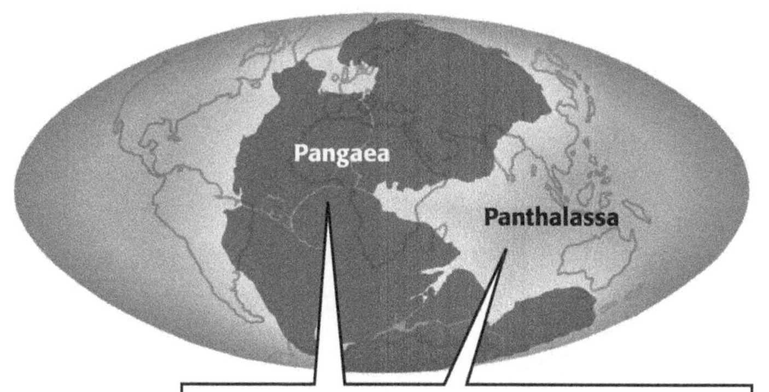

About 245 million years ago, all of the continents were joined into one large land mass called *Pangaea*. It was surrounded by a huge sea called *Panthalassa*, which is Greek for "all seas."

TAKE A LOOK

9. Infer On the map, circle the part of Pangaea that probably became the continent of Africa.

MOVING CONTINENTS

In the 1960s and 1970s, new technology allowed geologists to learn more about the Earth's crust. They found that the crust is not one solid chunk of rock. Instead, it is broken into many pieces called *tectonic plates*. These plates move slowly over the Earth's surface. Some of the plates have continents on them. As the plates move, the continents are carried along. The theory of how the plates move is called the theory of **plate tectonics**. ☑

READING CHECK

10. Explain Why do continents move?

____

____

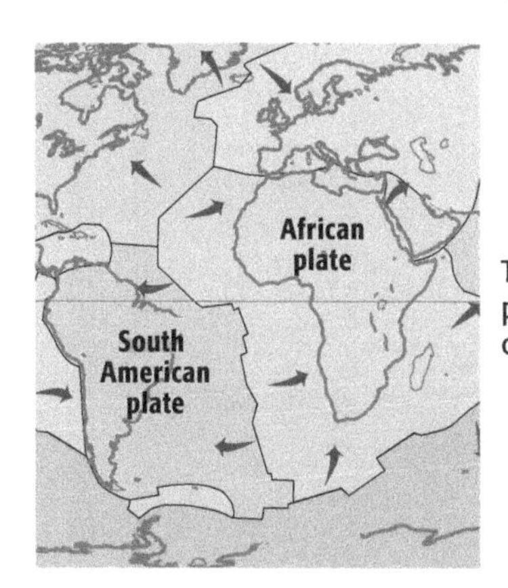

The continents ride on tectonic plates. The plates move slowly over the Earth's surface.

The theory of plate tectonics explains why fossils of tropical plants have been found in Antarctica. Millions of years ago, Antarctica was located near the equator. Over time, the plate containing Antarctica has moved to its current location.

EFFECTS OF PLATE MOTIONS ON LIVING THINGS

Sudden changes on Earth's surface, such as a meteorite impact, may cause mass extinctions. Slower changes, such as tectonic plate movements, give populations of organisms time to adapt. Geologists use fossils of organisms to study how populations have adapted to changes on Earth.

Name ______________________ Class ______________ Date ______________

Section 1 Review

NSES LS 1a, 3d, 5b, 5c

SECTION VOCABULARY

absolute dating any method of measuring the age of an event or object in years

extinct describes a species that has died out completely

fossil the trace or remains of an organism that lived long ago, most commonly preserved in sedimentary rock

geologic time scale the standard method used to divide the Earth's long natural history into manageable parts

plate tectonics the theory that explains how large pieces of the Earth's outermost layer, called tectonic plates, move and change shape

relative dating any method of determining whether an event or object is older or younger than other events or objects

1. Define Write your own definition for *geologic time scale.*

____

____

2. List Give three things that can cause mass extinctions.

____

____

____

3. Compare Which fossil in the rock layers below is probably the oldest? Explain your answer.

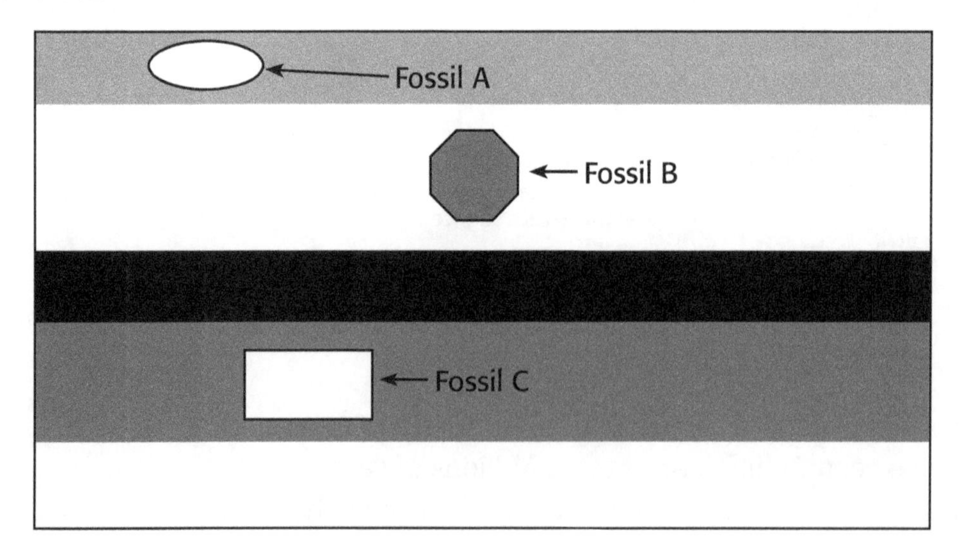

____

4. Explain Describe one way that a fossil can form.

____

____

____

Name ______________________ Class ______________ Date ______________

SECTION 2

Eras of the Geologic Time Scale

BEFORE YOU READ

After you read this section, you should be able to answer these questions:

- What kinds of organisms evolved during Precambrian time?
- What kinds of organisms evolved during the Paleozoic, Mesozoic, and Cenozoic eras?

National Science Education Standards

LS 1a, 1b, 3d, 5a, 5b, 5c

What Happened During Precambrian Time?

Remember that geologists divide the Earth's history into four main parts. These parts are Precambrian time, the Paleozoic era, the Mesozoic era, and the Cenozoic era. During each part of Earth's history, different species arose and evolved.

Precambrian time was the longest part of Earth's history. It lasted from the time the Earth formed, 4.6 billion years ago, until about 542 million years ago. Life on Earth began during this time.

Scientists think that the early Earth was very different than the Earth today. On the early Earth, the atmosphere contained very little oxygen. It was made mostly of carbon dioxide, water vapor, and nitrogen gas. In addition, volcanic eruptions, meteorite and comet impacts, and severe storms were common. There was no ozone layer, so more ultraviolet radiation reached the Earth's surface.

Scientists think that life developed from simple chemicals in the oceans and the atmosphere. According to this hypothesis, energy from radiation and storms caused these chemicals to react. The reactions produced more complex molecules. These complex molecules combined to produce structures such as cells. ☑

This is what the early Earth may have looked like. An artist drew this picture based on information from scientists.

STUDY TIP

Organize As you read this section, make a chart showing the kinds of organisms that evolved during each of the four main parts of Earth's history.

Math Focus

1. Calculate About how long did Precambrian time last? Round your answer to the nearest hundred million years.

____

READING CHECK

2. Identify How do scientists think complex molecules formed on the early Earth?

____

____

____

PHOTOSYNTHESIS AND OXYGEN

The first living things on Earth were probably *prokaryotes*, or single-celled organisms without nuclei. These early organisms did not need oxygen to survive. However, more than 3 billion years ago, organisms called *cyanobacteria* began to develop. Cyanobacteria perform photosynthesis. During *photosynthesis*, a living thing uses sunlight, water, and carbon dioxide to produce food and oxygen. ☑

READING CHECK

3. Define What is photosynthesis?

____

____

____

____

As the cyanobacteria carried out photosynthesis, they released oxygen into the atmosphere. This caused the amount of oxygen in the atmosphere to increase.

Some of the oxygen formed the ozone layer in the upper atmosphere. The ozone layer absorbs much of the ultraviolet radiation from the sun. Ultraviolet radiation is harmful to most living things. Before the ozone layer formed, life existed only in the oceans and underground. The new ozone layer reduced the radiation reaching Earth's surface. As a result, organisms were able to live on land.

TAKE A LOOK

4. Explain Where did the oxygen come from that formed the ozone layer?

____

____

____

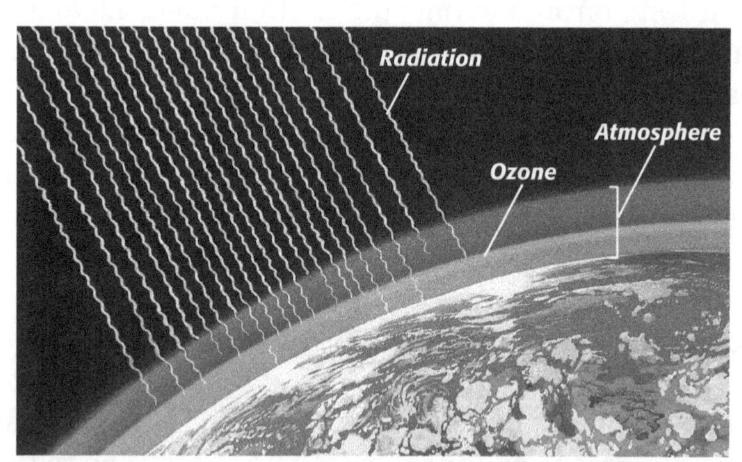

Cyanobacteria released oxygen into the atmosphere during photosynthesis. Some of this oxygen formed the ozone layer, which absorbs much of the harmful radiation from the sun.

MULTICELLULAR ORGANISMS

About 2 billion years ago, single-celled organisms that were larger and more complex than prokaryotes evolved. These organisms were *eukaryotes*. They had nuclei and other complex structures in their cells. These early eukaryotic cells probably evolved into organisms with many cells.

What Happened During the Paleozoic Era?

The **Paleozoic era** began about 542 million years ago. It ended about 251 million years ago. *Paleozoic* comes from Greek words meaning "ancient life." The Paleozoic era is the second-longest part of the Earth's history. However, it was less than 10% as long as Precambrian time.

Era	Period	Millions of years ago
Paleozoic era	Permian	299
	Carboniferous	359
	Devonian	416
	Silurian	444
	Ordovician	488
	Cambrian	542

TAKE A LOOK

5. List Give the six periods in the Paleozoic era in order from oldest to most recent.

____

____

____

____

____

____

LIFE IN THE PALEOZOIC ERA

Rocks from the Paleozoic era contain many fossils of *multicellular* organisms, or organisms with many cells. These organisms all evolved during the Paleozoic era. Most of the organisms that lived during the early Paleozoic era lived in the oceans. Some of the animals include sponges, corals, snails, clams, fishes, and sharks. ☑

READING CHECK

6. Identify Where did most animals live during the early Paleozoic?

____

During the Paleozoic era, plants and animals began to live on land. By the end of the era, forests of giant ferns, club mosses, and conifers covered much of the Earth. All of the major plant groups except for flowering plants developed during the Paleozoic. These plants provided food and shelter for animals.

Fossils indicate that crawling insects were some of the first animals to live on land. By the end of the Paleozoic, amphibians, reptiles, and winged insects had evolved. ☑

7. Identify What kinds of animals were probably the first to live on land?

____

THE END OF THE PALEOZOIC

The end of the Paleozoic is marked by the largest known mass extinction. During this extinction, as many as 90% of marine species died out. Scientists are not sure what caused this extinction.

Era	Period	Millions of years ago
Mesozoic era	Cretaceous	146
	Jurassic	200
	Triassic	251

What Happened During the Mesozoic Era?

The **Mesozoic era** lasted from about 251 million years ago to about 65.5 million years ago. *Mesozoic* comes from Greek words that mean "middle life." Many species of reptiles evolved during the Mesozoic era. They became the dominant, or main, organisms on Earth. The Mesozoic era is sometimes called the "Age of Reptiles." ☑

8. Explain Why is the Mesozoic era sometimes called the "Age of Reptiles"?

____

____

____

LIFE IN THE MESOZOIC ERA

Dinosaurs are the most well-known reptiles that evolved during the Mesozoic. They dominated the Earth for about 150 million years. As you may know, there were many different species of dinosaurs. Some were taller than a 10-story building. Others were as small as modern chickens.

In addition to dinosaurs, the first birds and mammals evolved during the Mesozoic. In fact, many scientists think that birds evolved from some kinds of dinosaurs.

The most important land plants during the early Mesozoic era were conifers. These conifers formed huge forests. Flowering plants evolved later in the Mesozoic.

Critical Thinking

9. Infer Scientists sometimes call birds "living dinosaurs." Why do you think this is?

____

____

____

THE END OF THE MESOZOIC ERA

About 65.5 million years ago, dinosaurs and many other species became extinct. Most scientists think that this extinction was caused by the impact of a large meteorite or comet. ☑

The impact produced huge amounts of dust and ash. This blocked sunlight from reaching Earth's surface. It also caused global temperatures to drop for many years. Without sunlight, plants began to die. Plant-eating dinosaurs soon died out because they had nothing to eat. Then, meat-eating dinosaurs died out.

READING CHECK

10. Identify What do most scientists think caused the extinction of the dinosaurs?

____

What Has Happened During the Cenozoic Era?

The **Cenozoic era** began about 65.5 million years ago and continues today. *Cenozoic* comes from Greek words meaning "recent life." During the Cenozoic era, mammals have become the dominant organisms. Therefore, the Cenozoic era is sometimes called the "Age of Mammals."

Era	Period	Millions of years ago
Cenozoic era	Quaternary	1.8
	Tertiary	65.5

TAKE A LOOK

11. Identify How long ago did the Cenozoic era start?

____

Scientists have more information about the Cenozoic era than about any of the other parts of Earth's history. Fossils of organisms from this era formed more recently than fossils from earlier times. Therefore, most fossils of organisms from the Cenozoic era are near the surface. This makes them easier to find. In addition, the fossils from the Cenozoic era are less likely to have been destroyed by weathering and erosion. ☑

READING CHECK

12. Describe Give two reasons that scientists know more about the Cenozoic era than about other times in Earth's history.

____

____

____

____

____

____

LIFE IN THE CENOZOIC ERA

During the early parts of the Cenozoic era, most mammals were small. They lived in forests. Later in the era, larger mammals evolved. These included mastodons, saber-toothed cats, camels, and horses. Some had long legs for running. Some had special teeth for eating certain kinds of food. Some had large brains.

We are currently living in the Cenozoic era. Humans evolved during this era. The environment, landscapes, and organisms around us today are part of this era.

The climate has changed many times during the Cenozoic era. During some periods of time, the climate was much colder than it is today. These periods of time are known as *ice ages*. During ice ages, ice sheets and glaciers grew larger. Many organisms migrated toward the equator to survive the cold. Others adapted to the cold. Some species became extinct.

Name ______________________ Class ______________ Date ______________

Section 2 Review

NSES LS 1a, 1b, 3d, 5a, 5b, 5c

SECTION VOCABULARY

Cenozoic era the current geologic era, which began 65.5 million years ago; also called the Age of Mammals

Mesozoic era the geologic era that lasted from 251 million to 65.5 million years ago; also called the Age of Reptiles

Paleozoic era the geologic era that followed Precambrian time and that lasted from 542 million to 251 million years ago

Precambrian time the interval of time in the geologic time scale from Earth's formation to the beginning of the Paleozoic era, from 4.6 billion to 542 million years ago

1. **List** Give the four main parts of Earth's history in order from oldest to most recent.

____

____

2. **Identify** During which era did multicellular organisms evolve?

____

3. **Compare** Give three ways the early Earth was different from the Earth today.

____

____

____

4. **Describe** How did the Earth's atmosphere change because of cyanobacteria?

____

____

____

5. **Identify** What kind of event marks the end of both the Paleozoic era and the Mesozoic era?

____

6. **Infer** Why might birds and mammals have been more likely than reptiles to survive the events that caused the extinction of the dinosaurs?

____

____

7. **Describe** What happens during an ice age?

____

____

SECTION 3

Humans and Other Primates

BEFORE YOU READ

After you read this section, you should be able to answer these questions:

- What are two features of primates?
- How have hominids changed through time?

National Science Education Standards
LS 1a, 3d, 5a, 5b, 5c

How Are Humans Similar to Apes?

Humans, apes, and monkeys share many common features. In addition, scientists have found many fossils of organisms with features of both humans and apes. Therefore, scientists agree that humans, apes, and monkeys share a common ancestor. The species of humans, apes, and monkeys that are alive today all evolved from a single, earlier species. This species probably lived more than 45 million years ago.

Humans, monkeys, apes, and lemurs are all part of a group of mammals called **primates**. There are two features that primates share. First, a primate's eyes are located at the front of its head. Both eyes look in the same direction. This gives the primate *binocular*, or three-dimensional, vision.

A second feature primates share is flexible fingers. Almost all primates have *opposable thumbs*. This means that the thumb can move to touch each finger. It is not fixed in place like the toes of a dog or cat. Opposable thumbs help primates grip objects firmly.

STUDY TIP

Summarize As you read this section, make a timeline that shows how hominids have changed over time.

STANDARDS CHECK

LS 5a Millions of species of animals, plants, and microorganisms are alive today. Although different species might look dissimilar, the unity among organisms becomes apparent from an analysis of internal structures, the similarity of their chemical processes, and the evidence of common ancestry.

Word Help: structure
a whole that is built or put together from parts

Word Help: process
a set of steps, events, or changes

Word Help: evidence
information showing whether an idea or belief is true or valid

1. Identify What are two features that humans share with apes and monkeys?

____

____

____

A primate's eyes both point in the same direction. This gives the primate three-dimensional vision.

Primates have flexible fingers. Almost all primates, including humans and these orangutans, have opposable thumbs. Most primates also have opposable big toes, although humans do not.

THE FIRST PRIMATES

The ancestors of primates probably lived at the same time as the dinosaurs. They probably lived in trees and ate insects. They were small and looked similar to mice. The first primates evolved during the early Cenozoic era. About 45 million years ago, primates with larger brains evolved. These primates were the first to share features with monkeys, apes, and humans. ☑

2. Identify When did the first primates evolve?

____

____

APES AND CHIMPANZEES

The chimpanzee is a kind of ape. Scientists think that the chimpanzee is the closest living relative of humans. This does not mean that humans evolved from chimpanzees. It means that humans and chimpanzees share a more recent common ancestor than humans and other apes. Between 30 million and 6 million years ago, ancestors of humans, chimpanzees, and other apes began to evolve different features. ☑

READING CHECK

3. Identify What organism do scientists think is the closest living relative of humans?

____

HOMINIDS

Humans are part of a group of primates called hominids. The **hominid** family includes only humans and their human-like ancestors. Humans are the only species of hominids that are still living.

The main feature that separates hominids from other primates is bipedalism. *Bipedalism* means "walking on two feet." Humans are bipedal. Other primates are not bipedal. They mainly move around with all four limbs touching the ground.

The skeletons of humans and gorillas are similar. However, they are not exactly the same.

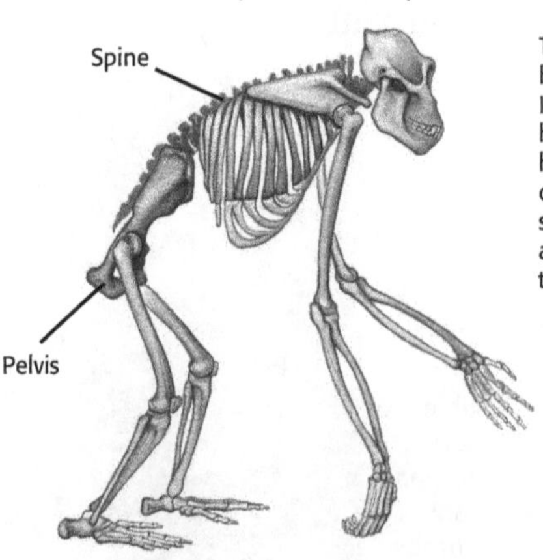

▲ The pelvis of this gorilla tilts the gorilla's upper body forward. A gorilla's spine is curved in a "C" shape. The gorilla's arms are long and touch the ground as it moves.

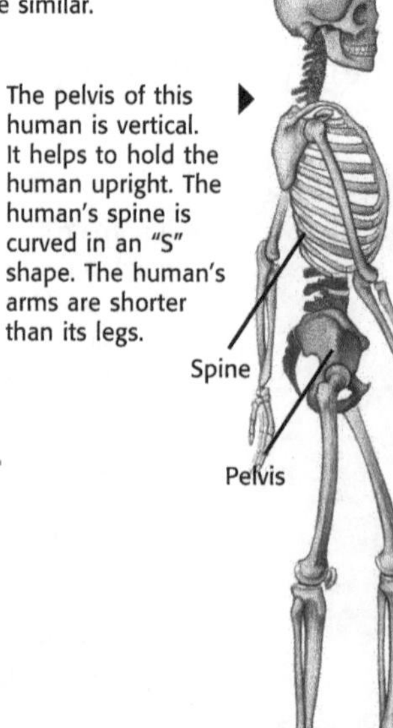

The pelvis of this human is vertical. It helps to hold the human upright. The human's spine is curved in an "S" shape. The human's arms are shorter than its legs. ▶

TAKE A LOOK

4. Compare How is the shape of a human spine different from that of a gorilla?

____

____

____

How Have Hominids Changed Through Time?

Scientists continue to find fossils of hominids. These fossils help scientists piece together the hominid family tree. As scientists find more fossils, they will better understand how modern humans evolved.

THE EARLIEST HOMINIDS

The earliest hominids had some humanlike features. Like all hominids, they could walk upright. They also had smaller teeth, flatter faces, and larger brains than earlier primates.

The oldest hominid fossils are more than 7 million years old. They have been found in Africa. Therefore, scientists think that hominids first evolved in Africa. ☑

READING CHECK

5. Explain Why do scientists think that hominids first evolved in Africa?

____

____

These are footprints from an early hominid. They are about 3.6 million years old. Anthropologist Mary Leaky discovered these footprints in Tanzania, Africa.

TAKE A LOOK

6. Apply Concepts Did the hominid that made these prints walk on two feet or four? Explain your answer.

____

____

Scientists have classified many early hominids as australopithecines. *Australopithecines* were apelike hominids. They had slightly larger brains than apes and may have used simple stone tools.

A VARIETY OF EARLY HOMINIDS

Many australopithecines and other types of hominids lived at the same time, just as many species of apes live today. Some australopithecines had slender bodies. They had humanlike jaws and teeth but small, apelike skulls. They probably lived in forests and grasslands. Scientists think that some of these australopithecines were the ancestors of modern humans. ☑

READING CHECK

7. Identify What type of hominid do scientists think was the ancestor of modern humans?

____

Some early hominids had large bodies and massive teeth and jaws. They had fairly small brains. They probably lived in tropical forests. Scientists think that these large-bodied hominids were probably not the ancestors of modern humans.

THE GROUP *HOMO*

About 2.4 million years ago, a new group of hominids evolved. These hominids were similar to the slender australopithecines, but were more humanlike. They had larger and more complex brains, rounder skulls, and flatter faces. They were probably scavengers, eating many different kinds of food. They may have migrated or changed the way they lived to adapt to climate change. ☑

READING CHECK

8. Compare How were the hominids that evolved 2.4 million years ago different from earlier australopithecines?

____

____

____

These new hominids were members of the group *Homo*, which includes modern humans. Fossil evidence shows that several different species of *Homo* may have lived at the same time on several continents. One of these species was *Homo habilis*, which evolved about 2.4 million years ago. Another species, *Homo erectus*, evolved about 1.8 million years ago. Members of the *Homo erectus* species could grow as tall as modern humans.

TAKE A LOOK

9. Explain How do scientists infer what early hominids looked like?

____

____

This is an artist's idea of what *Homo erectus* looked like. Artists and scientists work together to produce models like this. They used information from bones to infer what *Homo erectus* may have looked like.

When Did Modern Humans Evolve?

Until about 30,000 years ago, two types of hominids may have lived in the same areas at the same time. Both had large brains and made advanced tools, clothing, and art. Scientists think that one of these early hominids was the same species as modern humans.

Critical Thinking

10. Infer What kinds of evidence may scientists have used to determine that early hominids used tools and wore clothing?

NEANDERTHALS

One type of recent hominid is known as the *Neanderthal.* Neanderthals lived in Europe and western Asia. They may have evolved as early as 230,000 years ago. Neanderthals hunted large animals, made fires, and wore clothing. They may have cared for their sick and buried their dead. About 30,000 years ago, Neanderthals disappeared. Scientists do not know why they went extinct.

HOMO SAPIENS

Modern humans are members of the species **Homo sapiens**. The earliest members of *Homo sapiens* probably evolved in Africa 150,000 to 100,000 years ago. Some members of this species migrated out of Africa between 100,000 and 40,000 years ago. Compared to Neanderthals, *Homo sapiens* have smaller and flatter faces and more rounded skulls. *Homo sapiens* is the only species of hominid that is still alive. ☑

READING CHECK

11. Identify How many species of *Homo* are still alive?

Like modern humans, early *Homo sapiens* produced large amounts of art. They made sculptures, carvings, and paintings. Scientists have also found preserved villages and burial grounds from early *Homo sapiens.* These remains show that these early humans had an organized and complex society, like humans today.

This photo shows a museum recreation of early *Homo sapiens.*

Name ______________________ Class ______________ Date ____________

Section 3 Review

NSES **LS 1a, 3d, 5a, 5b, 5c**

SECTION VOCABULARY

hominid a type of primate characterized by bipedalism, relatively long lower limbs, and lack of a tail; examples include humans and their ancestors

Homo sapiens the species of hominids that includes modern humans and their closest ancestors and that first appeared about 100,000 to 150,000 years ago

primate a type of mammal characterized by opposable thumbs and binocular vision

1. Identify What feature separates hominids from other primates?

____

2. Identify When did humans, chimpanzees, and other apes begin to evolve different features?

____

3. Describe Give three features of the australopithecines that scientists think evolved into modern humans.

____

____

____

4. List Give three species of *Homo* and tell when each evolved.

____

____

____

5. Compare How are *Homo sapiens* different from Neanderthals? Give two ways.

____

____

6. Explain How do scientists know that many species of hominids once existed?

____

7. Infer What advantages could larger brains have given hominids? Give three examples.

____

____

____

Name ______________________ Class ______________ Date ______________

Sorting It All Out

BEFORE YOU READ

After you read this section, you should be able to answer these questions:

- What is classification?
- How do scientists classify organisms?
- How do scientists name groups of organisms?

National Science Education Standards
LS 5a

Why Do We Classify Things?

Imagine that you lived in a tropical rain forest and had to get your own food, shelter, and clothing from the forest. What would you need to know to survive? You would need to know which plants were safe to eat and which were not. You would need to know which animals you could eat and which ones could eat you. In other words, you would need to study the organisms around you and put them into useful groups. You would *classify* them.

Biologists use a *classification* system to group the millions of different organisms on Earth. **Classification** is putting things into groups based on characteristics the things share. Classification helps scientists answer several important questions:

- What are the defining characteristics of each species?
- When did the characteristics of a species evolve?
- What are the relationships between different species?

Organize As you read, make a diagram to show the eight-level system of organization.

Discuss With a partner, describe some items at home that you have put into groups. Explain why you grouped them and what characteristics you used.

How Do Scientists Classify Organisms?

What are some ways we can classify organisms? Perhaps we could group them by where they live or how they are useful to humans. Throughout history, people have classified organisms in many different ways.

In the 1700s, a Swedish scientist named Carolus Linnaeus created his own system. His system was based on the structure or characteristics of organisms. With his new system, Linnaeus founded modern taxonomy. **Taxonomy** is the science of describing, classifying, and naming organisms. ☑

READING CHECK

1. Explain How did Linnaeus classify organisms?

____

____

____

CLASSIFICATION TODAY

Taxonomists use an eight-level system to classify living things based on shared characteristics. Scientists also use shared characteristics to describe how closely related living things are.

The more characteristics organisms share, the more closely related they may be. For example, the platypus, brown bear, lion, and house cat are thought to be related because they share many characteristics. These animals all have hair and mammary glands, so they are grouped together as mammals. However, they can also be classified into more specific groups.

Critical Thinking

2. Infer What is the main difference between organisms that share many characteristics and organisms that do not?

____
____
____
____
____
____

BRANCHING DIAGRAMS

Shared characteristics can be shown in a *branching diagram.* Each characteristic on the branching diagram is shared by only the animals above it. The characteristics found higher on the diagram evolved more recently than the characteristics below them.

In the diagram below, all of the animals have hair and mammary glands. However, only the brown bear, lion, and house cat give birth to live young. More recent organisms are at the ends of branches high on the diagram. For example, according to the diagram, the house cat evolved more recently than the platypus. ☑

READING CHECK

3. Identify On a branching diagram, where would you see the characteristics that evolved most recently?

____

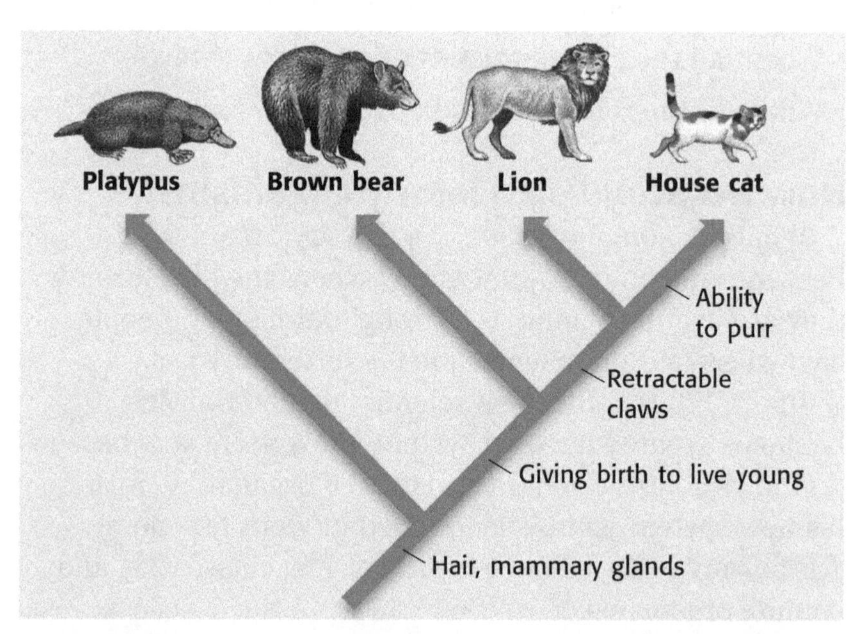

This branching diagram shows the similarities and differences between four kinds of mammals. The bottom of the diagram begins in the past, and the tips of the branches end in the present.

TAKE A LOOK

4. Identify According to the diagram, which organisms evolved before the lion? Circle these organisms.

What Are the Levels of Classification?

Scientists use shared characteristics to group organisms into eight levels of classification. At each level of classification, there are fewer organisms than in the level above. A domain is the largest, most general level of classification. Every living thing is classified into one of three domains.

Species is the smallest level of classification. A species is a group of organisms that can mate and produce fertile offspring. For example, dogs are all one species. They can mate with one another and have fertile offspring. The figure on the next page shows each of the eight levels of classification.

TWO-PART NAMES

We usually call organisms by common names. For example, "cat," "dog," and "human" are all common names. However, people who speak a language other than English have different names for a cat and dog. Sometimes, organisms are even called by different names in English. For example, cougar, mountain lion, and puma are three names for the same animal! ☑

Scientists need to be sure they are all talking about the same organism. They give organisms *scientific names.* Scientific names are the same in all languages. An organism has only one scientific name.

Scientific names are based on the system created by Linnaeus. He gave each kind of organism a two-part name. The first part of the name is the *genus*, and the second part is the *species*. All genus names begin with a capital letter. All species names begin with a lowercase letter. Both words in a scientific name are underlined or italicized. For example, the scientific name for the Asian elephant is *Elephas maximus.* ☑

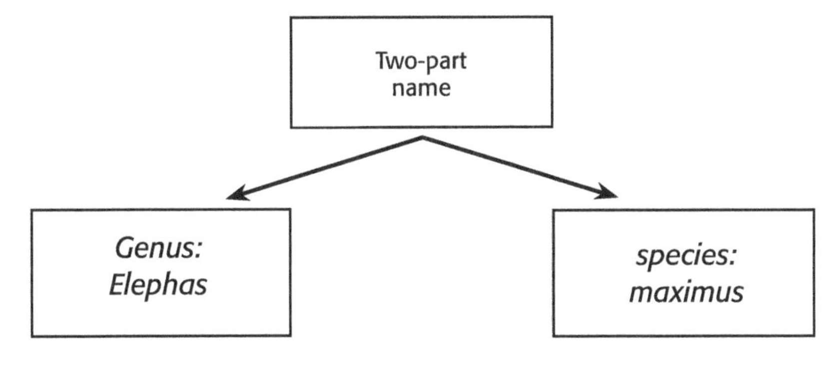

READING CHECK

5. List What are two problems with common names?

____

____

____

____

____

READING CHECK

6. Identify What are the two parts of a scientific name?

____

____

Levels of Classification of the House Cat

Kingdom Animalia: All animals are in the kingdom Animalia.

Phylum Chordata: All animals in the phylum Chordata have a hollow nerve cord. Most have a backbone.

Class Mammalia: Animals in the class Mammalia have a backbone. They also nurse their young.

Order Carnivora: Animals in the order Carnivora have a backbone and nurse their young. They also have special teeth for tearing meat.

Family Felidae: Animals in the family Felidae are cats. They have a backbone, nurse their young, have special teeth for tearing meat, and have retractable claws.

Genus *Felis:* Animals in the genus *Felis* share traits with other animals in the same family. However, these cats cannot roar; they can only purr.

Species *Felis catus*: The species *Felis catus* is the common house cat. The house cat shares traits with all of the organisms in the levels above the species level, but it also has unique traits.

TAKE A LOOK

7. Identify Which level contains organisms that are more closely related: a phylum or a class?

____

8. Describe How does the number of organisms change from the level of kingdom to the level of species?

____

____

____

What Is a Dichotomous Key?

What could you do if you found an organism that you did not recognize? You could use a special guide called a dichotomous key. A **dichotomous key** is set of paired statements that give descriptions of organisms. These statements let you rule out out certain species based on characteristics of your specimen. There are many dichotomous keys for many different kinds of organisms. You could even make your own!

In a dichotomous key, there are only two choices at each step. To use the key, you start with the first pair of statements. You choose the statement from the pair that describes the organism. At each step, the key may identify the organism or it may direct you to another pair of statements. By working through the statements in order, you can identify the organism.

Critical Thinking

9. Infer Why couldn't one single dichotomous key be used for all of the organisms on Earth?

____

____

____

____

____

Dichotomous Key to 10 Common Mammals in the Eastern United States

1. a. This mammal flies. Its "hand" forms a wing. **b.** This mammal does not fly. It's "hand" does not form a wing.	**little brown bat** **Go to step 2.**
2. a. This mammal has no hair on its tail. **b.** This mammal has hair on its tail.	**Go to step 3.** **Go to step 4.**
3. a. This mammal has a short, naked tail. **b.** This mammal has a long, naked tail.	**eastern mole** **Go to step 5.**
4. a. This mammal has a black mask across its face. **b.** This mammal does not have a black mask across its face.	**raccoon** **Go to step 6.**
5. a. This mammal has a tail that is flat and paddle shaped. **b.** This mammal has a tail that is not flat or paddle shaped.	**beaver** **opossum**
6. a. This mammal is brown and has a white underbelly. **b.** This mammal is not brown and does not have a white underbelly.	**Go to step 7.** **Go to step 8.**
7. a. This mammal has a long, furry tail that is black on the tip. **b.** This mammal has a long tail that has little fur.	**longtail weasel** **white-footed mouse**
8. a. This mammal is black and has a narrow white stripe on its forehead and broad white stripes on its back. **b.** This mammal is not black and does not have white stripes.	**striped skunk** **Go to step 9.**
9. a. This mammal has long ears and a short, cottony tail. **b.** This mammal has short ears and a medium-length tail.	**eastern cottontail** **woodchuck**

TAKE A LOOK

10. Identify Use this dichotomous key to identify the two animals shown.

____

____

Name ______________________ Class ______________ Date ______________

Section 1 Review

NSES LS 5a

SECTION VOCABULARY

classification the division of organisms into groups, or classes, based on specific characteristics **dichotomous key** an aid that is used to identify organisms and that consists of the answers to a series of questions	**taxonomy** the science of describing, naming, and classifying organisms

1. List Give the eight levels of classification from the largest to the smallest.

____

____

2. Identify According to the branching diagram below, which characteristic do ferns have that mosses do not?

____

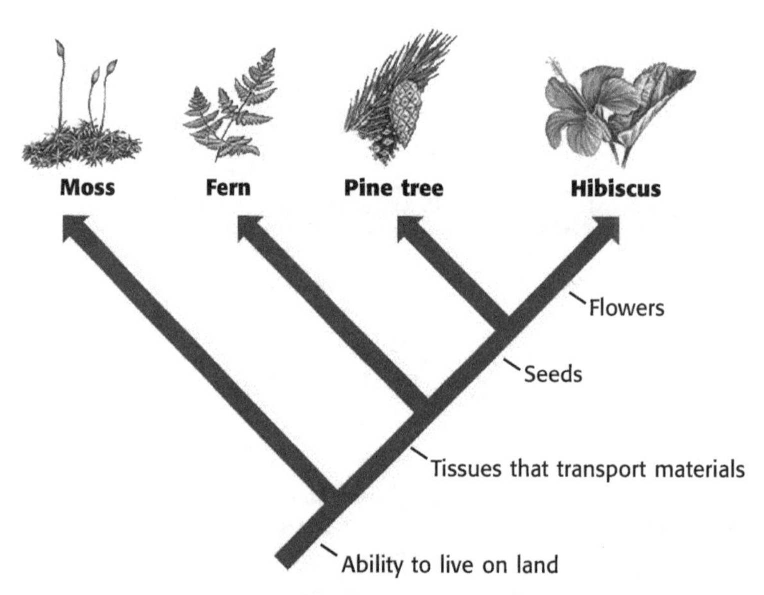

3. Analyze Which species in the diagram above is most similar to the hibiscus? Which is the least similar?

____

____

4. Identify What are the two parts of a scientific name?

____

5. Infer Could you use the dichotomous key in this section to identify a species of lizard? Explain your answer.

____

____

Name ______________________ Class ______________ Date ______________

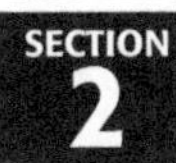

CHAPTER 9 Classification

SECTION 2

Domains and Kingdoms

BEFORE YOU READ

After you read this section, you should be able to answer these questions:

- How are prokaryotes classified?
- How are eukaryotes classified?

National Science Education Standards
LS 1f, 2a, 4b, 4c, 5b

How Do Scientists Classify Organisms?

For hundreds of years, all organisms were classified as either plants or animals. However, as more organisms were discovered, scientists found some organisms that did not fit well into these two kingdoms. Some animals, for example, had characteristics of both plants and animals.

What would you call an organism that is green and makes its own food? Is it a plant? What if the organism moved and could also eat other organisms? Plants generally do neither of these things. Is the organism a plant or an animal?

STUDY TIP
List As you read this section, make a list of the domains and kingdoms scientists use to classify organisms.

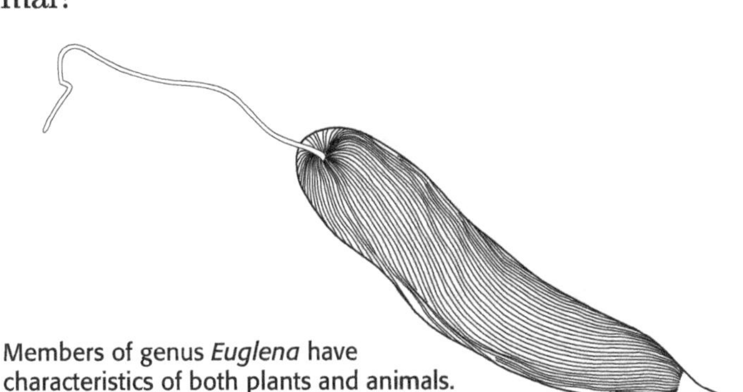

Members of genus *Euglena* have characteristics of both plants and animals.

The organism above belongs to genus *Euglena*. Its members show all of the characteristics just described. As scientists discovered organisms, such as *Euglena*, that didn't fit easily into existing groups, they created new ones. As they added kingdoms, scientists found that members of some kingdoms were closely related to members of other kingdoms. Today, scientists group kingdoms into *domains*.

All organisms on Earth are grouped into three domains. Two domains, Bacteria and Archaea, are made up of prokaryotes. The third domain, Eukarya, is made up of all the eukaryotes. Scientists are still working to describe the kingdoms in each of the three domains.

Critical Thinking

1. Apply Concepts In which domain would multicellular organisms be classified? Explain your answer.

____

____

____

____

How Are Prokaryotes Classified?

A prokaryote is a single-celled organism that does not have a nucleus. Prokaryotes are the oldest group of organisms on Earth. They make up two domains: Archaea and Bacteria.

DOMAIN ARCHAEA

Domain **Archaea** is made up of prokaryotes. The cell walls and cell membranes of archaea are made of different substances than those of other prokaryotes. Many archaea can live in extreme environments where other organisms could not survive. Some archaea can also be found in more moderate environments, such as the ocean. ☑

READING CHECK

2. Compare How are members of Archaea different from other prokaryotes?

____

____

____

____

____

DOMAIN BACTERIA

All bacteria belong to domain **Bacteria**. Bacteria can be found in the air, in soil, in water, and even on and inside the human body!

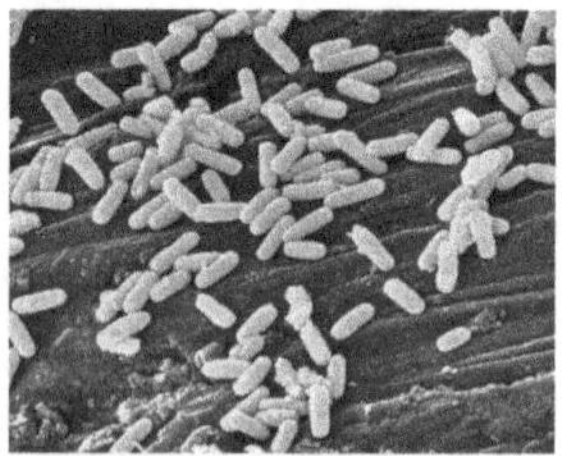

We often think of bacteria as bad, but not all bacteria are harmful. One kind of bacterium changes milk into yogurt. *Escherichia coli* is a bacterium that lives in human intestines. It helps break down undigested food and produces vitamin K. Some bacteria do cause diseases, such as pneumonia. However, other bacteria make chemicals that can help us fight bacteria that cause disease. ☑

READING CHECK

3. Explain Are all bacteria harmful? Explain your answer.

____

____

____

____

The Grand Prismatic Spring in Yellowstone National Park contains water that is about 90°C (190°F). Most organisms would die in such a hot environment.

TAKE A LOOK

4. Apply Concepts What kind of prokaryotes do you think could live in this spring? Explain your answer.

____

____

____

____

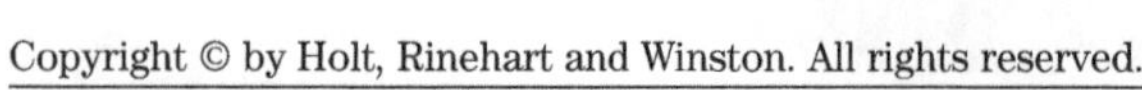

How Are Eukaryotes Classified?

Organisms that have cells with membrane-bound organelles and a nucleus are called *eukaryotes*. All eukaryotes belong to domain **Eukarya**. Domain Eukarya includes the following kingdoms: Protista, Fungi, Plantae, and Animalia.

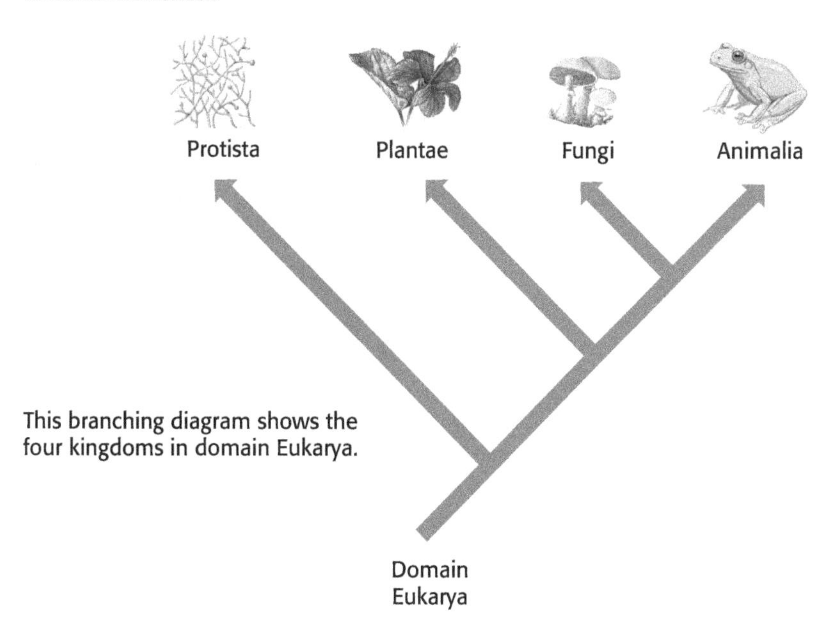

This branching diagram shows the four kingdoms in domain Eukarya.

STANDARDS CHECK

LS 5a Biological evolution accounts for the diversity of species developed through gradual processes over many generations. Species acquire many of their unique characteristics through biological adaptation, which involves the selection of naturally occurring variations in populations. Biological adaptations include changes in structures, behaviors, or physiology that enhance survival and reproductive success in a particular environment.

Word Help: selection
the process of choosing

Word Help: variation
a difference in the form or function

5. Identify Based on the branching diagram, which two kingdoms in Eukarya evolved most recently? How do you know?

____

____

____

____

____

____

KINGDOM PROTISTA

Members of kingdom **Protista** are either single-celled or simple multicellular organisms. They are commonly called *protists*. Scientists think that the first protists evolved from ancient bacteria about 2 billion years ago. Much later, plants, fungi, and animals evolved from ancient protists.

Kingdom Protista contains many different kinds of organisms. Some, such as *Paramecium*, resemble animals. They are called *protozoa*. Plantlike protists are called *algae*. Some algae, such as phytoplankton, are single cells. Others, such as kelp, are multicellular. Multicellular slime molds also belong to kingdom Protista.

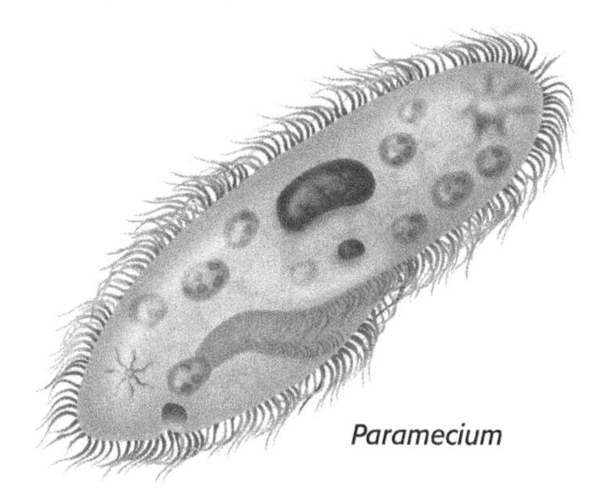

Paramecium

KINGDOM FUNGI

Molds and mushrooms are members of kingdom **Fungi**. Some fungi (singular, *fungus*) are unicellular. That is, they are single-celled organisms. Yeast is unicellular. Most other fungi are multicellular. Unlike plants, fungi do not perform photosynthesis. However, they also do not eat food, as animals do. Instead, fungi use digestive juices to break down materials in the environment and absorb them. ☑

READING CHECK

6. Describe How do fungi get food?

____

____

____

____

Amanita is a poisonous fungus. You should never eat wild fungi.

KINGDOM PLANTAE

Although plants differ in size and appearance, most people can easily identify the members of kingdom Plantae. Kingdom **Plantae** contains organisms that are eukaryotic, have cell walls, and make food by photosynthesis. Most plants need sunlight to carry out photosynthesis. Therefore, plants must live in places where light can reach.

The food that plants make is important for the plants and also for other organisms. Many animals, fungi, protists, and bacteria get nutrients from plants. When they digest the plant material, they get the energy stored by the plant. Plants also provide homes for other organisms.

Math Focus

7. Calculate The average student's arms extend about 1.3 m. How many students would have to join hands to form a human chain around a giant sequoia?

____

The giant sequoia is one of the largest members of kingdom Plantae. A giant sequoia can measure 30 m around the base and grow more than 91 m tall!

KINGDOM ANIMALIA

Kingdom **Animalia** contains complex, multicellular organisms. Organisms in kingdom Animalia are commonly called *animals*. The following are some characteristics of animals:

- Their cells do not have cell walls.
- They are able to move from place to place.
- They have sense organs that help them react quickly to their environment.

TAKE A LOOK

8. Identify Which animal characteristic from above can be seen in this bald eagle?

____

STRANGE ORGANISMS

Some organisms are not easy to classify. For example, some plants can eat other organisms to get nutrition as animals do. Some protists use photosynthesis as plants do but also move around as animals do.

Red Cup Sponge

Critical Thinking

9. Apply Concepts To get nutrients, a Venus' flytrap uses photosynthesis and traps and digests insects. Its cells have cell walls. Into which kingdom would you place this organism? Explain your answer.

____

____

____

____

____

What kind of organism is this red cup sponge? It does not have sense organs and cannot move for most of its life. Because of this, scientists once classified sponges as plants. However, sponges cannot make their own food as plants do. They must eat other organisms to get nutrients. Today, scientists classify sponges as animals. Sponges are usually considered the simplest animals.

Name ______________________ Class ______________ Date ______________

Section 2 Review

NSES LS 1f, 2a, 4b, 4c, 5b

SECTION VOCABULARY

Animalia a kingdom made up of complex, multicellular organisms that lack cell walls, can usually move around, and quickly respond to their environment

Archaea in a modern taxonomic system, a domain made up of prokaryotes (most of which are known to live in extreme environments) that are distinguished from other prokaryotes by differences in their genetics and in the makeup of their cell wall; this domain aligns with the traditional kingdom Archaebacteria

Bacteria in a modern taxonomic system, a domain made up of prokaryotes that usually have a cell wall and that usually reproduce by cell division. This domain aligns with the traditional kingdom Eubacteria

Eukarya in a modern taxonomic system, a domain made up of all eukaryotes; this domain aligns with the traditional kingdoms Protista, Fungi, Plantae, and Animalia

Fungi a kingdom made up of nongreen, eukaryotic organisms that have no means of movement, reproduce by using spores, and get food by breaking down substances in their surroundings and absorbing the nutrients

Plantae a kingdom made up of complex, multicellular organisms that are usually green, have cell walls made of cellulose, cannot move around, and use the sun's energy to make sugar by photosynthesis

Protista a kingdom of mostly one-celled eukaryotic organisms that are different from plants, animals, bacteria, and fungi

1. **Compare** What is one major difference between domain Eukarya and domains Bacteria and Archaea?

____

____

2. **Explain** Why do scientists continue to add new kingdoms to their system of classification?

____

____

3. **Analyze Methods** Why do you think Linnaeus did not include classifications for archaea and bacteria?

____

____

____

4. **Apply Concepts** Based on its characteristics described at the beginning of this section, in which kingdom would you classify *Euglena*?

____

Name ______________________ Class ______________ Date ______________

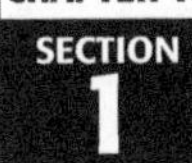

CHAPTER 10 Bacteria and Viruses

SECTION 1

Bacteria and Archaea

BEFORE YOU READ

After you read this section, you should be able to answer these questions:

- What are bacteria and archaea?
- What are the characteristics of bacteria?
- How do archaea and bacteria differ?

National Science Education Standards
LS 1b, 1c, 2a, 3a, 3b, 4b

What Are Bacteria and Archaea?

Organisms are grouped by traits they have in common. All living things can be grouped into one of three domains: Bacteria, Archaea, or Eukarya.

All organisms in domain Eukarya are eukaryotes. Each cell of a *eukaryote* has a nucleus and membrane-bound organelles. All organisms in domains Bacteria and Archaea are prokaryotes. A **prokaryote** is an organism that is single-celled and has no nucleus.

Although many prokaryotes live in groups, they are single organisms that can move, get food, and make copies of themselves. Most prokaryotes are very small and cannot be seen without a microscope. However, you can see some very large bacteria with your naked eye. ☑

STUDY TIP

Underline Use colored pencils to underline the characteristics of bacteria in red, characteristics of archaea in blue, and characteristics shared by both in green.

READING CHECK

1. Identify What are two characteristics of prokaryotes?

____

____

____

____

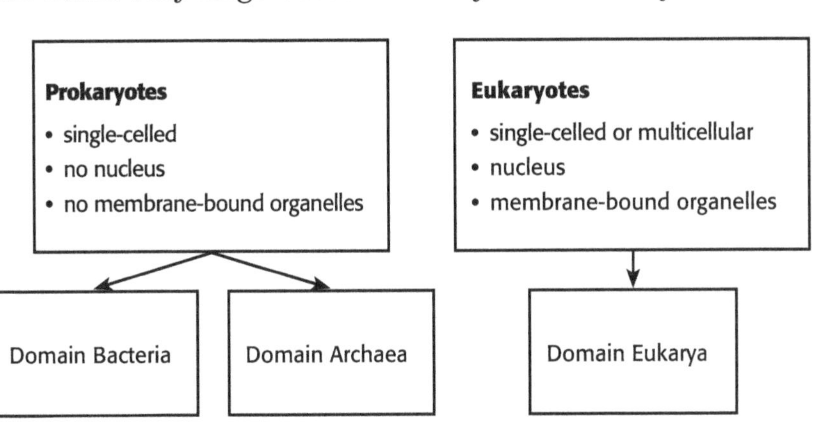

What Characteristics Do Archaea and Bacteria Share?

NO NUCLEUS

Prokaryotes do not store their DNA in a nucleus as eukaryotes do. Their DNA is stored as a circular loop inside the cell. ☑

READING CHECK

2. Describe What does the DNA of a prokaryote look like?

____

____

REPRODUCTION

Prokaryotes copy themselves, or reproduce, by a process called binary fission. **Binary fission** is reproduction in which a single-celled organism splits into two single-celled organisms. Before a prokaryote can reproduce, it must make a copy of its loop of DNA. After the cell splits, the two new cells are identical to the original cell. ☑

READING CHECK

3. Identify How do prokaryotes reproduce?

____

____

Binary Fission

❶ The cell grows.

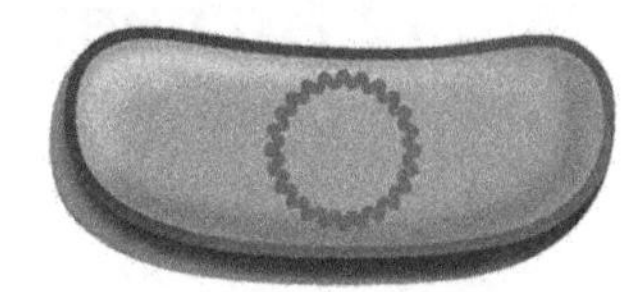

❷ The cell makes a copy of its DNA. Both copies attach to the cell membrane.

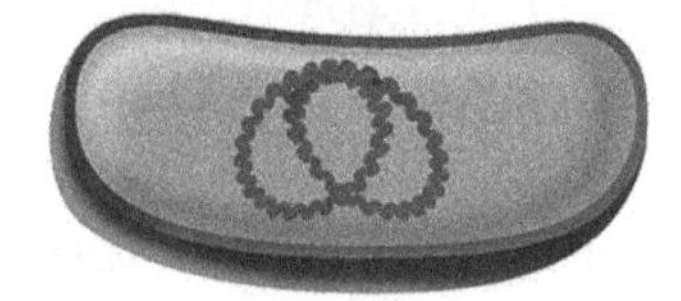

❸ The DNA and its copy separate as the cell grows larger.

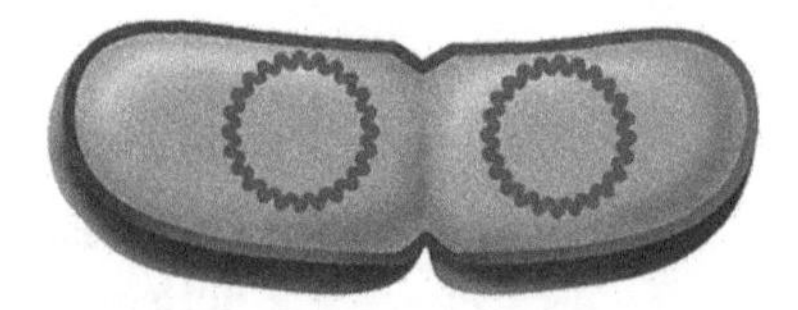

❹ The cells separate. Each new cell has a copy of the DNA.

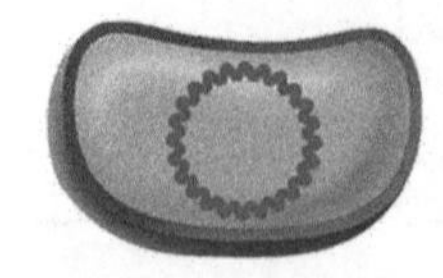
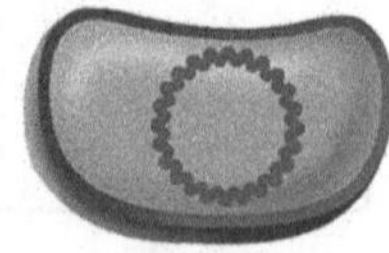

TAKE A LOOK

4. Describe After binary fission, how do the two cells compare to the original cell?

____

____

What Are Some Characteristics of Bacteria?

Most of the prokaryotes that scientists have found are bacteria. Domain Bacteria has more individual members than domains Archaea and Eukarya combined have. Bacteria can be found almost everywhere.

SHAPE

Most bacteria are one of three shapes: bacilli, cocci, and spirilla. *Bacilli* are rod-shaped. *Cocci* are spherical. *Spirilla* are long and spiral-shaped. Different shapes help bacteria survive. Most bacteria have a stiff cell wall that gives them their shape. ☑

Some bacteria have hairlike parts called *flagella* (singular, *flagellum*). A flagellum works like a tail to push a bacterium through fluids.

READING CHECK

5. List What are three common shapes of bacteria?

____

____

____

The Most Common Shapes of Bacteria

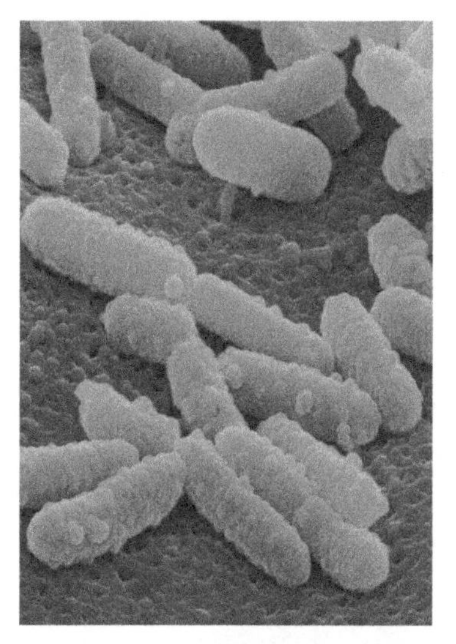

Bacilli are rod shaped. They have a large surface area, which helps them take in nutrients. However, a large surface area causes them to dry out quickly.

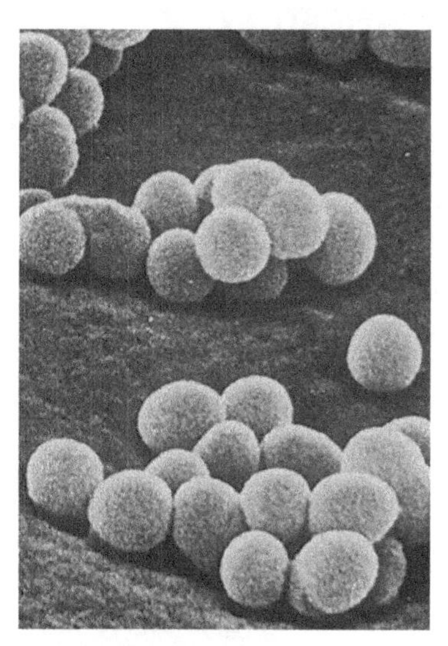

Cocci are spherical. They do not dry out as quickly as rod-shaped bacteria.

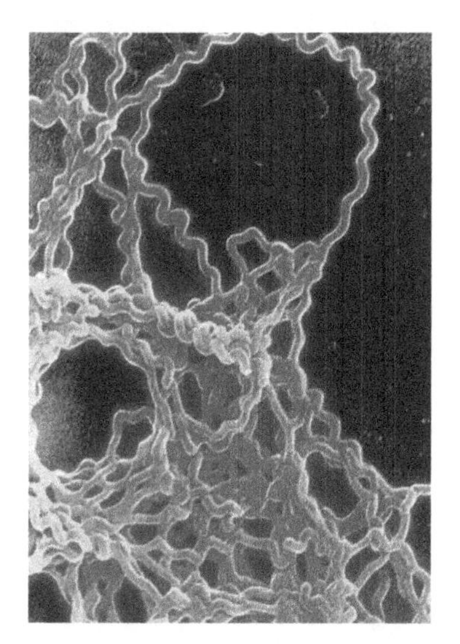

Spirilla are long and spiral-shaped. They have flagella at both ends. These tail-like structures help the bacteria move through fluids.

TAKE A LOOK

6. Compare What advantage do cocci have over bacilli?

____

____

____

7. Compare What advantage do bacilli have over cocci?

____

____

____

____

ENDOSPORES

Most bacteria do well in warm, moist places. Some species of bacteria die in dry and cold surroundings. However, some bacteria form endospores to survive these kinds of conditions. An **endospore** is a thick, protective covering that forms around the DNA of a bacterium. The endospore protects the DNA from changes in the environment. When conditions are good, the endospores break open, and the bacteria begin to grow.

STANDARDS CHECK

LS 4b Populations of organisms can be categorized by the functions they serve in an ecosystem. Plants and some microorganisms are producers—they make their own food. All animals, including humans, are consumers, which obtain their food by eating other organisms. Decomposers, primarily bacteria and fungi, are consumers that use waste materials and dead organisms for food. Food webs identify the relationship among producers, consumers, and decomposers in an ecosystem.

Word Help: categorized to put into groups or classes

8. List Name three roles bacteria can play in an ecosystem.

____

____

____

CLASSIFICATION

Scientists can classify bacteria by the way the bacteria get food. There are three ways for bacteria to get food: consume it, decompose it, or produce it.

- *Consumers* eat other organisms.
- *Decomposers* eat dead organisms or waste.
- *Producers* make their own food. Some bacteria can make food using the energy from sunlight.

Decomposers, such as the ones helping to decay this leaf, return nutrients to the soil. This allows other living things to use those nutrients.

CYANOBACTERIA

Cyanobacteria are producers. These bacteria have a green pigment called *chlorophyll*. Chlorophyll traps the energy from the sun. The cell uses this energy to make food.

Some scientists think that billions of years ago, bacteria similar to cyanobacteria began to live inside larger cells. According to this hypothesis, the bacteria made food for itself and the larger cells. In return, the larger cells protected the bacteria. This relationship may have led to the first plant cells on Earth.

How Do Archaea Differ from Bacteria?

Like bacteria, archaea are prokaryotes. However, archaea are different from bacteria. For example, not all archaea have cell walls. When they do have them, the cell walls are made of different materials than the cell walls of bacteria. ☑

There are three types of archaea: heat lovers, salt lovers, and methane makers. *Heat lovers* live in hot ocean vents and hot springs. They usually live in water that is 60°C to 80°C. However, they have been found in living in water as hot as 250°C.

Salt lovers live where there are high levels of salt, such as the Dead Sea. *Methane makers* give off methane gas. Methane makers often live in swamps. They can also live inside animal intestines.

READING CHECK

9. Identify What is one difference between archaea and bacteria?

____

TAKE A LOOK

10. Identify What type of archaea do you think would live in this swamp?

____

HARSH ENVIRONMENTS

Although bacteria can be found almost anywhere, archaea can live in places where even bacteria cannot survive. For example, many archaea live in places with little or no oxygen. Many can also survive very high temperatures and pressures.

Scientists have found archaea in the hot springs at Yellowstone National Park and beneath 430 m of ice in Antarctica. Archaea have even been found 8 km below the surface of the Earth! Even though they can be found in harsh environments, many archaea also live in more moderate environments, such as the ocean.

Critical Thinking

11. Infer What kind of prokaryote would most likely be found near vents at the bottom of the ocean with extremely high temperatures? Explain your answer.

____

Name ______________________ Class ______________ Date ______________

Section 1 Review

NSES LS 1b, 1c, 2a, 3a, 3b, 4b

SECTION VOCABULARY

binary fission a form of asexual reproduction in single-celled organisms by which one cell divides into two cells of the same size

endospore a thick-walled protective spore that forms inside a bacterial cell and resists harsh conditions

prokaryote an organism that consists of a single cell that does not have a nucleus

1. **List** What are the three domains that include all living things?

____

2. **Compare** Fill in the Venn Diagram to compare bacteria and archaea.

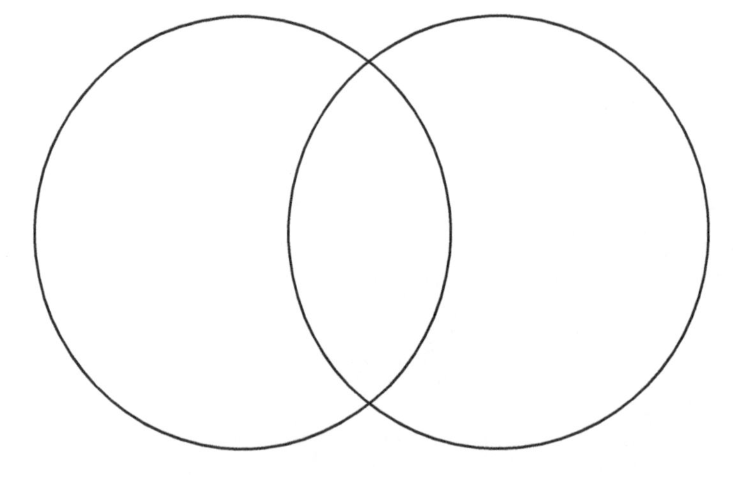

3. **Describe** How do some scientists think the first plants appeared on Earth?

____

____

____

____

4. **List** Name three kinds of archaea.

____

____

5. **Infer** Do you think it would be possible to find bacteria and archaea living in the same environment? Explain your answer.

____

____

____

____

Name ______________________ Class ______________ Date ______________

CHAPTER 10 Bacteria and Viruses

SECTION 2

Bacteria's Role in the World

BEFORE YOU READ

After you read this section, you should be able to answer these questions:

- How are some bacteria helpful?
- How are some bacteria harmful?

National Science Education Standards
LS 1f

Are All Bacteria Harmful?

Bacteria are everywhere. They live in our water, our food, and our bodies. Some bacteria cause disease. However, most of the types of bacteria are helpful to organisms and the environment. Some bacteria move nitrogen throughout the environment. Other bacteria help recycle dead animals and plants. Still other bacteria are used to help scientists make medicines. Bacteria help us every day.

STUDY TIP

Outline As you read, make an outline of this section. Use the header questions to help you make your outline.

Say It

Discuss Does it surprise you to learn that not all bacteria are harmful? What were some things you used to believe about bacteria? With a partner, talk about how your view of bacteria has changed.

How Are Bacteria Helpful to Plants?

Plants need nitrogen to live. Although nitrogen makes up about 78% of the air, plants cannot use this nitrogen directly. They need to take in a different form of nitrogen. Bacteria in the soil take nitrogen from the air and change it into a form that plants can use. This process is called *nitrogen fixation.*

Bacteria's Role in the Nitrogen Cycle

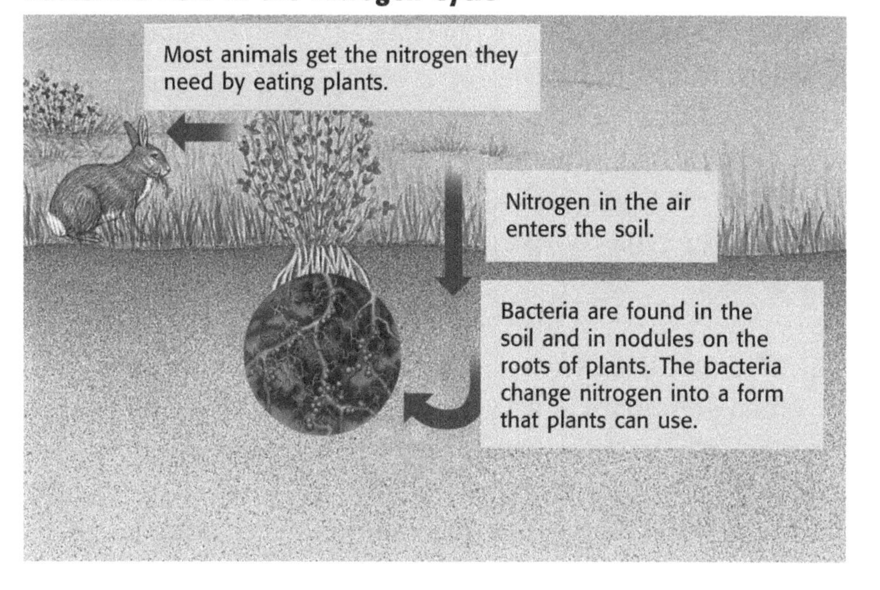

TAKE A LOOK

1. Explain How do most animals get the nitrogen that they need?

____

____

____

How Are Bacteria Helpful to the Environment?

Bacteria are useful for recycling nutrients and cleaning up pollution. In nature, bacteria break down dead plants and animals. Breaking down dead matter helps recycle the nutrients so they can be used by other organisms.

Some bacteria are used to fight pollution.This process is called bioremediation. **Bioremediation** means using microorganisms, such as bacteria, to clean up harmful chemicals. Bioremediation is used to clean up oil spills and waste from cities, farms, and industries.

Math Focus

2. Calculate An ounce (1 oz) is equal to about 28 g. If 1 g of soil contains 25 billion bacteria, how many bacteria are in 1 oz of soil?

____

How Are Bacteria Helpful to People?

MAKING FOOD

Many of the foods you eat are made with the help of bacteria. Bacteria are commonly used in dairy products. Every time you eat cheese, yogurt, buttermilk, or sour cream, you are also eating bacteria. These products are made using milk and bacteria. Bacteria change the sugar in milk, called *lactose*, into *lactic acid.* Lactic acid adds flavor to the food and preserves it.

Bacteria are used to make many kinds of food.

MAKING MEDICINES

What's the best way to fight bacteria that cause disease? Would you believe that the answer is to use other bacteria? An **antibiotic** is a medicine that can kill bacteria and other microorganisms. Many antibiotics are made by bacteria.

READING CHECK

3. Explain What does genetic engineering let scientists do?

____

____

____

____

GENETIC ENGINEERING

When scientists change an organism's DNA, it is called *genetic engineering*. For example, scientists have put genes from different organisms into bacteria. The added DNA gives the bacterium instructions for making different proteins. Genetic engineering lets scientists make products that are hard to find in nature. ☑

Genes from the *Xenopus* frog were used to produce the first genetically engineered bacteria.

Scientists have used genetic engineering to produce insulin. The human body needs *insulin* to break down and use sugar. People who have a disease called *diabetes* cannot make enough insulin. In the 1970s, scientists found a way to put genes into bacteria so that the bacteria would produce human insulin.

Critical Thinking

4. Infer Why do you think it is helpful to engineer bacteria to produce insulin?

____

____

____

____

____

How Can Bacteria Be Harmful?

Some bacteria can be harmful to people and other organisms. Bacteria that cause disease are called **pathogenic bacteria**. Pathogenic bacteria get inside a host organism and take nutrients from the organism's cells. Pathogenic bacteria can harm the organism. Today, people protect themselves from some bacterial diseases by vaccination. Many bacterial infections can be treated with antibiotics. ☑

READING CHECK

5. Identify What can be used to treat bacterial infections?

____

Bacteria can cause disease in other organisms as well as people. Bacteria can rot or discolor a plant and its fruit. To stop this, plants are sometimes treated with antibiotics. Scientists also try to grow plants that have been genetically engineered to resist bacteria that cause disease.

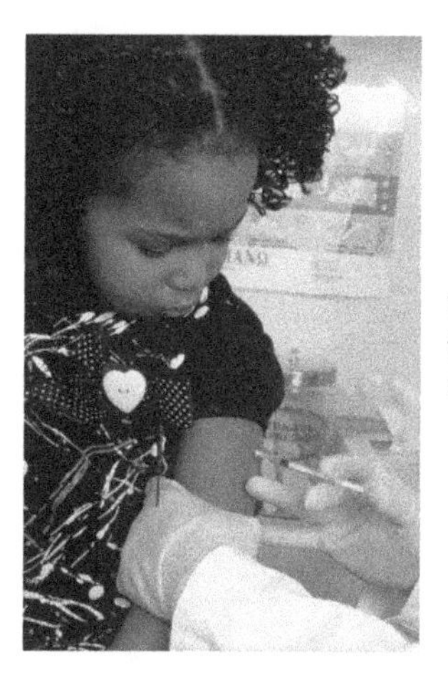

Vaccines can protect you from bacterial diseases such as tetanus and diphtheria.

Name ______________________ Class ______________ Date ______________

Section 2 Review

NSES LS 1f

SECTION VOCABULARY

antibiotic medicine used to kill bacteria and other organisms **bioremediation** the biological treatment of hazardous waste by living organisms	**pathogenic bacteria** bacteria that cause disease

1. **Define** What is nitrogen fixation?

____

____

2. **Complete** Fill in the process chart for the nitrogen cycle. Be sure to describe what is happening during each step.

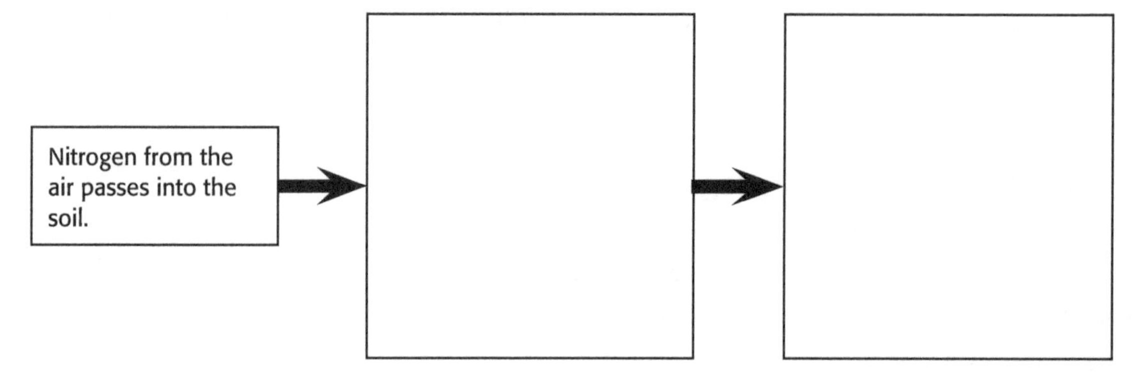

3. **Explain** What are two ways that bacteria are helpful to other living things?

____

____

____

4. **Explain** What is genetic engineering?

____

____

____

5. **Explain** How do pathogenic bacteria harm an organism?

____

____

6. **Identify Relationships** Legumes, which include peanuts and beans, are good nitrogen-fixers. Legumes are also a good source of amino acids. What chemical element would you expect to find in amino acids?

____

Name ______________________ Class ______________ Date ______________

CHAPTER 10 Bacteria and Viruses

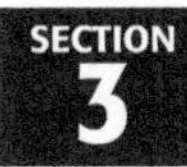

SECTION 3 Viruses

BEFORE YOU READ

After you read this section, you should be able to answer these questions:

- What is a virus?
- How does a virus survive?
- How do viruses make more of themselves?

National Science Education Standards
LS 1f, 2a

What Is a Virus?

Most people have either had chickenpox or seen someone with the disease. Chickenpox is caused by a virus. A **virus** is a tiny particle that gets inside a cell and usually kills it. Many viruses cause diseases, such as the common cold, the flu, and acquired immune deficiency syndrome (AIDS). Viruses are smaller than bacteria and can only be seen with a microscope. Viruses can also change quickly. These traits make it hard for scientists to fight viruses. ☑

STUDY TIP

Compare As you read, make a Venn Diagram to compare viruses and bacteria.

READING CHECK

1. Explain Why is it difficult for scientists to fight viruses?

____

____

____

Are Viruses Living?

Like living things, viruses have protein and genetic material. However, viruses are not alive. A virus cannot eat, grow, or reproduce like a living thing. For a virus to function, it needs to get inside a living cell. Viruses use cells as hosts. A **host** is a living thing that a virus lives on or in. A virus uses a host cell to make more viruses.

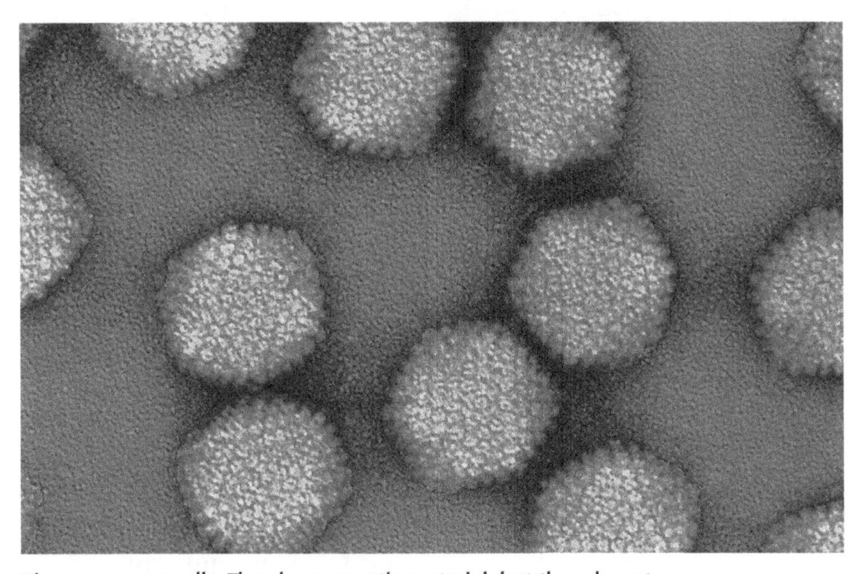

Viruses are not cells. They have genetic material, but they do not have cytoplasm or organelles.

STANDARDS CHECK

LS 2a Reproduction is a characteristic of all living systems; because no living organism lives forever, reproduction is essential to the continuation of every species. Some organisms reproduce asexually. Others reproduce sexually.

2. List Give three reasons that viruses are not considered living things.

____

____

____

____

____

What Is the Structure of a Virus?

Every virus is made up of genetic material inside a protein coat. The protein coat protects the genetic material and helps the virus enter a cell.

How Are Viruses Grouped?

Viruses can be grouped in several ways. These include shape and type of genetic material. The genetic material in viruses is either RNA or DNA. RNA is made of only one strand of nucleotides. DNA is made of two strands of nucleotides. The viruses that cause chickenpox and warts have DNA. The viruses that cause colds, flu, and AIDS have RNA. ☑

READING CHECK

3. Identify What two kinds of genetic material can viruses have?

____

____

The Basic Shapes of Viruses

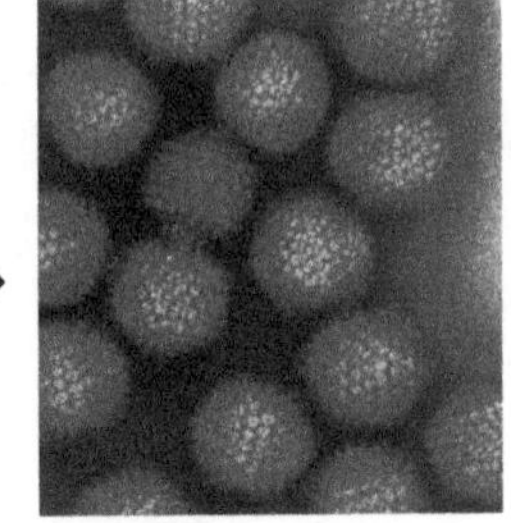

Crystals
The polio virus is shaped like the crystals shown here.

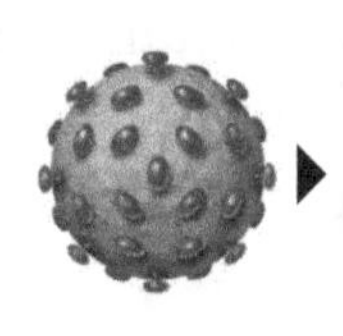

Spheres
Influenza viruses look like spheres. HIV, the virus that causes AIDS, also has this structure.

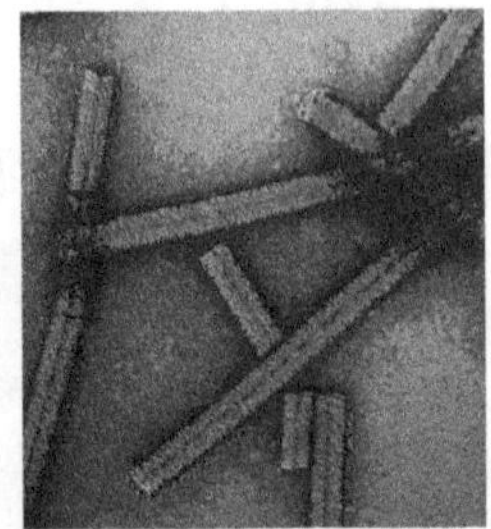

Cylinders
The tobacco mosaic virus is shaped like a cylinder and attacks tobacco plants.

Spacecraft
One group of viruses attacks only bacteria. Many of these look almost like spacecraft.

TAKE A LOOK

4. Identify What shape are viruses that attack bacteria?

____

How Do Viruses Make More Viruses?

THE LYTIC CYCLE

Like living things, viruses make more of themselves. However, viruses cannot reproduce on their own. They attack living cells and turn them into virus factories. This process is called the *lytic cycle.* ☑

READING CHECK

5. Identify Name one way viruses are like living things.

____

____

____

The Lytic Cycle

❶ The virus finds and joins itself to a host cell.

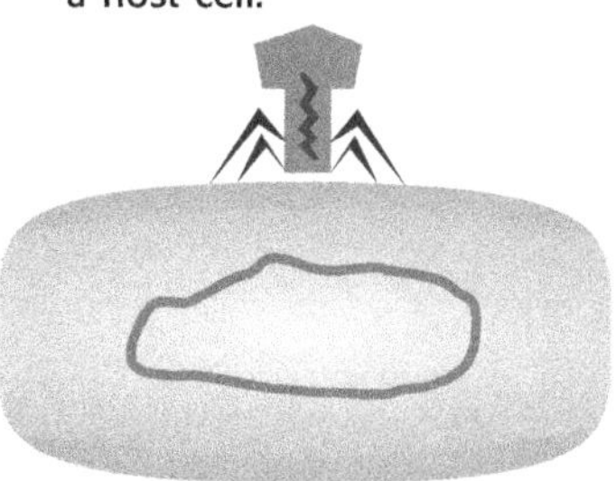

❷ The virus enters the cell, or the virus's genetic material is injected into the cell.

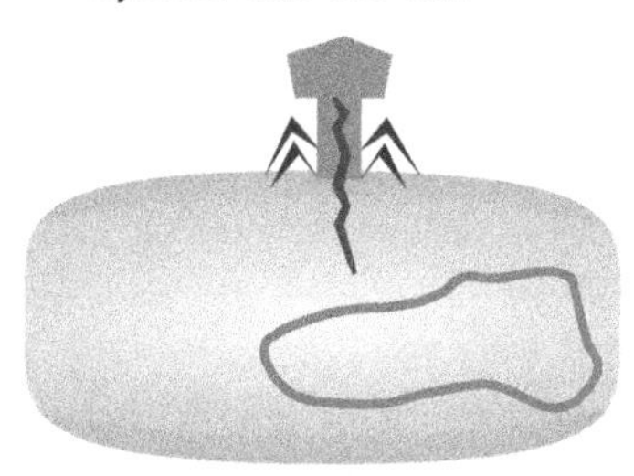

❸ Once the virus's genes are inside, they take control of the host cell and turn it into a virus factory.

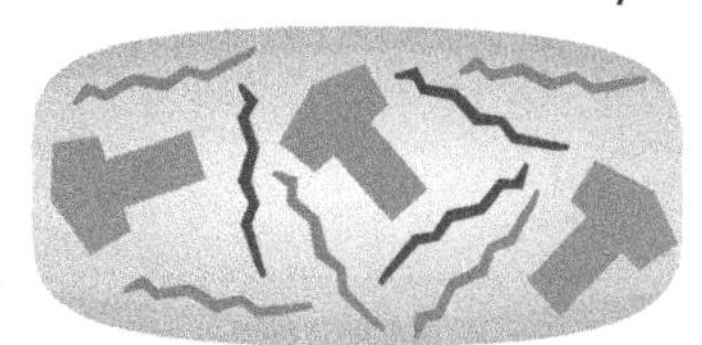

❹ The host cell dies when the new viruses break out of it. The new viruses look for host cells, and the cycle continues.

Math Focus

6. Calculate If you enlarged an average virus 600,000 times, it would be about the size of a small pea. How tall would you be if you were enlarged 600,000 times?

____

TAKE A LOOK

7. Explain What happens to a host cell when the new viruses are released?

____

THE LYSOGENIC CYCLE

Some viruses don't go right into the lytic cycle. These viruses put their genetic material inside a host cell, but new viruses are not made. As the host cells divides, the genetic material of the virus is passed to the new cells. This is called the *lysogenic cycle*. The host cells can carry the genes of the virus for a long time. When the genes do become active, the lytic cycle begins.

How Can You Stop Viruses?

Antibiotics cannot destroy viruses. However, scientists have made some medicines called *antiviral medications*. These medicines stop viruses from reproducing in their host. Most diseases caused by viruses do not have cures. It is best to try to stop a virus from entering your body. Washing your hands helps you avoid some viruses. The vaccinations some children get also help to prevent viral infections.

Critical Thinking

8. Infer Why shouldn't you take antibiotics to treat a cold?

____

____

____

____

Name ______________________ Class ______________ Date ______________

Section 3 Review

NSES LS 1f, 2a

SECTION VOCABULARY

host an organism from which a parasite takes food or shelter	**virus** a microscopic particle that gets inside a cell and often destroys the cell

1. **Compare** How are viruses similar to living things?

____

____

2. **Describe** What is the structure of a virus?

____

____

3. **List** List four shapes that viruses may have.

____

4. **Summarize** Complete the process chart to show the steps of the lytic cycle.

The virus attaches to a host cell.
↓
↓
↓
The new viruses break out of the host cell, and the host cell dies. The new viruses look for hosts.

5. **Explain** Why do viruses need hosts?

____

____

5. **Compare** How is the lysogenic cycle different from the lytic cycle?

____

____

____

____

6. **List** Name two ways to prevent a viral infection.

____

____

SECTION 1

Protists

BEFORE YOU READ

After you read this section, you should be able to answer these questions:

- How are protists different from other organisms?
- How do protists get food?
- How do protists reproduce?

National Science Education Standards

LS 1b, 1c, 1f, 2a, 4b, 5a

What Are Protists?

Protists are members of the kingdom Protista. They are a very diverse group of organisms. Most are single-celled, but some have many cells. Some are single-celled but live in colonies. Members of the kingdom Protista have many different ways of getting food, reproducing, and moving.

Protists are very different from plants, animals, and fungi. They are also very different from each other. Scientists group protists together because they do not fit into any other kingdom. However, they do share a few characteristics:

- They are *eukaryotic* (each of their cells has a nucleus).
- They are less complex than other eukaryotic organisms.
- They do not have specialized tissues.

STUDY TIP

List As you read, make a table showing the ways different protists can obtain food and reproduce.

Critical Thinking

1. Infer Are protists prokaryotes or eukaryotes? Explain your answer.

____

____

____

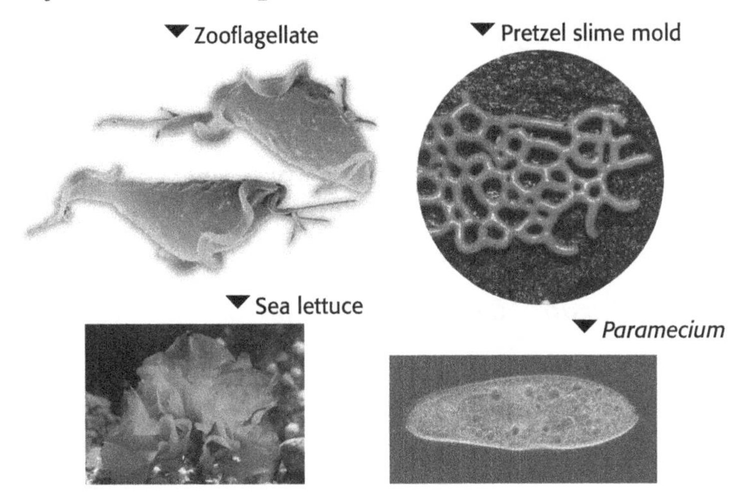

Protists have many different shapes.

How Do Protists Get Food?

Some protists can make their own food. Other protists need to get their food from other organisms or from the environment. Some protists use more than one method of getting food.

AUTOTROPHS

Some protists are producers because they make their own food. The cells of these protists have structures called *chloroplasts*, which capture energy from the sun. Protists use this energy to produce food in a process called *photosynthesis*. Plants use the same process to make their own food. Producers are also known as *autotrophs*. *Autos* is Greek for "self." ☑

2. Define What is an autotroph?

____

____

HETEROTROPHS

Some protists cannot make their own food. They are called **heterotrophs**. *Heteros* is Greek for "different." There are several ways that protist heterotrophs can obtain food.

- *Consumers* eat other organisms, such as bacteria, yeasts, or other protists.
- *Decomposers* break down dead organic matter.
- **Parasites** are organisms that feed on or invade other living things. The organism a parasite invades is called a **host**. Parasites usually harm their hosts. ☑

READING CHECK

3. List What are the three kinds of protist heterotrophs?

____

____

____

How Do Protists Reproduce?

Like all living things, protists reproduce. Some protists reproduce asexually, and some reproduce sexually. Some protists reproduce asexually at one stage in their life cycle and sexually at another stage.

ASEXUAL REPRODUCTION

Most protists reproduce asexually. In asexual reproduction, only one parent is needed to make offspring. These offspring are identical to the parent.

- In *binary fission*, a single-celled protist divides into two cells.
- In *multiple fission*, a single-celled protist divides into more than two cells. Each new cell is a single-celled protist.

Math Focus

4. Calculate If seven individuals of genus Euglena reproduce at one time, how many individuals will result?

____

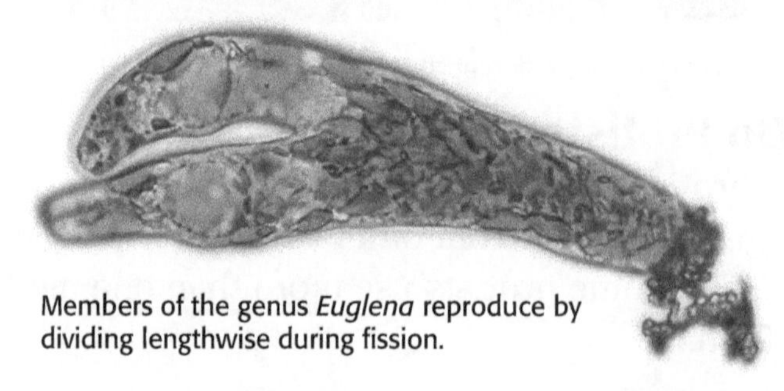

Members of the genus *Euglena* reproduce by dividing lengthwise during fission.

SEXUAL REPRODUCTION

Sexual reproduction requires two parents. The offspring have new combinations of genetic material. The two protists below are reproducing by conjugation. During *conjugation*, two organisms join and exchange genetic material. Then they divide to produce four new protists.

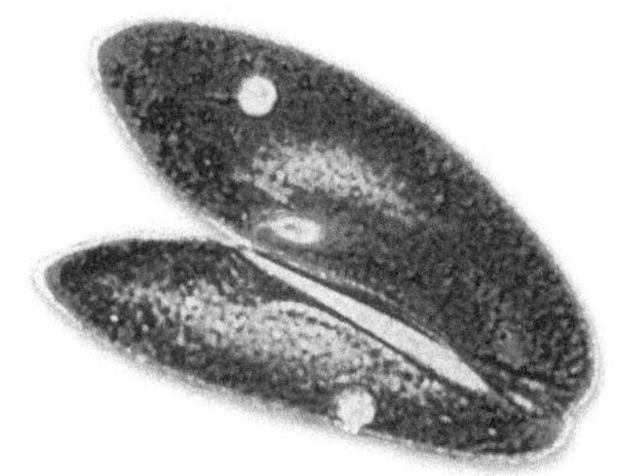

Protists of the genus *Paramecium* can reproduce by conjugation, a kind of sexual reproduction.

Some protists can reproduce both sexually and asexually. For example, some can reproduce asexually when environmental conditions are good. When conditions become difficult, such as when there is little food, the protists reproduce sexually. This allows the offspring to get new combinations of traits that might be helpful under difficult conditions.

What Is a Reproductive Cycle?

Some protists have complex reproductive cycles. These protists may change forms many times. For example, *Plasmodium vivax* is one of the protists that cause the disease malaria. *P. vivax* is a parasite. It needs both humans and mosquitoes to reproduce. Its reproductive cycle is shown below.

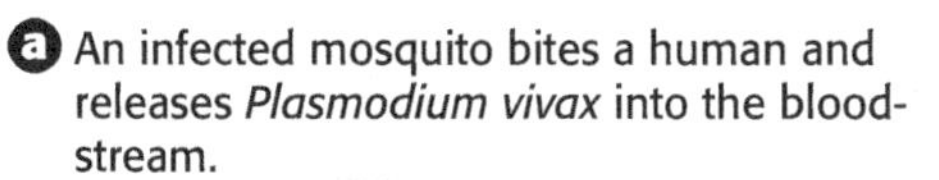

Critical Thinking

5. Apply Concepts Are the offspring produced by conjugation genetically identical to their parents? Explain your answer.

____

____

____

____

____

____

STANDARDS CHECK

LS 2a Reproduction is a characteristic of all living systems; because no living organism lives forever, reproduction is essential to the continuation of every species. Some organisms reproduce asexually. Others reproduce sexually.

6. Explain How does sexual reproduction during times of difficult conditions help protist offspring?

____

____

____

____

____

TAKE A LOOK

7. Identify Inside the human body, where does *P. vivax* reproduce?

____

Name ______________________ Class ______________ Date ______________

Section 1 Review

NSES LS 1b, 1c, 1f, 2a, 4b, 5a

SECTION VOCABULARY

heterotroph an organism that gets food by eating other organisms or their byproducts and that cannot make organic compounds from inorganic materials

host an organism from which a parasite takes food or shelter

parasite an organism that feeds on an organism of another species (the host) and that usually harms the host; the host never benefits from the presence of the parasite

protist an organism that belongs to the kingdom Protista

1. List Name three traits protists have in common.

____

____

____

2. Identify What kind of reproduction results in offspring that are different from their parents? What kind of reproduction results in offspring that are identical to their parents?

____

____

3. Apply Concepts Why are mosquito control programs used to prevent the spread of malaria?

____

____

____

4. List Name four ways protists obtain food.

____

____

____

____

5. Explain Some scientists think kingdom Protista should be divided into several kingdoms. What do you think is the reason for this?

____

____

____

SECTION 2

Kinds of Protists

BEFORE YOU READ

After you read this section, you should be able to answer these questions:

- How are protists classified?
- What structures do protists use to move?

National Science Education Standards
LS 1a, 1b, 1f, 3a, 5a

How Do Scientists Classify Protists?

Protists are hard to classify, or group, because they are a very diverse group of organisms. Scientists are still learning how protists are related to each other. One way scientists classify protists is by their shared traits. Using this method, scientists classify protists into three groups: producers, heterotrophs that can move, and heterotrophs that cannot move. However, these groups do not show how protists are related to each other.

STUDY TIP

Organize As you read this section, make a table that describes each type of protist, how it gets food, and if and how it moves around.

Which Protists Are Producers?

Protist producers are known as **algae** (singular, *alga*). Like plants, algae have chloroplasts. Recall that chloroplasts contain a green pigment called chlorophyll that captures energy from the sun. During photosynthesis, protists use this energy, carbon dioxide, and water to make food. Most algae also have other pigments that give them different colors. ☑

Almost all algae live in water. Some are single-celled, and some are made of many cells. Many-celled algae are called *seaweeds*. They usually live in shallow water and can grow to be many meters long.

Free-floating, single-celled algae are called **phytoplankton**. They are too small to be seen without a microscope. They usually float near the surface of the water. They provide food and oxygen for other organisms.

READING CHECK

1. Identify What are protist producers called?

____

Critical Thinking

2. Compare What is the difference between seaweeds and phytoplankton?

____

____

____

____

____

Diatoms are a kind of plankton.

RED ALGAE

Most of the world's seaweeds are red algae. Their cells contain chlorophyll, but a red pigment gives them their color. Most red algae live in tropical oceans. They can thrive in very deep water. Their red pigment allows them to absorb the light that filters into deep water. ☑

READING CHECK

3. Explain What is the function of red pigment in red algae?

____

____

____

GREEN ALGAE

Green algae are the most diverse group of protist producers. They are green because chlorophyll is their main pigment. Most live in water or moist soil, but they can also live in melting snow, on tree trunks, and inside other organisms. Many are single-celled. Some have many cells and can grow to be eight meters long. Some individual cells live in groups called *colonies*.

Volvox is a green alga that grows in round colonies.

TAKE A LOOK

4. Infer Is *Volvox* single-celled or multicellular? Explain your answer.

____

____

____

BROWN ALGAE

Most of the seaweeds found in cool climates are brown algae. These algae have chlorophyll and a yellow-brown pigment. Brown algae live in oceans. They attach to rocks or form large floating groups called *beds*. ☑

Brown algae can grow to be very large. Some grow 60 m in just one season. That is about as long as 20 cars! Only a brown alga's top is exposed to sunlight. This part of the alga makes food through photosynthesis. The food is carried to parts of the algae that are too deep in the water to get sunlight.

READING CHECK

5. Identify In what type of climate are brown algae likely to be found?

____

DIATOMS

Diatoms make up a large percentage of phytoplankton. Most are single-celled protists found in both salt water and fresh water. They get their energy from photosynthesis. Diatoms are enclosed in thin, two-part shells. The shell is made of a glass-like substance called *silica*. ☑

READING CHECK

6. Identify What substance makes up the thin shells of diatoms?

____

DINOFLAGELLATES

Most dinoflagellates are single-celled. They generally live in salt water, but a few species live in fresh water, or even in snow. Dinoflagellates spin through the water using whiplike structures called *flagella* (singular, *flagellum*). Most dinoflagellates get their energy from photosynthesis. However, a few are consumers, decomposers, or parasites.

EUGLENOIDS

Euglenoids are single-celled protists. They move through the water using flagella. Most have two flagella, one long and one short. They cannot see, but they have eyespots that sense light. A special structure called a *contractile vacuole* removes excess water from the cell.

Euglenoids do not fit well into any one protist group because they can get their food in several ways. Many euglenoids are producers. When there is not enough light to make food, they can be heterotrophs. Some euglenoids are full-time consumers or decomposers. ☑

READING CHECK

7. Explain Why don't euglenoids fit well into any one group of protists?

____

____

____

Structure of Euglenoids

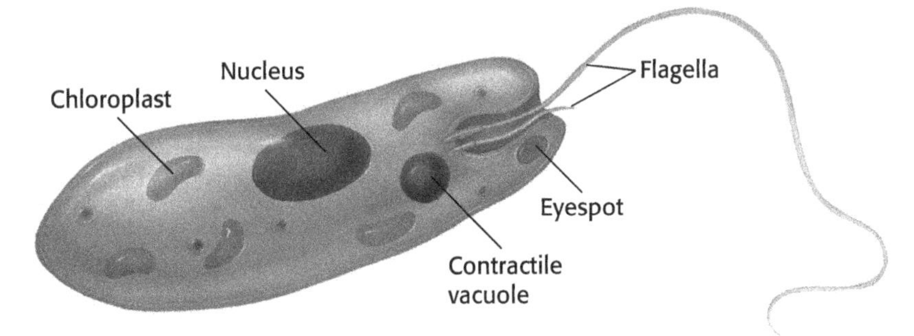

Which Heterotrophs Can Move?

Heterotrophic protists that can move are sometimes called *protozoans*. They are usually single-celled consumers or parasites. They move using special structures, such as pseudopodia, flagella, or cilia. ☑

READING CHECK

8. Identify What is another name for heterotrophic protists that can move?

____

AMOEBAS

Amoebas are soft, jelly-like protozoans. Many amoebas are found in fresh water, salt water, and soil. Amoebas have contractile vacuoles to get rid of excess water. They move and catch food using pseudopodia (singular, *pseudopod*). *Pseudopodia* means "false feet." The figure below shows how an amoeba uses pseudopodia to move.

How Amoebas Move

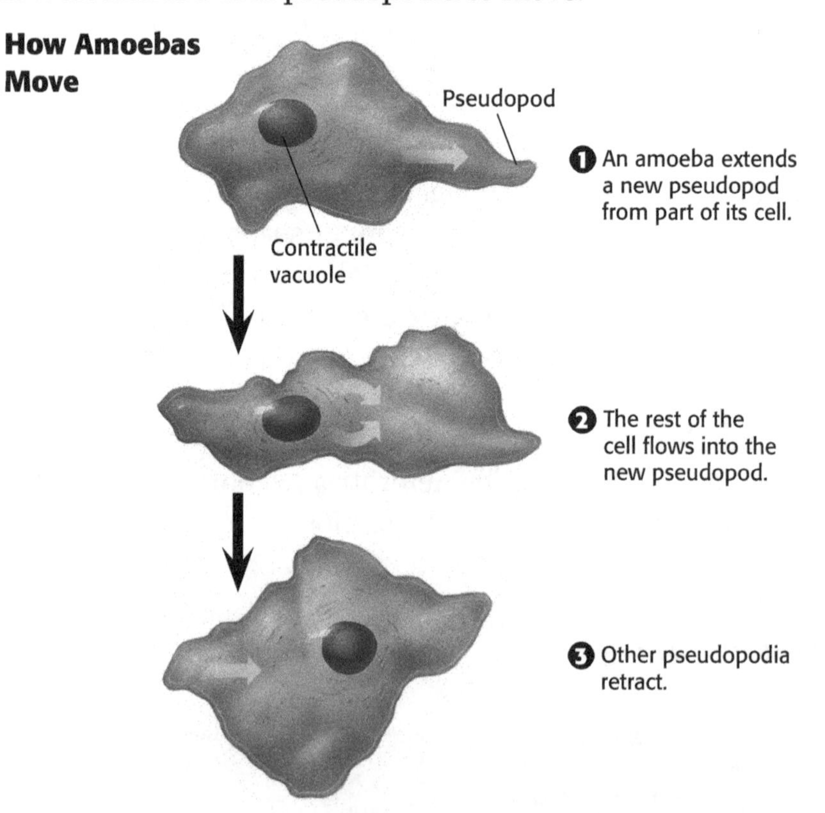

TAKE A LOOK

9. Infer Are pseudopodia fixed structures? Explain your answer.

____

____

____

____

____

Some amoebas are consumers that eat bacteria and smaller protists. Some are parasites that get their food by invading other organisms. These include the amoebas that live in human intestines and cause a painful disease called *amoebic dysentery*.

Amoebas also use pseudopodia to catch food. When an amoeba senses food, it moves toward it using its pseudopodia. It surrounds the food with pseudopodia to form a *food vacuole*. Enzymes move into the vacuole to digest food. The digested food then passes into the amoeba.

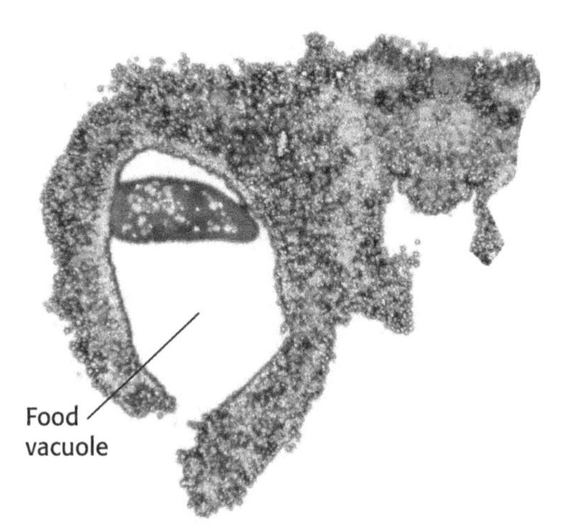

An amoeba uses pseudopodia to catch food.

SHELLED AMOEBA-LIKE PROTISTS

Some amoeba-like protists have an outer shell. They move by poking pseudopodia out of pores in the shells. Some, such as *foraminiferans*, have snail-like shells. Others, such as *radiolarian*, look like glass ornaments. ☑

READING CHECK

10. Identify What are two kinds of shelled amoeba-like protists?

____

____

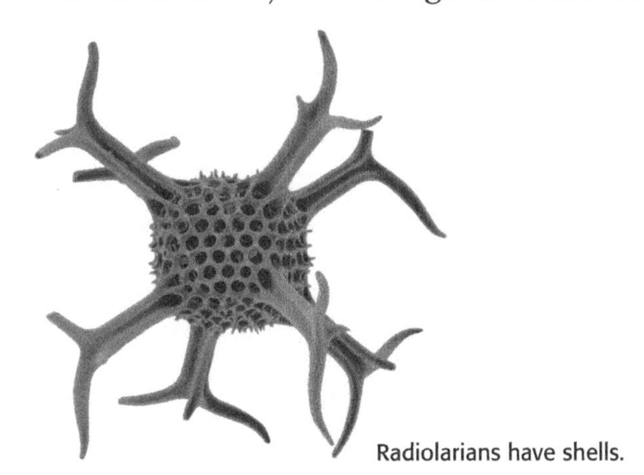

Radiolarians have shells.

ZOOFLAGELLATES

Zooflagellates are heterotrophic protists that move by waving flagella back and forth. Some live in water. Others live in the bodies of other organisms.

Some zooflagellates are parasites that cause disease. People who drink water containing the zooflagellate *Giardia lamblia* can get severe stomach cramps.

Some zooflagellates live in mutualism with other organisms. In *mutualism*, two different organisms live closely together and help each other survive. The zooflagellate shown on the next page lives in the gut of termites. The organisms help each other. The zooflagellate helps the termite digest wood, and the termite gives the protist food and a place to live. ☑

READING CHECK

11. Define What is mutualism?

____

____

____

____

Structure of Flagellates

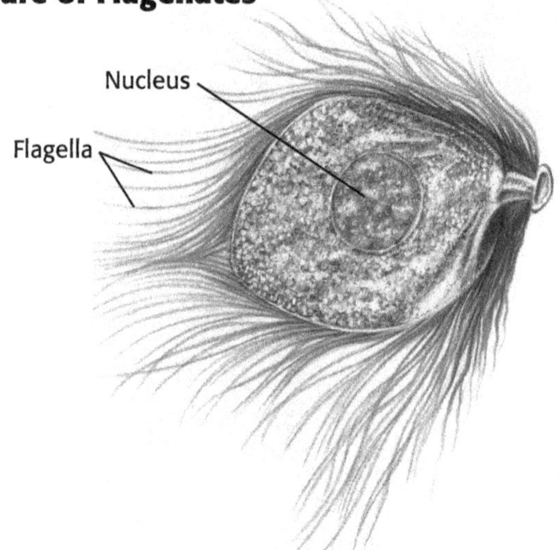

TAKE A LOOK

12. Apply Concepts Is this organism a prokaryote or a eukaryote? Explain your answer.

____

____

CILIATES

Ciliates are complex protists. They have hundreds of tiny, hairlike structures called *cilia*. The cilia beat back and forth very quickly to move the ciliate forward. Cilia also sweep food toward the ciliate's food passageway.

A *Paramecium* is a kind of ciliate. It has several important features:

- a large nucleus, called a *macronucleus*, that controls the cell's functions
- a smaller nucleus, called a *micronucleus*, that can pass genes to another *Paramecium* during sexual reproduction
- a food vacuole, where enzymes digest food
- an anal pore, where food waste is removed from the cell
- a contractile vacuole to remove excess water ☑

READING CHECK

13. Identify What structure in a *Paramecium* is used to exchange genes with another *Paramecium* during sexual reproduction?

____

Structure of Paramecium

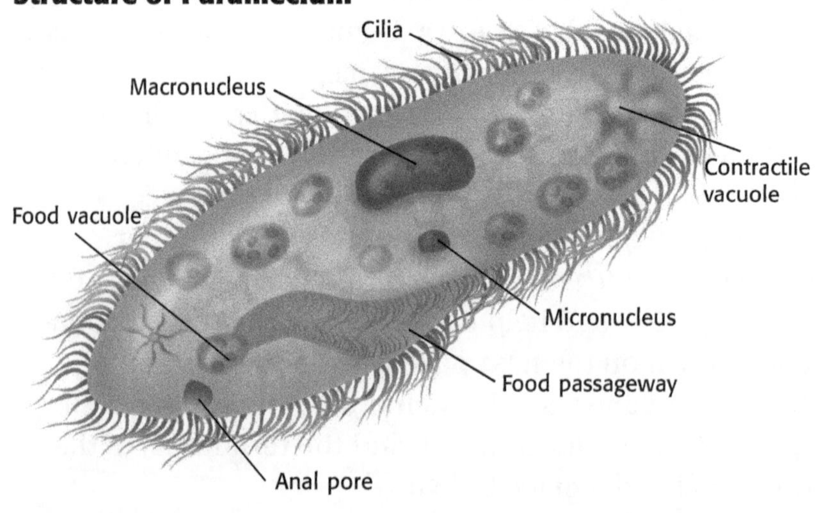

Which Heterotrophs Cannot Move?

Not all protist heterotrophs can move. Some of these are parasites. Others in this group only move at certain phases in their life cycles.

SPORE-FORMING PROTISTS

Many spore-forming protists are parasites. They absorb nutrients from their hosts. Spore-forming protists have complicated life cycles that usually include two or more hosts. For example, the spore-forming protists that cause malaria use mosquitoes and humans as hosts.

WATER MOLDS

Most water molds are small, single-celled protists. They live in water, moist soil, or other organisms. Some water molds are decomposers. Many water molds are parasites. Their hosts can be living plants, animals, algae, or fungi.

SLIME MOLDS

Slime molds live in cool, moist places. They look like colorful globs of slime. At certain phases of their life cycles, slime molds can move using pseudopodia. Some live as a giant cell with many nuclei at one stage of life. Some live as single-celled organisms. Slime molds eat bacteria and yeast. They also surround bits of rotting matter and digest them.

When water or nutrients are hard to find, slime molds grow stalk-like structures with round knobs on top. The knobs are called *sporangia*, and they contain spores. *Spores* are small reproductive cells covered by a thick cell wall. The spores can survive for a long time without water or nutrients. When conditions improve, the spores develop into new slime molds. ☑

The sporangia of a slime mold contain spores.

Critical Thinking

14. Identify What is one characteristic that slime molds, at certain phases of their life cycles, share with amoebas?

____

____

READING CHECK

15. Explain How can spores help a species of slime mold survive difficult conditions?

____

____

____

____

Name ______________________ Class ______________ Date ______________

Section 2 Review

NSES LS 1a, 1b, 1f, 3a, 5a

SECTION VOCABULARY

algae eukaryotic organisms that convert the sun's energy into food through photosynthesis but that do not have roots, stems, or leaves (singular, alga)	**phytoplankton** the microscopic, photosynthetic organisms that float near the surface of marine or fresh water

1. List Name three kinds of protists that use flagella to move.

____

____

2. Organize Fill in the Venn Diagram below to organize the different kinds of protists based on how they get food. Remember that some protists can get food in more than one way.

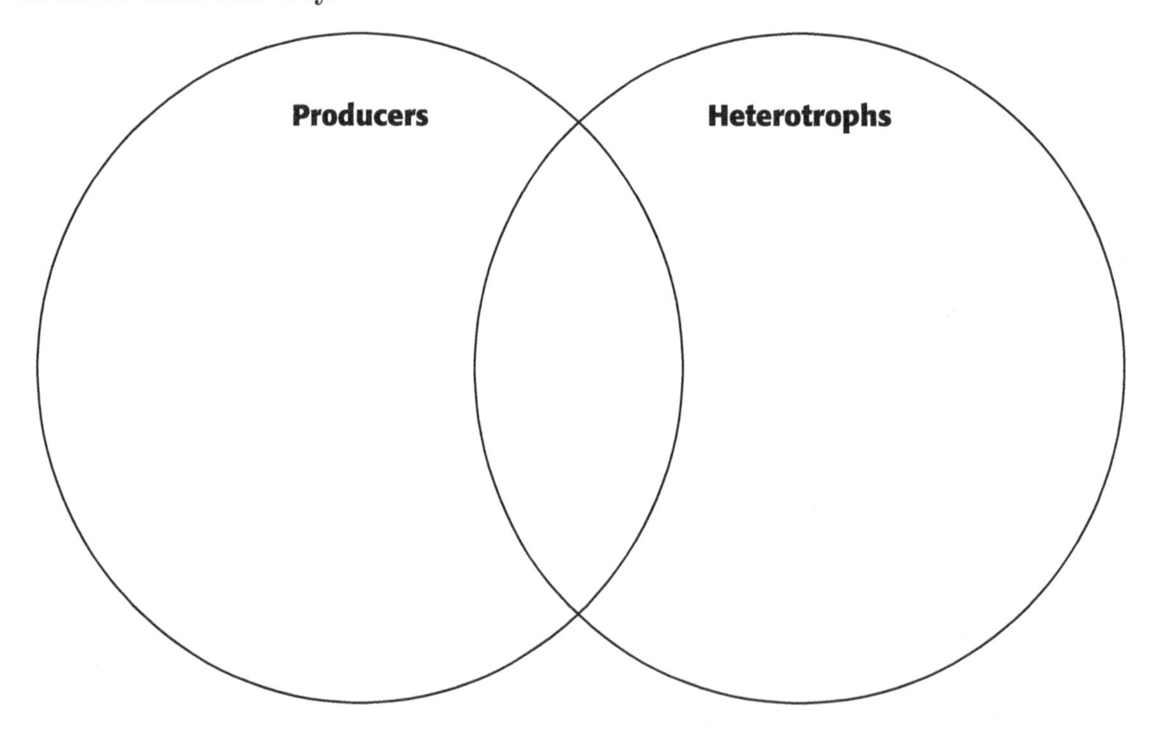

3. List Give two examples of each of the following: protist producers, heterotrophs that can move, and heterotrophs that cannot move.

____

____

____

4. Identify What are two ways amoebas use pseudopodia?

____

____

Name ______________________ Class ______________ Date ______________

CHAPTER 11 Protists and Fungi

SECTION 3

Fungi

BEFORE YOU READ

After you read this section, you should be able to answer these questions:

- What are the characteristics of fungi?
- What are the four groups of fungi?

National Science Education Standards

LS 1a, 1b, 1d, 1f, 2a, 4b, 5a

What Are Fungi?

Fungi (singular, *fungus*) are everywhere. The mushrooms on pizza are a type of fungus. The yeast used to make bread is also a fungus. If you have ever had athlete's foot, you can thank a fungus for that, too.

Fungi are so different from other organisms that they are placed in their own kingdom. There are many different shapes, sizes, and colors of fungi, but they have several characteristics in common.

- They are eukaryotic. (Their cells have nuclei.)
- They are heterotrophs. (They cannot make their own food.)
- Their cells have rigid cell walls.
- They have no chlorophyll.
- They produce reproductive cells called spores.

Compare Make a chart showing how fungi in each group reproduce.

Critical Thinking

1. Compare What are two ways that fungi differ from plants?

____

____

____

▼ Straight coral fungus

▲ Bird's nest fungus

Describe Where have you seen fungi before? What did they look like? With a partner, describe any fungi you have seen.

STANDARDS CHECK

LS 4b Populations of organisms can be categorized by the functions they serve in an ecosystem. Plants and some microorganimsms are producers—they make their own food. All animals, including humans, are consumers, which obtain their food by eating other organisms. Decomposers, primarily bacteria and fungi, are consumers that use waste materials and dead organisms for food. Food webs identify the relationship among producers, consumers, and decomposers in an ecosystem.

Word Help: categorize
to put into groups or classes

Word Help: function
use or purpose

2. List What are three ways that fungi can get food?

____

____

____

Some fungi are single cells, but most are made of many cells. Many-celled fungi are made up of chains of cells called *hyphae* (singular, *hypha*). A **hypha** is a thread-like filament. The cells in these filaments have openings in their cell walls. These openings allow cytoplasm to move between cells.

Most of the hyphae that make up a fungus grow together to form a twisted mass called the **mycelium**. The mycelium is generally the largest part of the fungus. It grows underground.

How Do Fungi Get Nutrients?

Fungi are heterotrophs, which means they feed on other organisms. Unlike other heterotrophs, however, fungi cannot catch or surround food. They must live on or near their food supply. They secrete digestive juices onto food and absorb the nutrients from the dissolved food.

Most fungi are decomposers. They feed on dead plant or animal matter. Some fungi are parasites. Some live in mutualism with other organisms. For example, some fungi grow on or in the roots of a plant. This relationship is called a *mycorrhiza*. The plant provides nutrients to the fungus. The fungus helps the plant absorb minerals and protects it from some diseases.

How Do Fungi Reproduce?

Reproduction in fungi can be asexual or sexual. Many fungi reproduce using spores. **Spores** are small reproductive cells or structures that can grow into a new fungus. They can be the formed by sexual or asexual reproduction. Many spores have thick cell walls that protect them from harsh environments.

ASEXUAL REPRODUCTION

Asexual reproduction can occur in several ways. Some fungi produce spores. The spores are light and easily spread by the wind. Asexual reproduction can also happen when hyphae break apart. Each new piece can become a new fungus. Single-celled fungi called yeasts reproduce through a process called *budding*. In budding, a new cell pinches off from an existing cell. ☑

READING CHECK

3. List What are three types of asexual reproduction in fungi?

____

____

____

SEXUAL REPRODUCTION

In sexual reproduction, special structures form to make sex cells. The sex cells join to produce sexual spores that grow into a new fungus.

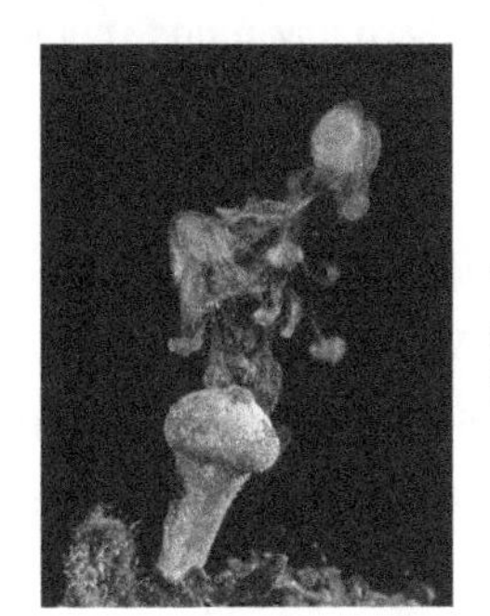

This puffball is releasing sexual spores that can grow into new fungi.

What Are the Four Kinds of Fungi?

Scientists classify fungi based on the shape of the fungi and the way they reproduce. There are four main groups of fungi: threadlike fungi, sac fungi, club fungi, and imperfect fungi.

THREADLIKE FUNGI

Have you ever seen fuzzy black mold growing on bread? A **mold** is a shapeless, fuzzy fungus. Most threadlike fungi live in the soil and are decomposers. However, some are parasites.

Threadlike fungi can reproduce asexually. Structures called *sporangia* (singular, *sporangium*) produce spores. When sporangia break open, they release the spores into the air. ☑

READING CHECK

4. Define What are sporangia?

____

____

____

Threadlike fungi can also reproduce sexually. Hyphae from two different individuals can join together and grow into specialized sporangia. These sporangia can survive in the cold and with little water. When conditions improve, these specialized sporangia release spores that can grow into new fungi.

These groups of sporangia are magnified. Each tiny, round sporangium contains thousands of spores.

TAKE A LOOK

5. Circle On the figure, circle one sporangium.

SAC FUNGI

Sac fungi are the largest group of fungi. Members of this group include yeasts, mildew, truffles, and morels. Sac fungi can reproduce both asexually and sexually. Usually they use asexual reproduction. When they do reproduce sexually, they form a sac called an *ascus*. Sexual spores develop inside the ascus. ☑

6. Identify How do sac fungi usually reproduce?

____

Most sac fungi are multicellular. However, yeasts are single-celled sac fungi. When yeasts reproduce asexually, they use a process called budding. In budding, a new cell pinches off from an existing cell.

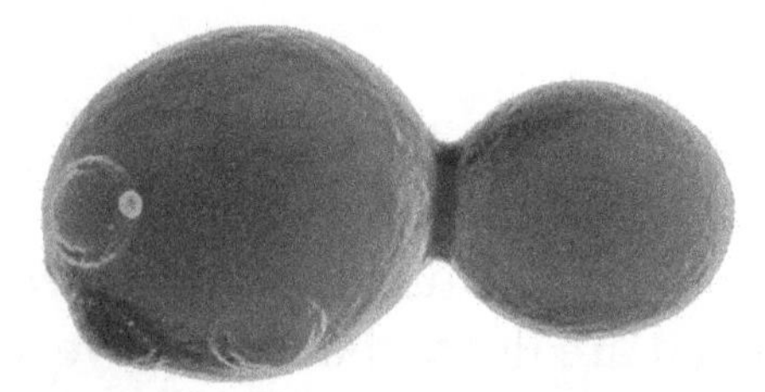

Yeasts are the only fungi that reproduce by budding.

Yeasts use sugar as food and produce carbon dioxide and alcohol as waste. Because of this, humans use yeasts to make products such as alcohol and bread. Bread dough rises because of trapped bubbles of waste carbon dioxide from the yeast.

Some sac fungi are used to make vitamins and antibiotics. Others cause plant diseases. One of these diseases, Dutch elm disease, has killed millions of elm trees. Humans can use some sac fungi, such as morels and truffles, as food. ☑

7. List Name three ways humans use sac fungi.

____

____

____

Morels are only part of a larger sac fungus. They are the reproductive part of a fungus that lives under the soil.

CLUB FUNGI

Mushrooms are the most familiar club fungi. Other club fungi include bracket fungi, smuts, rusts, and puffballs. Bracket fungi grow on wood and form small shelves or brackets. Smuts and rusts are common plant parasites that can attack corn and wheat.

Investigate Use your school's media center to research a bracket, rust, or smut club fungus. Learn about where the fungus lives and how it gets nutrients. Describe this fungus to your class.

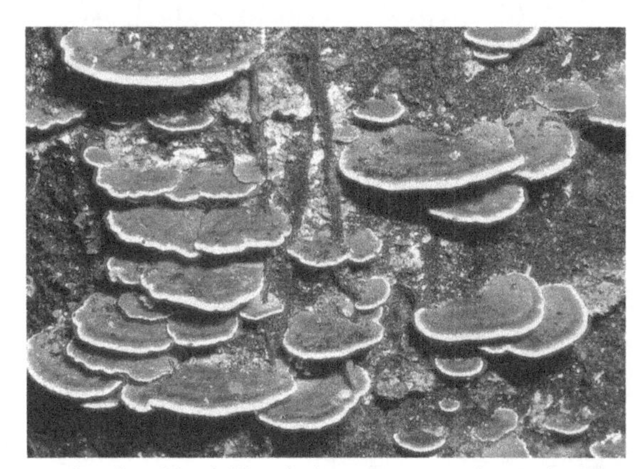
Bracket fungi look like shelves on trees. The underside of the bracket contains spores,

Club fungi reproduce sexually. They grow special hyphae that form clublike structures called *basidia* (singular, *basidium*). Sexual spores develop on the basidia. The spore-producing part of the organism is the only part of a mushroom that is above the ground. Most of the organism is underground. ☑

The most familiar mushrooms are called *gill fungi*. The basidia of these mushrooms develop in structures called *gills*, under the mushroom cap. Some gill fungi are the edible mushrooms sold in supermarket. However, other gill fungi are extremely poisonous.

READING CHECK

8. Identify What part of a mushroom grows above the ground?

____

____

TAKE A LOOK

9. Infer Does this picture show the largest part of the fungus? Explain your answer.

____

____

____

Witch's hat fungus is a gill fungus.

IMPERFECT FUNGI

This group includes all of the fungi that do not fit into the other groups. Imperfect fungi do not reproduce sexually. This group includes some fungi that are harmful to humans and some that are useful.

Most imperfect fungi are parasites that cause diseases in plants and animals. One kind of imperfect fungus causes a skin disease called athlete's foot. Another kind of imperfect fungus produces aflatoxin. *Aflatoxin* is a poison that can cause cancer. Some imperfect fungi are used to make medicines, including the antibiotic penicillin. Other imperfect fungi are used to produce cheeses and soy sauce.

STANDARDS CHECK

LS 1f Disease is the breakdown in structures or functions of an organism. Some diseases are the result of intrinsic failures of the system. Others are the result of damage by infection by other organisms.

10. Identify Name two diseases that can be caused by an imperfect fungus.

____

____

The fungus *Penicillium* produces a substance that kills certain bacteria.

What Are Lichens?

A **lichen** is the combination of a fungus and an alga. The alga lives inside the cell walls of the fungus. This creates a mutualistic relationship. In a mututalistic relationship, both organisms benefit from living closely with another species. Even though lichens are made of two different organisms, scientists give lichens their own scientific names. This is because the two organisms together function like one organism. ☑

READING CHECK

11. Identify What two kinds of organisms make up a lichen?

____

____

These are some of the many types of lichens.

Critical Thinking

12. Infer How do you think a fungus and an alga evolved to live together?

The fungus and alga that make up a lichen benefit each other. Unlike fungi alone, lichens are producers. The algae in the lichens produce food through photosynthesis. Algae alone would quickly dry out in a dry environment. Fungi, however, have protective walls that keep moisture inside the lichen. This allows lichens to live in dry environments.

Lichens need only air, light, and minerals to grow. They can grow in very dry and very cold environments. They can even grow on rocks. Lichens are important to other organisms because they can help form soil. As they grow on a rock, lichens produce acid. The acid breaks down the rock. Soil forms from bits of rock and dead lichens. Once soil forms, plants can move into the area. Animals that eat the plants can also move into the area. ☑

Lichens are sensitive to air pollution. Because of this, they are good *ecological indicators*. This means that if lichens start to die off in an area, there may be something wrong in the environment.

READING CHECK

13. Explain How do lichens make soil?

Name ______________________ Class ______________ Date ______________

Section 3 Review

NSES **LS 1a, 1b, 1d, 1f, 2a, 4b, 5a**

SECTION VOCABULARY

fungus an organism whose cells have nuclei, rigid cell walls, and no chlorophyll and that belongs to the kingdom Fungi

hypha a nonreproductive filament of a fungus

lichen a mass of fungal and algal cells that grow together in a symbiotic relationship and that are usually found on rocks or trees

mold in biology, a fungus that looks like wool or cotton

mycelium the mass of fungal filaments, or hyphae, that forms the body of a fungus

spore a reproductive cell or multicellular structure that is resistant to stressful environmental conditions and that can develop into an adult without fusing with another cell

1. List What are the four groups of fungi?

____

____

2. Explain How does a mycorrhiza help both the plant and the fungus?

____

____

3. Identify Relationships How are a hypha and a mycelium related?

____

4. Identify What part of a club fungus grows above the ground? What part grows below the ground?

____

____

5. Explain What is the function of sporangia?

____

6. Compare How are lichens different from fungi?

____

____

____

7. Compare What is the difference between a lichen and a mycorrhiza?

____

____

8. Identify Which group of fungi forms basidia during sexual reproduction?

____

CHAPTER 12 Introduction to Plants

SECTION 1

What Is a Plant?

BEFORE YOU READ

After you read this section, you should be able to answer these questions:

- What characteristics do all plants share?
- What are two differences between plant cells and animal cells?

National Science Education Standards

LS 1a, 2b, 4c, 5a

What Are the Characteristics of Plants?

A plant is an organism that uses sunlight to make food. Trees, grasses, ferns, cactuses, and dandelions are all types of plants. Plants can look very different, but they all share four characteristics.

STUDY TIP

Organize As you read, make a diagram to show the major groups of plants. Be sure to include the characteristics of each group.

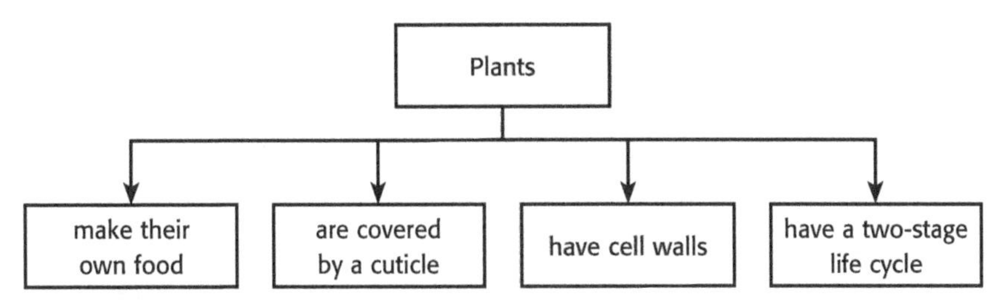

PHOTOSYNTHESIS

Plants make their own food from carbon dioxide, water, and energy from sunlight. This process is called *photosynthesis*. Photosynthesis takes place in special organelles called *chloroplasts*. Inside the chloroplasts, a green pigment called *chlorophyll* collects energy from the sun for photosynthesis. Chlorophyll is what makes most plants look green. Animal cells do not have chloroplasts. ☑

READING CHECK

1. Define What is chlorophyll?

____

____

____

____

CUTICLES

Every plant has a cuticle that covers and protects it. A *cuticle* is a waxy layer that coats a plant's leaves and stem. The cuticle keeps plants from drying out by keeping water inside the plant.

CELL WALLS

How do plants stay upright? They do not have skeletons, as many animals do. Instead, each plant cell is surrounded by a stiff cell wall. The cell wall is outside the cell membrane. Cell walls support and protect the plant cell. Animal cells do not have cell walls.

TAKE A LOOK

2. Identify What structure in a plant cell stores water?

____

____

3. Identify Where is chlorophyll found?

____

Structures in a Plant Cell

Large central vacuole A vacuole stores water and helps support the cell.

Chloroplast Chloroplasts contain chlorophyll. Chlorophyll captures energy from the sun. Plants use this energy to make food.

Cell membrane The cell membrane surrounds a plant cell and lies under the cell wall.

Cell wall The cell wall surrounds the cell membrane. It supports and protects the plant cell.

TWO-STAGE LIFE CYCLE

Many organisms, including plants, produce offspring when a sperm joins with an egg. This is called *sexual reproduction*. In animals, sexual reproduction happens in every generation. However, plants do not produce sperm and eggs in every generation.

Instead, plants have a two-stage life cycle. This means that they need two generations to produce eggs and sperm. In the *sporophyte* stage, a plant makes spores. A spore is a cell that can divide and grow into a new plant. This new plant is called a *gametophyte*. In the gametophyte stage, the plants produce sperm and eggs. The sperm and eggs must join to produce a new sporophyte. ☑

READING CHECK

4. List What are the two stages of the plant life cycle?

____

____

What Are the Main Groups of Plants?

There are two main groups of plants: vascular and nonvascular. A **vascular plant** has specialized vascular tissues. *Vascular tissues* move water and nutrients from one part of a plant to another. A **nonvascular plant** does not have vascular tissues to move water and nutrients.

The Main Groups of Plants

Nonvascular Plants	Vascular Plants		
Mosses, liverworts, and hornworts	Seedless plants	Seed plants	
	Ferns, horsetails, and club mosses	Nonflowering	Flowering
		Gymnosperms	Angiosperms

NONVASCULAR PLANTS

Nonvascular plants depend on diffusion to move water and nutrients through the plant. In *diffusion*, water and nutrients move through a cell membrane and into a cell. Each cell must get water and nutrients from the environment or a cell that is close by.

Nonvascular plants can rely on diffusion because they are small. If a nonvascular plant were large, not all of its cells would get enough water and nutrients. Most nonvascular plants live in damp areas, so each of their cells is close to water.

Critical Thinking

5. Apply Concepts Do you think a sunflower is a gymnosperm or an angiosperm? Explain your answer.

____

____

____

VASCULAR PLANTS

Many of the plants we see in gardens and forests are vascular plants. Vascular plants are divided into two groups: seedless plants and seed plants. Seed plants are divided into two more groups—flowing and nonflowering. Nonflowering seed plants, such as pine trees, are called **gymnosperms**. Flowering seed plants, such as magnolias, are called **angiosperms**.

STANDARDS CHECK

LS 5a Millions of species of animals, plants, and microorganisms are alive today. Although different species might look dissimilar, the unity among organisms becomes apparent from an analysis of internal structures, the similarity of their chemical processes, and the evidence of common ancestry.

Word Help: evidence information showing whether an idea is true or valid

6. List Give three reasons scientists think plants evolved from an ancient green algae.

____

____

____

____

____

____

How Did Plants Evolve?

What would you see if you traveled back in time about 440 million years? The Earth would be a strange, bare place. There would be no plants on land. Where did plants come from?

The green alga in the figure below may look like a plant, but it is not. However, it does share some characteristics with plants. Both algae and plants have the same kind of chlorophyll and make their food by photosynthesis. Like plants, algae also have a two-stage life cycle. Scientists think these similarities mean that plants evolved from a species of green algae millions of years ago.

A modern green alga and plants, such as ferns, share several characteristics. Because of this, scientists think that both types of organisms shared an ancient ancestor.

Section 1 Review

NSES LS 1a, 2b, 4c, 5a

SECTION VOCABULARY

angiosperm a flowering plant that produces seeds within a fruit

gymnosperm a woody, vascular seed plant whose seeds are not enclosed by an ovary or fruit

nonvascular plant the three groups of plants (liverworts, hornworts, and mosses) that lack specialized conducting tissues and true roots, stems, and leaves

vascular plant a plant that has specialized tissues that conduct materials from one part of the plant to another

1. Explain What are the two main differences between a plant cell and an animal cell?

____

____

____

____

2. Organize Fill in each box in the figure below with one of the main characteristics of plants.

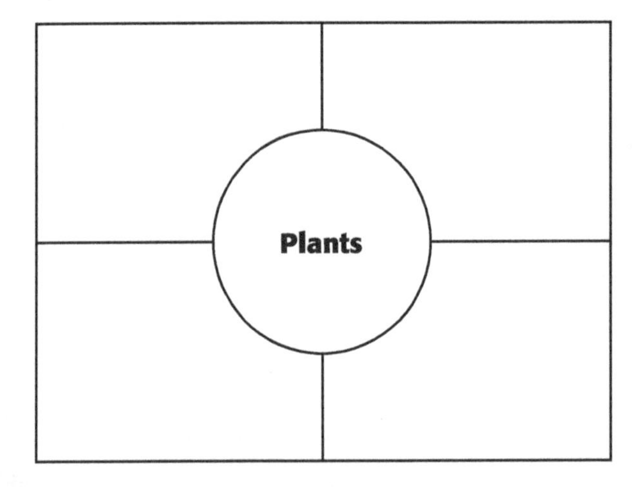

3. Predict What would happen to a plant if its chloroplasts stopped working? Explain your answer.

____

____

____

4. Compare What is the main difference between vascular and nonvascular plants?

____

____

____

Name ______________________ Class ______________ Date ______________

CHAPTER 12 Introduction to Plants

SECTION 2

Seedless Plants

BEFORE YOU READ

After you read this section, you should be able to answer these questions:

- What are the differences between seedless vascular plants and nonvascular plants?
- How can plants reproduce without seeds?

National Science Education Standards

LS 1a, 1c, 2b, 5c

What Are Seedless Plants?

When you think of plants, you probably think of plants that make seeds, such as flowers and trees. However, there are many plants that don't make seeds.

Remember that plants are divided into two main groups: nonvascular plants and vascular plants. All nonvascular plants are seedless, and some vascular plants are seedless, as well.

STUDY TIP

Organize As you read this section, make a chart that compares vascular plants and nonvascular plants.

What Are the Features of Nonvascular Plants?

Mosses, liverworts, and hornworts are types of nonvascular plants. Remember that nonvascular plants do not have vascular tissue to deliver water and nutrients. Instead, each plant cell gets water and nutrients directly from the environment or from a nearby cell. Therefore, nonvascular plants usually live in places that are damp.

Nonvascular plants do not have true stems, roots, or leaves. However, they do have features that help them to get water and stay in place. For example, a **rhizoid** is a rootlike structure that holds some nonvascular plants in place. Rhizoids also help plants get water and nutrients. ☑

Nonvascular plants
• have no vascular tissue
• have no true roots, stems, leaves, or seeds
• are usually small
• live in damp places

Critical Thinking

1. Apply Concepts Why wouldn't you expect to see nonvascular plants in the desert?

____

____

____

____

READING CHECK

2. List What are two functions of the rhizoid?

____

____

____

REPRODUCTION IN NONVASCULAR PLANTS

Like all plants, nonvascular plants have a two-stage life cycle. They have a sporophyte generation, which produces spores, and a gametophyte generation, which produces eggs and sperm. Sperm from these plants need water so they can swim to the eggs. Nonvascular plants can also reproduce asexually, that is, without eggs and sperm.

Moss Life Cycle

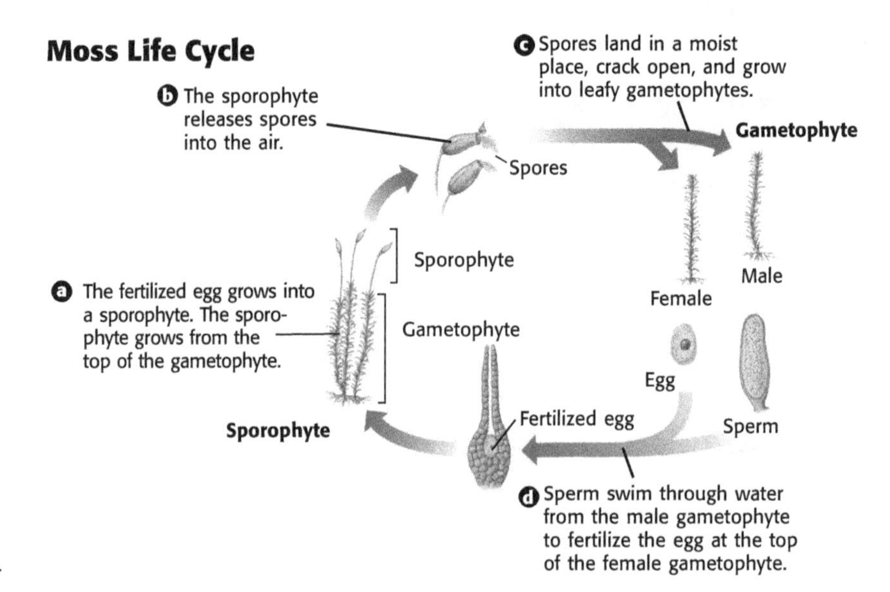

TAKE A LOOK

3. Identify Are the male and female gametophytes separate plants or part of the same plant?

____

IMPORTANCE OF NONVASCULAR PLANTS

Nonvascular plants are usually the first plants to live in a new environment, such as newly exposed rock. When these plants die, they break down and help form a thin layer of soil. Then plants that need soil in order to grow can move into these areas.

Some nonvascular plants are important as food or nesting material for animals. A nonvascular plant called peat moss is important to humans. When it turns to peat, it can be burned as a fuel.

Critical Thinking

4. Apply Concepts Why do you think nonvascular plants can be the first plants to grow in a new environment?

____

____

____

____

What Are the Features of Seedless Vascular Plants?

Vascular plants have specialized tissues that carry water and nutrients to all their cells. These tissues generally make seedless vascular plants larger than nonvascular plants. Because they have tissues to move water, vascular plants do not have to live in places that are damp. ☑

Many seedless vascular plants, such as ferns, have a structure called a rhizome. The **rhizome** is an underground stem that produces new leaves and roots.

READING CHECK

5. Explain How do the cells of a seedless vascular plant get water?

____

____

____

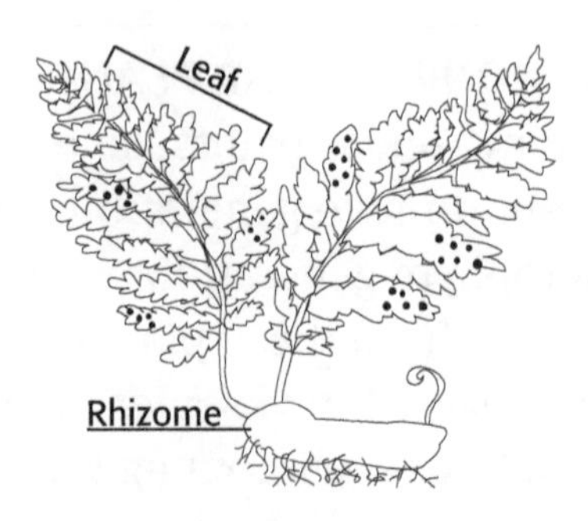

REPRODUCTION IN SEEDLESS VASCULAR PLANTS

The life cycles of vascular plants and nonvascular plants are similar. Sperm and eggs are produced in gametophytes. They join to form a sporophyte. The sporophyte produces spores. The spores are released. They grow into new gametophytes. ☑

Seedless vascular plants can also reproduce asexually in two ways. New plants can branch off from older plants. Pieces of a plant can fall off and begin to grow as new plants.

READING CHECK

6. Identify What do spores grow into?

____

Fern Life Cycle

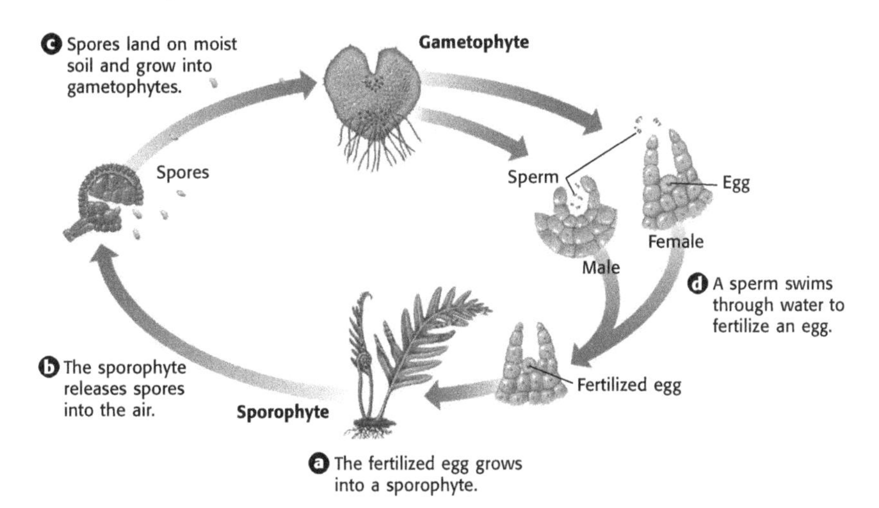

TAKE A LOOK

7. Apply Concepts Does this figure show sexual or asexual reproduction? Explain your answer.

____

____

____

____

IMPORTANCE OF SEEDLESS VASCULAR PLANTS

Did you know that seedless vascular plants that lived 300 million years ago are important to people today? After these ancient ferns, horsetails, and club mosses died, they formed coal and oil. Coal and oil are fossil fuels that people remove from Earth's crust to use for energy. They are called *fossil fuels* because they formed from plants (or animals) that lived long ago. ☑

Another way seedless vascular plants are important is they help make and preserve soil. Seedless vascular plants help form new soil when they die and break down. Their roots can make the soil deeper, which allows other plants to grow. Their roots also help prevent soil from washing away.

Many seedless vascular plants are used by humans. Ferns and some club mosses are popular houseplants. Horsetails are used in some shampoos and skincare products.

READING CHECK

8. Explain Where does coal come from?

____

____

____

____

____

Name ______________________ Class ______________ Date ______________

Section 2 Review

NSES LS 1a, 1c, 2b, 5c

SECTION VOCABULARY

rhizoid a rootlike structure in nonvascular plants that holds the plants in place and helps plants get water and nutrients	**rhizome** a horizontal underground stem that produces new leaves, shoots, and roots

1. Compare What are two differences between a rhizoid and a rhizome?

____

____

____

____

2. Explain In which generation does sexual reproduction occur? Explain your answer.

____

____

3. Compare Use a Venn Diagram to compare vascular and nonvascular plants.

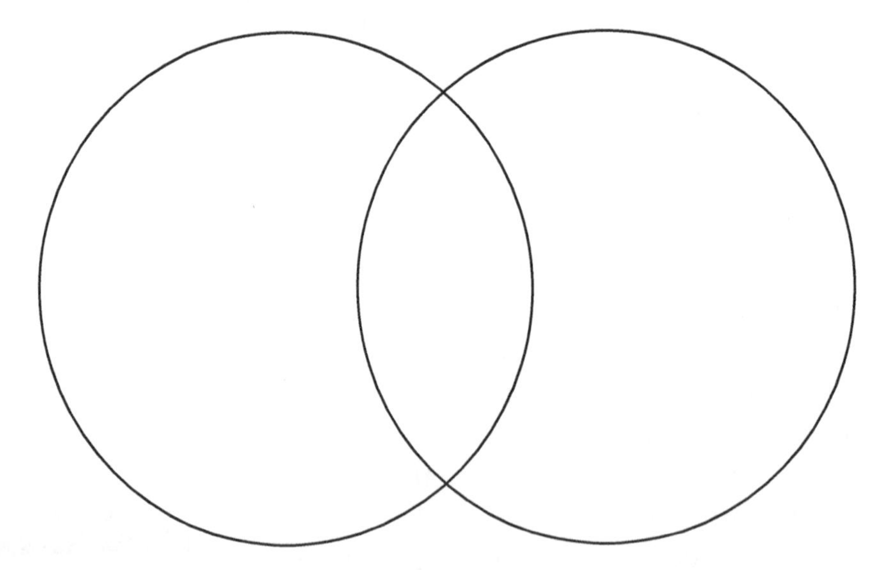

4. Describe What are two ways in which seedless nonvascular plants reproduce asexually?

____

____

5. Apply Concepts Nonvascular plants are usually very small. How does their structure limit their size?

____

____

Name ______________________ Class ______________ Date ______________

CHAPTER 12 Introduction to Plants

SECTION 3

Seed Plants

BEFORE YOU READ

After you read this section, you should be able to answer these questions:

- How are seed plants different from seedless plants?
- What are the parts of a seed?
- How do gymnosperms and angiosperms reproduce?

National Science Education Standards
LS 1a, 1d, 2b, 2c, 2d, 4b, 4c, 4d, 5b

What Are Seed Plants?

Think about the seed plants that you use during the day. You probably use dozens of seed plants, including the food you eat and the paper you write on. Seed plants include trees, such as oaks and pine trees, as well as flowers, such as roses and dandelions. Seed plants are one of the two main groups of vascular plants.

STUDY TIP

Organize As you read this section, make cards showing the parts of the life cycle of seed plants. Practice arranging the cards in the correct sequence.

Like all plants, seed plants have a two-stage life cycle. However, seed plants differ from seedless plants, as shown below.

Seedless plants	Seed plants
They do not produce seeds.	They produce seeds.
The gametophyte grows as an independent plant.	The gametophyte lives inside the sporophyte.
Sperm need water to swim to the eggs.	Sperm are carried to the eggs by pollen.

TAKE A LOOK

1. Compare How do the gametophytes of seedless plants differ from those of seed plants?

____

Seed plants do not depend on moist habitats for reproduction, the way seedless plants do. Because of this, seed plants can live in many more places than seedless plants can. Seed plants are the most common plants on Earth today.

What Is a Seed?

A *seed* is a structure that feeds and protects a young plant. It forms after fertilization, when a sperm and an egg join. A seed has the following three main parts:

- a young plant, or sporophyte
- *cotyledons*, early leaves that provide food for the young plant
- a seed coat that covers and protects the young plant ☑

READING CHECK

2. Identify What process must occur before a seed can develop?

____

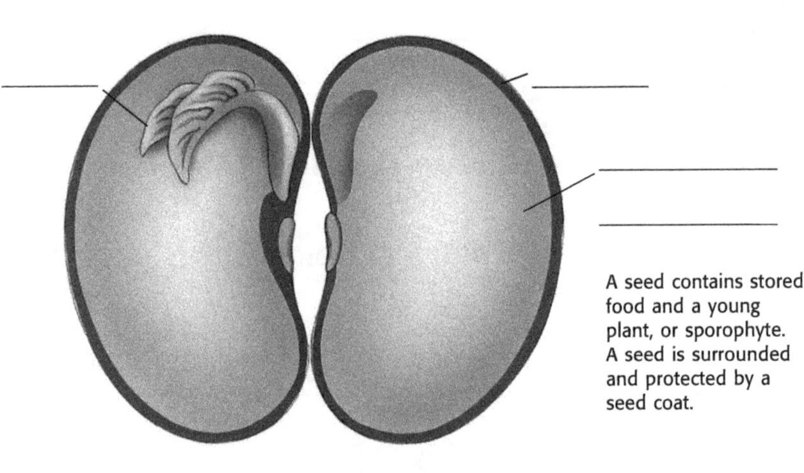

A seed contains stored food and a young plant, or sporophyte. A seed is surrounded and protected by a seed coat.

TAKE A LOOK

3. Label Label the parts of a seed with these terms: young plant, seed coat, cotyledon.

ADVANTAGES OF HAVING SEEDS

Seeds have some advantages over spores. For example, when the young plant inside a seed begins to grow, it uses the food stored in the seed. In contrast, the spores of seedless plants don't have stored food to help a new plant grow. Therefore, they will live only if they start growing when and where there are enough resources available.

Another advantage is that seeds can be spread by animals. The spores of seedless plants are usually spread by wind. Animals often spread seeds more efficiently than the wind spreads spores. Therefore, seeds that are spread by animals are more likely to find a good place to grow.

STANDARDS CHECK

LS 5b Biological evolution accounts for the diversity of species developed through gradual processes over many generations. Species acquire many of their unique characteristics through biological adaptation, which involves the selection of naturally occurring variations in populations. Biological adaptations include changes in structures, behaviors, or physiology that enhance survival and reproductive success in a particular environment.

Word Help: diversity
variety

Word Help: process
a set of steps, events, or changes

Word Help: survival
the continuing to live or exist

4. Identify What are two advantages seeds have over spores?

What Kinds of Plants Have Seeds?

Seed plants are divided into two main groups: gymnosperms and angiosperms. *Gymnosperms* are nonflowering plants, and *angiosperms* are flowering plants.

GYMNOSPERMS

Gymnosperms are seed plants that do not have flowers or fruits. They include plants such as pine trees and redwood trees. Many gymnosperms are evergreen, which means that they keep their leaves all year. Gymnosperm seeds usually develop in a cone, such as a pine cone.

REPRODUCTION IN GYMNOSPERMS

The most well-known gymnosperms are the conifers. Conifers are evergreen trees and shrubs, such as pines, spruces, and firs, that make cones to reproduce. They have male cones and female cones. Spores in male cones develop into male gametophytes, and spores in female cones develop into female gametophytes. The gametophytes produce sperm and eggs.

A **pollen** grain contains the tiny male gametophyte. The wind carries pollen from the male cones to the female cones. This movement of pollen to the female cones is called **pollination**. Pollination is part of sexual reproduction in plants. ☑

After pollination, sperm fertilize the eggs in the female cones. A fertilized egg develops into a new sporophyte inside a seed. Eventually, the seeds fall from the cone. If the conditions are right, the seeds will grow into plants.

5. Explain How is gymnosperm pollen carried from one plant to another?

____

The Life Cycle of a Pine Tree

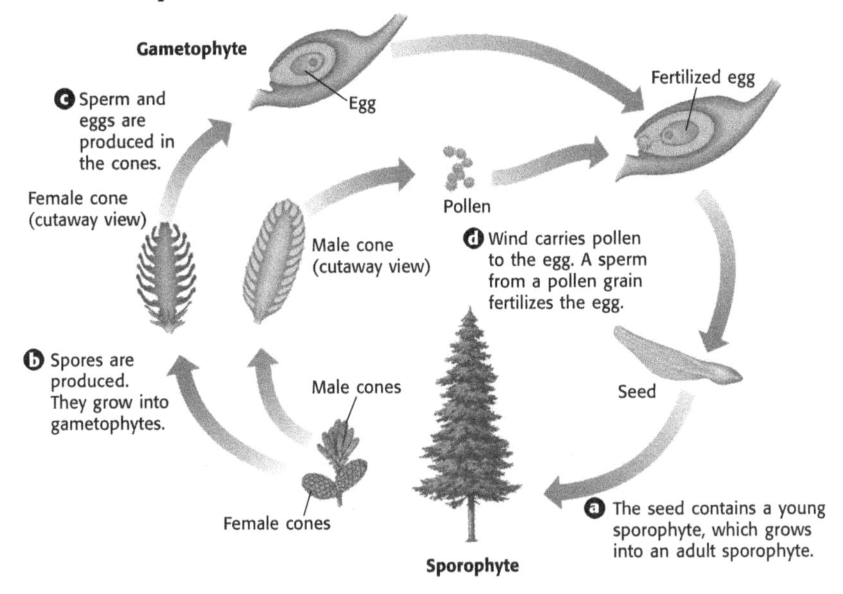

TAKE A LOOK

6. Explain Does this picture show an example of sexual or asexual reproduction? Explain.

____

____

____

____

IMPORTANCE OF GYMNOSPERMS

Gymnosperms are used to make many products, such as medicines, building materials, and household products. Some conifers produce a drug used to fight cancer. Many trees are cut so that their wood can be used to build homes and furniture. Pine trees make a sticky substance called resin. Resin can be used to make soap, paint, and ink.

What Are Angiosperms?

Angiosperms are seed plants that produce flowers and fruit. Maple trees, daisies, and blackberries are all examples of angiosperms. There are more angiosperms on Earth than any other kind of plant. They can be found in almost every land ecosystem, including grasslands, deserts, and forests.

Math Focus

7. Calculate Percentages More than 265,000 species of plants have been discovered. About 235,000 of those species are angiosperms. What percentage of plants are angiosperms?

____

TWO KINDS OF ANGIOSPERMS

There are two kinds of angiosperms: monocots and dicots. These plants are grouped based on how many cotyledons, or seed leaves, the seeds have. *Monocots* have seeds with one cotyledon. Grasses, orchids, palms, and lilies are all monocots.

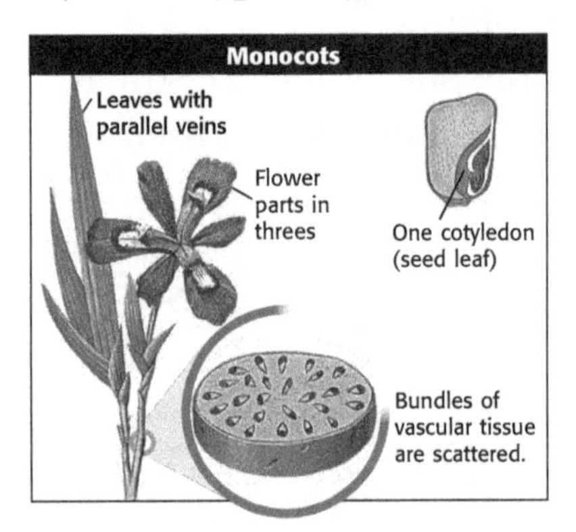

Dicots have seeds with two cotyledons. Roses, sunflowers, peanuts, and peas are all dicots.

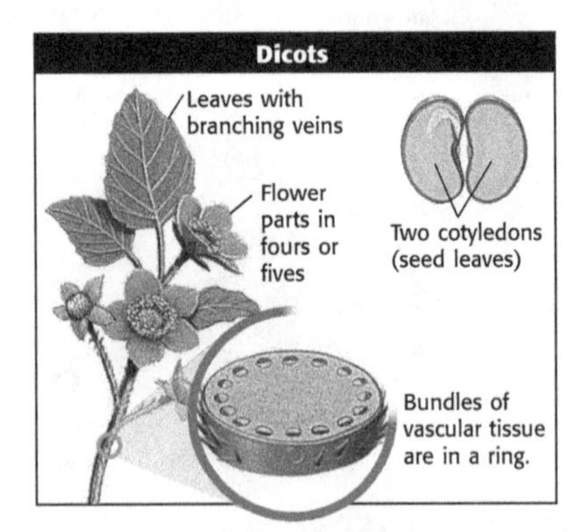

Monocots	Dicots
	flower parts in fours or fives
leaves with parallel veins	
	two cotyledons
bundles of vascular tissue scattered	

TAKE A LOOK

8. Complete Fill in the table to show the differences between monocots and dicots.

REPRODUCTION IN ANGIOSPERMS

In angiosperms, pollination takes place in flowers. Some angiosperms depend on the wind for pollination. Others rely on animals such as bees and birds to carry pollen from flower to flower.

Angiosperm seeds develop inside fruits. Some fruits and seeds, like those of a dandelion, are made to help the wind carry them. Other fruits, such as blackberries, attract animals that eat them. The animals drop the seeds in new places, where they can grow into plants. Some fruits, such as burrs, travel by sticking to animal fur. ☑

9. Identify Where do angiosperm seeds develop?

____

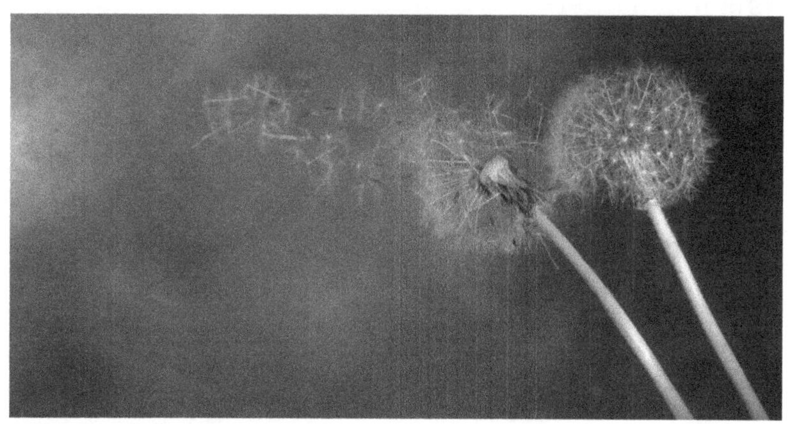

Each of the fluffy structures on this dandelion is actually a fruit. Each of the fruits contains a seed.

TAKE A LOOK

10. Identify How are the fruits of this dandelion spread?

____

IMPORTANCE OF ANGIOSPERMS

Like many other plants, flowering plants provide food for animals. A mouse that eats seeds and berries uses flowering plants directly as food. An owl that eats a field mouse uses flowering plants indirectly as food. Flowering plants can also provide food for the animals that pollinate them.

People use flowering plants, too. Major food crops, such as corn, wheat, and rice, come from flowering plants. Many flowering trees, such as oak trees, can be used for building materials. Plants such as cotton and flax are used to make clothing and rope. Flowering plants are also used to make medicines, rubber, and perfume oils.

Describe Think of all the products you used today that came from angiosperms. Describe to the class five items you used in some way and what kind of angiosperm they came from.

Name _______________ Class _______________ Date _______________

Section 3 Review

NSES LS 1a, 1d, 2b, 2c, 2d, 4b, 4c, 4d, 5b

SECTION VOCABULARY

pollen the tiny granules that contain the male gametophyte of seed plants	**pollination** the transfer of pollen from the male reproductive structures to the female reproductive structures of seed plants

1. Compare How are the gametophytes of seed plants different from the gametophytes of seedless plants?

____

____

____

2. Describe What happens during pollination?

____

____

3. Compare Use a Venn Diagram to compare gymnosperms and angiosperms.

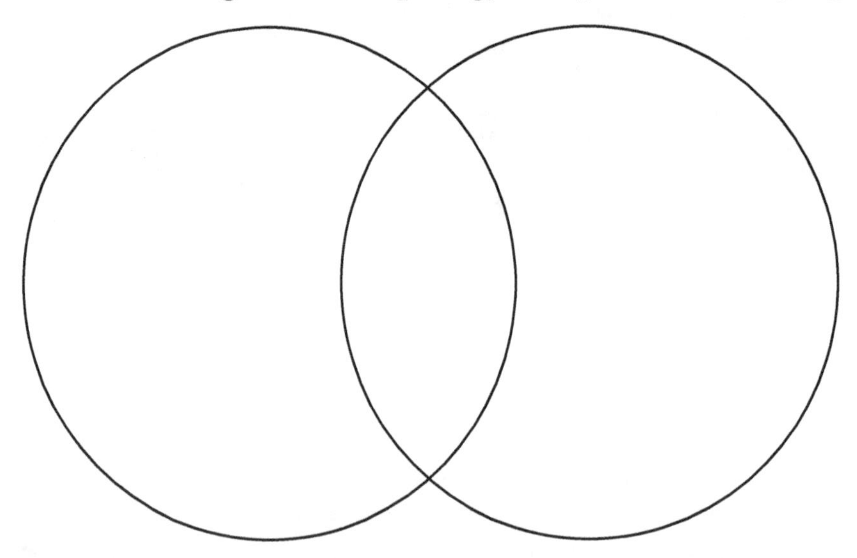

4. List What are the three main parts of a seed? What does each part do?

____

____

____

5. Identify In what structure do gymnosperm seeds usually develop?

____

6. Identify What two structures are unique to angiosperms?

____

Name ______________________ Class ______________ Date ______________

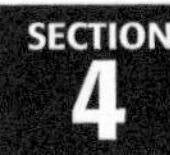

SECTION 4 Structures of Seed Plants

BEFORE YOU READ

After you read this section, you should be able to answer these questions:

- What are the functions of roots and stems?
- What is the function of leaves?
- What is the function of a flower?

National Science Education Standards
LS 1a, 1d, 2b, 3d, 4c, 5b

What Structures Are Found in a Seed Plant?

Remember that seed plants include trees, such as oaks and pine trees, as well as flowers, such as roses and dandelions. Seed plants are one of the two main groups of vascular plants.

You have different body systems that carry out many functions. Plants have systems too. Vascular plants have a root system, a shoot system, and a reproductive system. A plant's root and shoot systems help the plant to get water and nutrients. Roots are often found underground. Shoots include stems and leaves. They are usually found above ground. ☑

STUDY TIP

List As you read this section, make a chart listing the structures of seed plants and their functions.

READING CHECK

1. Identify What are the three main parts of a seed plant?

____

____

____

Onion

Dandelion

Carrots

The roots of plants absorb and store water and nutrients.

STANDARDS CHECK

LS 1a Living systems at all levels of organization demonstrate the complementary nature of structure and function. Important levels of organization for structure and function include cells, organs, tissues, organ systems, whole organisms, and ecosystems.

Word Help: structure
the way in which a whole is put together

Word Help: function
use or purpose

2. Describe What are the functions of xylem and phloem?

____

____

____

____

____

VASCULAR TISSUE

Like all vascular plants, seed plants have specialized tissues that move water and nutrients through the plant. There are two kinds of vascular tissue: xylem and phloem. **Xylem** moves water and minerals from the roots to the shoots. **Phloem** moves food molecules to all parts of the plant. The vascular tissues in the roots and shoots are connected.

What Are Roots?

Roots are organs that have three main functions:

- to absorb water and nutrients from the soil
- to hold plants in the soil
- to store extra food made in the leaves

Roots have several structures that help them do these jobs. The *epidermis* is a layer of cells that covers the outside of the root, like skin. Some cells of the epidermis, called *root hairs*, stick out from the root. These hairs expose more cells to water and minerals in the soil. This helps the root absorb more of these materials.

Roots grow longer at their tips. A *root cap* is a group of cells found at the tip of a root. These cells produce a slimy substance. This helps the root push through the soil as it grows.

Critical Thinking

3. Apply Concepts What do you think happens to water and minerals right after they are absorbed by roots?

____

____

____

The Parts of a Root

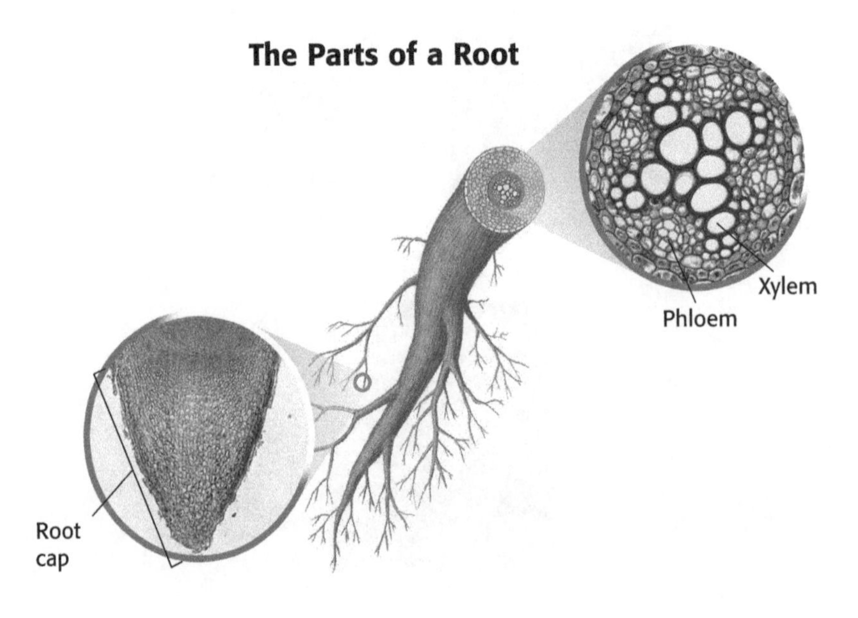

TAKE A LOOK

4. Identify Where is the vascular tissue located in this root?

____

____

TYPES OF ROOT SYSTEMS

There are two kinds of root systems: taproot systems and fibrous root systems. A *taproot system* has one main root, or taproot, that grows downward. Many smaller roots branch from the taproot. Taproots can reach water deep underground. Carrots are plants that have taproot systems.

A *fibrous root system* has several roots that spread out from the base of a plant's stem. The roots are usually the same size. Fibrous roots usually get water from near the soil surface. Many grasses have fibrous root systems.

What Are Stems?

A stem is an organ that connects a plant's roots to its leaves and reproductive structures. A stem does the following jobs: ☑

- Stems support the plant body. Leaves are arranged along stems so that each leaf can get sunlight.
- Stems hold up reproductive structures such as flowers. This helps bees and other pollinators find the flowers.
- Stems carry materials between the root system and the leaves and reproductive structures. Xylem carries water and minerals from the roots to the rest of the plant. Phloem carries the food made in the leaves to roots and other parts of the plant.
- Some stems store materials. For example, the stems of cactuses can store water.

READING CHECK

5. Define What is a stem?

____

____

____

____

HERBACEOUS STEMS

There are two different types of stems: herbaceous and woody. *Herbaceous* stems are thin, soft, and flexible. Flowers, such as daisies and clover, have herbaceous stems. Many crops, such as tomatoes, corn, and beans, also have herbaceous stems.

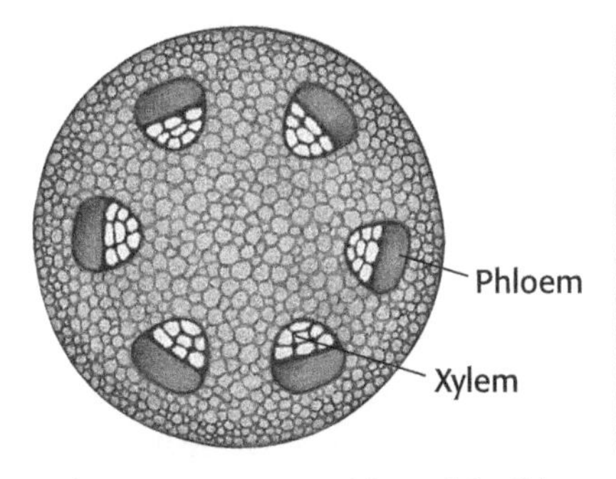

Herbaceous stems are thin and flexible

TAKE A LOOK

6. Compare Examine this figure and the pictures of woody stems on the next page. How are herbaceous and woody stems similar?

____

____

Critical Thinking

7. Infer How do you think growth rings can be used to tell how old a tree is?

____

____

____

____

____

____

____

WOODY STEMS

Other plants have woody stems. *Woody* stems are stiff and are often covered by bark. Trees and shrubs have woody stems. The trunk of a tree is actually its stem!

Trees or shrubs that live in areas with cold winters grow mostly during the spring and summer. During the winter, these plants are *dormant*. This means they are not growing or reproducing. Plants that live in areas with wet and dry seasons are dormant during the dry season.

When a growing season starts, the plant produces large xylem cells. These large cells appear as a light-colored ring when the plant stem is cut. In the fall, right before the dormant period, the plant produces smaller xylem cells. The smaller cells produce a dark ring in the stem. A ring of dark cells surrounding a ring of light cells makes up a *growth ring*. The number of growth rings can show how old the tree is.

TAKE A LOOK

8. Compare How are herbaceous and woody stems different?

____

____

____

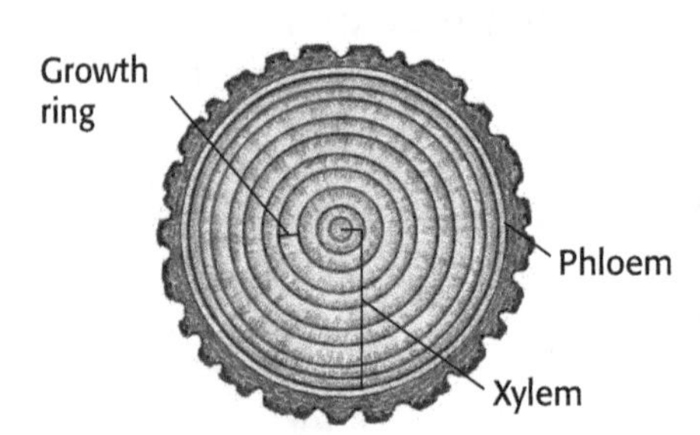

Woody stems are usually thick and stiff.

What Are Leaves?

FUNCTION OF LEAVES

Leaves are organs, too. The main function of leaves is to make food for the plant. The leaves are where most photosynthesis happens. Chloroplasts in the leaf cells trap energy from sunlight. The leaves also absorb carbon dioxide from the air. They use this energy, carbon dioxide, and water to make food. ☑

READING CHECK

9. Identify What is the main function of a leaf?

____

All leaf structures are related to the leaf's main job, photosynthesis. A *cuticle* covers the surfaces of the leaf. It prevents the leaf from losing water. The *epidermis* is a single layer of cells beneath the cuticle. Tiny openings in the epidermis, called *stomata* (singular, *stoma*), let carbon dioxide enter the leaf. *Guard cells* open and close the stomata.

Structure of a Leaf

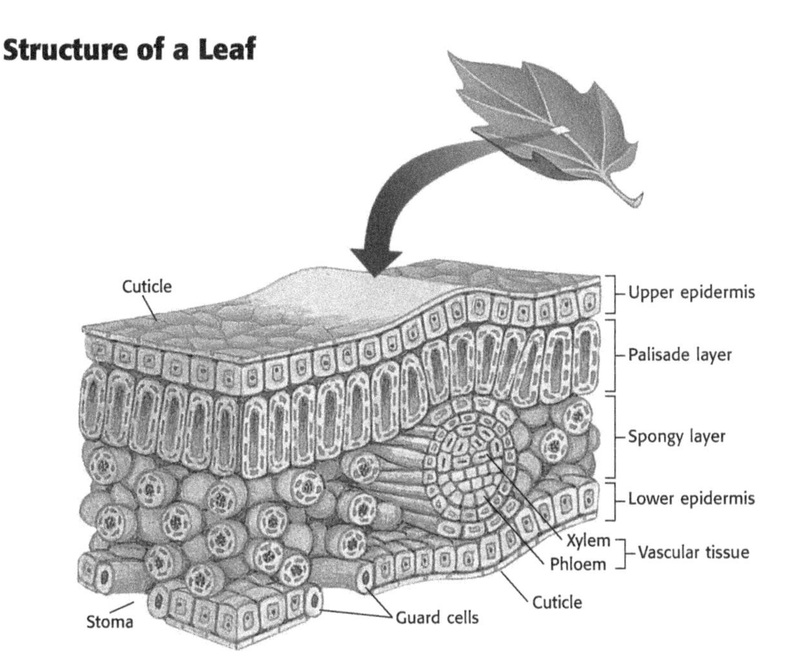

TAKE A LOOK

10. Explain Is this plant vascular or nonvascular? Explain your answer.

____

____

____

LEAF LAYERS

Most photosynthesis takes place in the two layers in the middle of the leaf. The upper layer, called the *palisade layer*, contains many chloroplasts. Sunlight is captured in this layer. The lower layer, called the *spongy layer*, has spaces between the cells, where carbon dioxide can move. The spongy layer also has the vascular tissues that bring water to the leaves and move food away.

LEAF SHAPES

Different kinds of plants can have different shaped leaves. Leaves may be round, narrow, heart-shaped, or fan-shaped. Leaves can also be different sizes. The raffia palm has leaves that may be six times longer than you are tall! Duckweed is a tiny plant that lives in water. Its leaves are so small that several of them could fit on your fingernail. Some leaves, such as those of poison ivy below, can be made of several leaflets.

Describe Some people are allergic to poison ivy. They can get a rash from touching its leaves. Some other plants can be poisonous to eat. Are there any other plants you know of that can be poisonous to touch or eat? Describe some of these plants to a partner.

This is one poison ivy leaf. It is made up of three leaflets

What Are Flowers?

All plants have reproductive structures. In angiosperms, or flowering plants, flowers are the reproductive structures. Flowers produce eggs and sperm for sexual reproduction. ☑

READING CHECK

11. Identify For which group of plants are flowers the reproductive structures?

____

PARTS OF A FLOWER

There are four basic parts of a flower: sepals, petals, stamens, and one or more pistils. These parts are often arranged in rings, one inside the other. However, not all flowers have every part.

Different species of flowering plants can have different flower types. Flowers with all four parts are called *perfect flowers*. Flowers that have stamens but no pistils are male. Flowers that have pistils but no stamens are female.

TAKE A LOOK

12. Label As you read, fill in the missing labels on the diagram.

13. Identify What two parts make up the stamen?

____

____

14. Identify What three parts make up the pistil?

____

____

____

Parts of a Flower

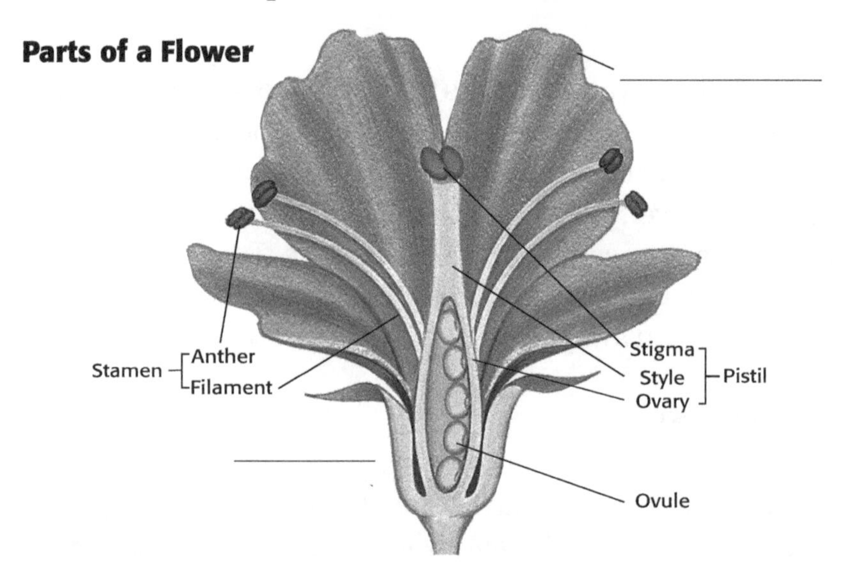

SEPALS

Sepals are leaves that make up the outer ring of flower parts. They are often green like leaves, but they may have other colors. Sepals protect and cover the flower while it is still a bud. When the flower begins to open, the sepals fold back, so the petals can be seen.

PETALS

Petals are leaflike parts of a flower. They make up the next ring inside of the sepals. Petals are sometimes brightly colored, like the petals of poppy flowers or roses. Many plants need animals to help spread their pollen. These colors help attract insects and other animals.

STAMENS

A **stamen** is the male reproductive structure of a flower. Structures on the stamen called *anthers* produce pollen. Pollen contains the male gametophyte, which produces sperm. The anther rests on a thin stalk called a *filament.* ☑

READING CHECK

15. Identify What is the male reproductive structure of a flower?

____

PISTILS

A **pistil** is the female reproductive structure. The tip of the pistil is called the *stigma*. The long, thin part of the pistil is called the *style*. The rounded base of the pistil is called the **ovary**. The ovary contains one or more ovules. Each ovule contains an egg.

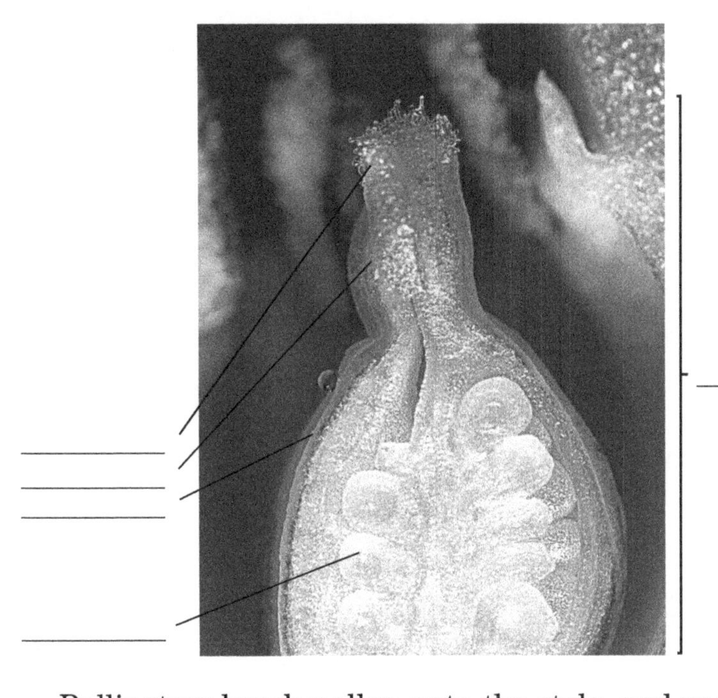

TAKE A LOOK

16. Label Label the female reproductive structures in this picture.

Pollinators brush pollen onto the style, and sperm from inside the pollen travel down the style to the ovary. One sperm can fertilize the egg of one ovule. After fertilization, an ovule develops into a seed. The ovary surrounding the ovule develops into a fruit.

IMPORTANCE OF FLOWERS

Flowers are important to plants because they help plants reproduce. They are also important to animals, such as insects and bats, that use parts of flowers for food. Humans also use flowers. Some flowers, such as broccoli and cauliflower, can be eaten. Others, such as chamomile, are used to make tea. Flowers are also used in perfumes, lotions, and shampoos.

Discuss What is your favorite flower? Have you ever seen any unusual flowers in nature? In groups of two or three, discuss your experiences with flowers.

Name ______________________ Class ______________ Date ______________

Section 4 Review

NSES LS 1a, 1d, 2b, 3d, 4c, 5b

SECTION VOCABULARY

ovary in flowering plants, the lower part of a pistil that produces eggs in ovules

petal one of the usually brightly colored, leaf-shaped parts that make up one of the rings of a flower

phloem the tissue that conducts food in vascular plants

pistil the female reproductive part of a flower that produces seeds and consists of an ovary, style, and stigma

sepal in a flower, one of the outermost rings of modified leaves that protect the flower bud

stamen the male reproductive structure of a flower that produces pollen and consists of an anther at the tip of a filament

xylem the type of tissue in vascular plants that provides support and conducts water and nutrients from the roots

1. Label Label the parts of this flower.

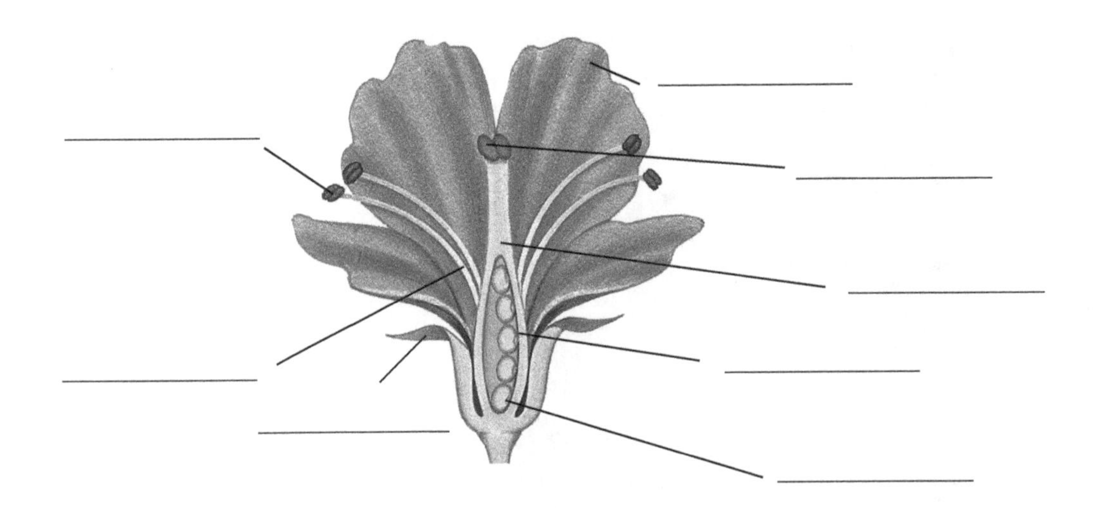

2. Compare How do taproot and fibrous root systems differ?

____

____

____

3. Describe What are the three functions of a stem?

____

____

____

4. List What are the four main organs of a flowering seed plant?

____

____

Name ______________________ Class ______________ Date ______________

CHAPTER 13 Plant Processes

SECTION 1

Photosynthesis

BEFORE YOU READ

After you read this section, you should be able to answer these questions:

- How do plants make food?
- How do plants get energy from food?
- How do plants exchange gases with the environment?

National Science Education Standards

LS 1a, 1c, 3c, 4c

What Is Photosynthesis?

Many organisms, including humans, have to eat to get energy. Plants, however, are able to make their own food. Plants make their food by a process called **photosynthesis**. During photosynthesis, plants use carbon dioxide, water, and energy from sunlight to make sugars.

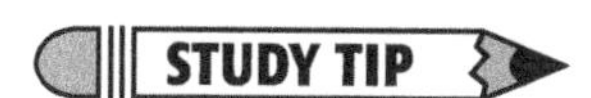

STUDY TIP

Outline As you read, outline the steps of photosynthesis. Use the questions in the section titles to help you make your outline.

How Do Plants Get Energy from Sunlight?

Plant cells have organelles called *chloroplasts*. Chloroplasts capture the energy from sunlight. Inside a chloroplast, membranes called *grana* contain chlorophyll. **Chlorophyll** is a green pigment that absorbs light energy. Many plants look green because chlorophyll reflects the green wavelengths of light. ☑

READING CHECK

1. Define What is chlorophyll?

____

____

____

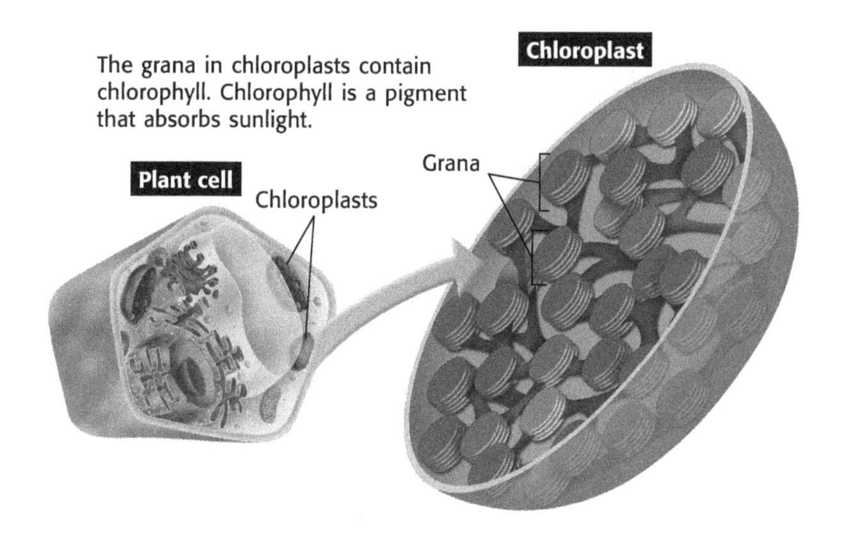

The grana in chloroplasts contain chlorophyll. Chlorophyll is a pigment that absorbs sunlight.

TAKE A LOOK

2. Identify Where is chlorophyll found in a plant cell?

____

____

How Do Plants Make Sugar?

During photosynthesis, plants take in water and carbon dioxide and absorb light energy. Plants use the light energy captured by chlorophyll to help form glucose molecules. *Glucose* is the sugar that plants use for food. In addition to producing sugar, plants give off oxygen during photosynthesis. ☑

READING CHECK

3. Identify What are two products of photosynthesis?

____

____

The following chemical equation summarizes photosynthesis:

$$6CO_2 + 6H_2O \xrightarrow{\text{light energy}} C_6H_{12}O_6 + 6O_2$$

(carbon dioxide) (water) (glucose) (oxygen)

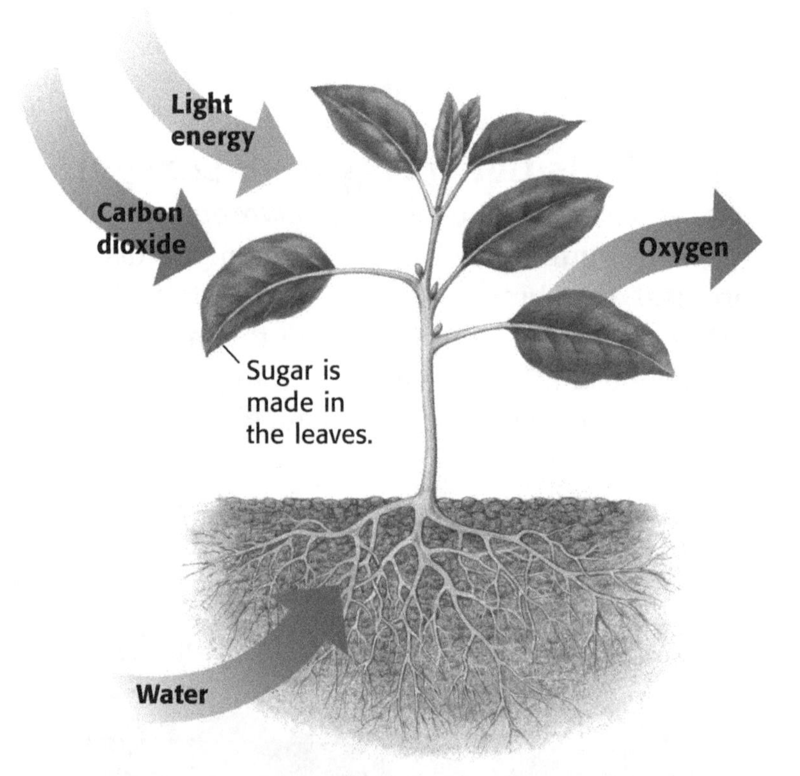

During photosynthesis, plants take in carbon dioxide and water and absorb light energy. They make sugar and release oxygen.

STANDARDS CHECK

LS 1c Cells carry on the many functions needed to sustain life. They grow and divide, thereby producing more cells. This requires that they take in nutrients, which they use to provide energy for the work cells do and to make the materials that a cell or an organism needs.

4. Identify Which cell structures release the energy stored in sugar?

____

How Do Plants Get Energy from Sugar?

Glucose molecules store energy. To use this energy, a plant cell needs its mitochondria to break down the glucose. This process of breaking down food molecules to get energy is called **cellular respiration**. During cellular respiration, cells use oxygen to break down food molecules. Like all cells, plant cells then use the energy from food to do work.

How Does a Plant Take in the Gases It Needs?

Plants take in carbon dioxide and give off oxygen. These gases move into and out of the leaf through openings called **stomata** (singular, *stoma*). Stomata allow gases to move through the plant's *cuticle*, the waxy layer that prevents water loss. Each stoma is surrounded by two guard cells. The guard cells act like double doors by opening and closing a stoma.

Water vapor also moves out of the leaf through stomata. The loss of water from leaves is called **transpiration**. Stomata open to allow carbon dioxide to enter a leaf. They close to prevent too much water loss.

Critical Thinking

5. Predict What do you think would happen if a plant had no stomata?

____

____

____

____

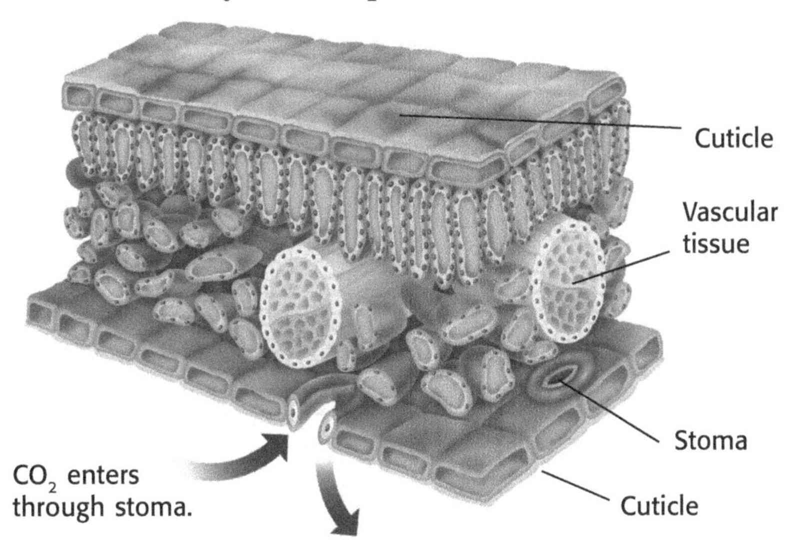

TAKE A LOOK

6. Identify Circle the guard cells in this picture. What is their function?

____

____

Why Is Photosynthesis Important?

Plants, along with many bacteria and protists that also use photosynthesis, form the bases of most food chains on Earth. During photosynthesis, plants store light energy as chemical energy. Animals get this energy when they eat plants. Other animals get energy from plants indirectly. They eat the animals that eat plants. Most organisms could not survive without photosynthetic organisms. ☑

Photosynthesis is also important because it produces oxygen. Recall that cellular respiration requires oxygen to break down food. Most organisms, including plants and animals, depend on cellular respiration to get energy from their food. Without the oxygen produced during photosynthesis, most organisms could not survive.

Say It

Describe Think of all the ways in which photosynthesis is important to you. Describe to the class three ways you depend on photosynthesis.

READING CHECK

7. Complete During photosynthesis, plants store light energy as

____

________________________________.

Name ______________________ Class ______________ Date ______________

Section 1 Review

NSES LS 1a, 1c, 3c, 4c

SECTION VOCABULARY

cellular respiration the process by which cells use oxygen to produce energy from food

chlorophyll a green pigment that captures light energy for photosynthesis

photosynthesis the process by which plants, algae, and some bacteria use sunlight, carbon dioxide, and water to make food

stoma one of many openings in a leaf or a stem of a plant that enable gas exchange to occur (plural, *stomata*)

transpiration the process by which plants release water vapor into the air through stomata; also the release of water vapor into the air by other organisms

1. **Explain** Why does chlorophyll look green?

____

____

2. **Identify** What is the role of mitochondria in plants? In what process do they take part?

____

____

3. **Compare** Complete the chart below to show the relationship between photosynthesis and cellular respiration.

Photosynthesis	Cellular respiration
	Cells break down food to provide energy.
Oxygen is produced.	

4. **Identify** What two structures in plant leaves help prevent the loss of water?

____

5. **Explain** Why are photosynthetic organisms, such as plants, so important to life on Earth?

____

____

____

6. **Explain** If plants need to take in carbon dioxide, why don't they keep their stomata open all the time?

____

Name ______________________ Class ______________ Date ______________

CHAPTER 13 Plant Processes

SECTION 2

Reproduction of Flowering Plants

BEFORE YOU READ

After you read this section, you should be able to answer these questions:

- What are pollination and fertilization?
- How do seeds and fruits form?
- How can flowering plants reproduce asexually?

National Science Education Standards

LS 1a, 2a, 2b, 2d, 5b

What Are Pollination and Fertilization?

Flowering plants are most noticeable to us when they are in bloom. As flowers bloom, they surround us with bright colors and sweet fragrances. However, flowers are not just for us to enjoy. They are the structures for sexual reproduction in flowering plants. Pollination and fertilization take place in flowers.

Summarize As you read, write out or draw the steps of pollination and fertilization.

Pollination and Fertilization

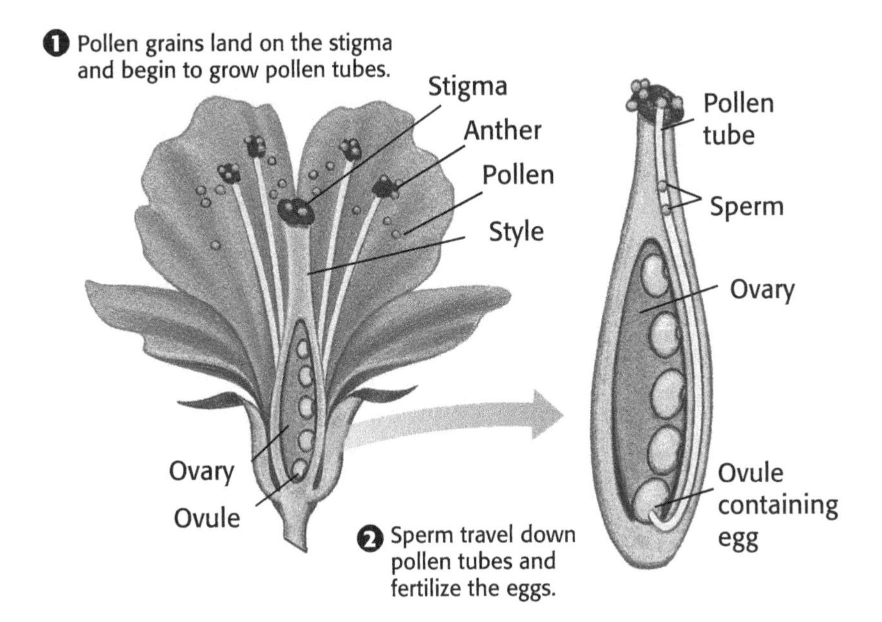

TAKE A LOOK

1. Identify Circle the part of the flower where pollination occurs.

2. Identify Draw an arrow to show where fertilization will take place.

Sexual reproduction begins in flowers when wind or animals move pollen from one flower to another. *Pollination* occurs when pollen from an anther lands on a stigma. Each pollen grain grows a tube through the style to the ovary. The ovary has ovules, each of which contains an egg. *Fertilization* occurs when a sperm joins with the egg inside an ovule.

STANDARDS CHECK

LS 2a Reproduction is a characteristic of all living systems; because no living organism lives forever, reproduction is essential to the continuation of every species. Some organisms reproduce asexually. Others reproduce sexually.

3. Explain Where do seeds and fruits come from?

____

____

____

____

____

What Happens After Fertilization?

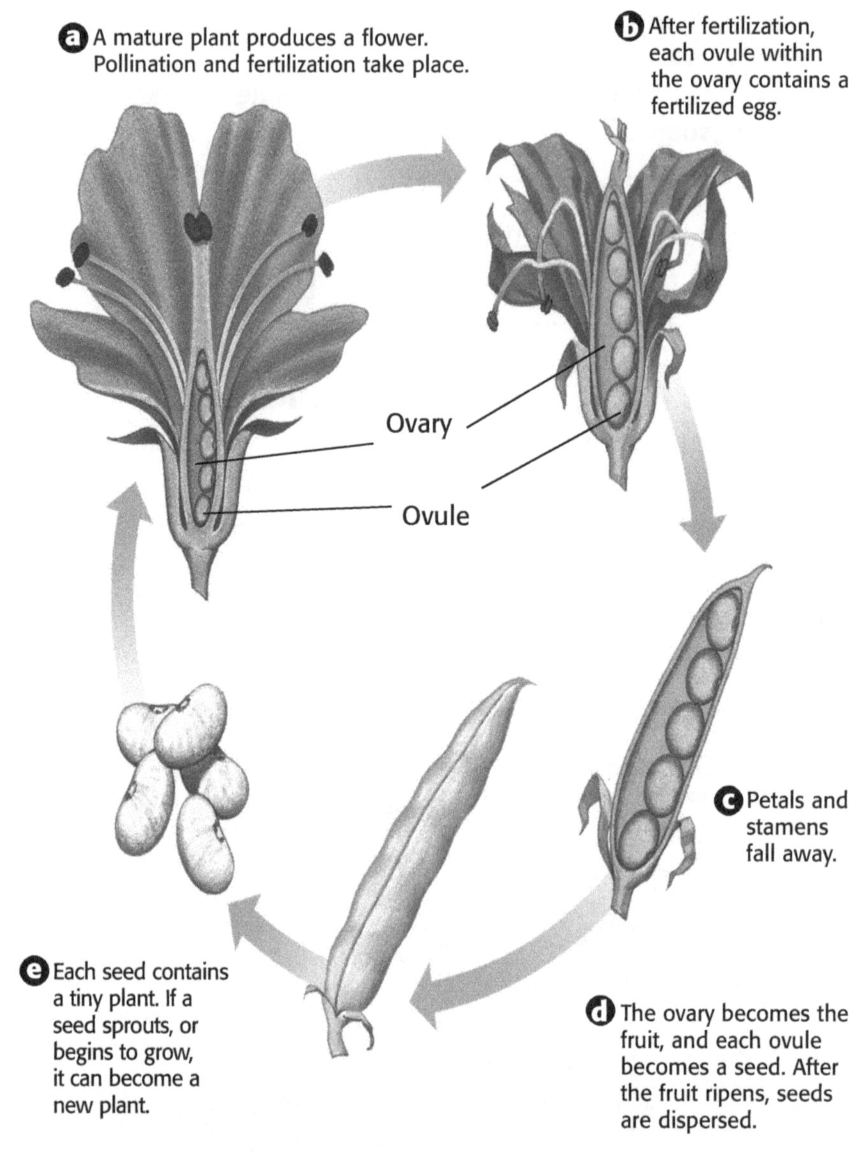

TAKE A LOOK

4. Identify In step C, circle the structures that will become seeds.

THE FUNCTIONS OF FRUITS

When people think of fruit, they often think of apples or bananas. However, many things we call vegetables, such as tomatoes or green beans, are also fruits! A fruit is the ovary of the flower that has grown larger.

Fruits have two major functions. They protect seeds while the seeds develop. Fruits also help a plant spread its seeds to new environments. For example, an animal might eat a fruit and drop the seeds far from the parent plant. Fruits such as burrs spread when they get caught in an animal's fur. Other fruits are carried to new places by the wind or even by water. ☑

READING CHECK

5. List What are two functions of a fruit?

____

____

____

How Do Seeds Grow into New Plants?

The new plant inside a seed, called the *embryo*, stops growing once the seed is fully developed. However, the seed might not sprout right away. To sprout, most seeds need water, air, and warm temperatures. A seed might become **dormant**, or inactive, if the conditions are not right for a new plant to grow. For example, if the environment is too cold or too dry, a young plant will not survive. ☑

Dormant seeds often survive for long periods of time during droughts or freezing weather. Some seeds actually need extreme conditions, such as cold winters or forest fires, to *germinate*, or sprout.

READING CHECK

6. Explain Why would a seed become dormant?

____

____

____

Seeds grow into new plants. First, the roots begin to grow. Then, the shoots grow up through the soil.

TAKE A LOOK

7. Identify Which part of a new plant grows first?

____

How Else Can Flowering Plants Reproduce?

Flowering plants can also reproduce asexually, or without flowers. In asexual reproduction, sperm and eggs do not join. A new plant grows from a plant part such as a root or stem. These plant parts include plantlets, tubers, and runners.

Three Structures for Asexual Reproduction

Kalanchoe produces plantlets along the edges of their leaves. The plantlets will fall off and take root in the soil.

A potato is a tuber, or underground stem. The "eyes" of potatoes are buds that can grow into new plants.

The strawberry plant produces runners, or stems that grow along the ground. Buds along the runners take root and grow into new plants.

Critical Thinking

8. Infer When would asexual reproduction be important for the survival of a flowering plant?

____

____

____

____

____

____

Name ______________________ Class ______________ Date ______________

Section 2 Review

NSES LS 1a, 2a, 2b, 2d, 5b

SECTION VOCABULARY

dormant describes the inactive state of a seed or other plant part when conditions are unfavorable to growth	

1. Apply Concepts Is fertilization part of asexual reproduction or sexual reproduction? Explain your answer.

____

____

2. Compare What is the difference between pollination and fertilization?

____

____

3. Summarize Complete the Process Chart below to summarize how sexual reproduction produces new plants.

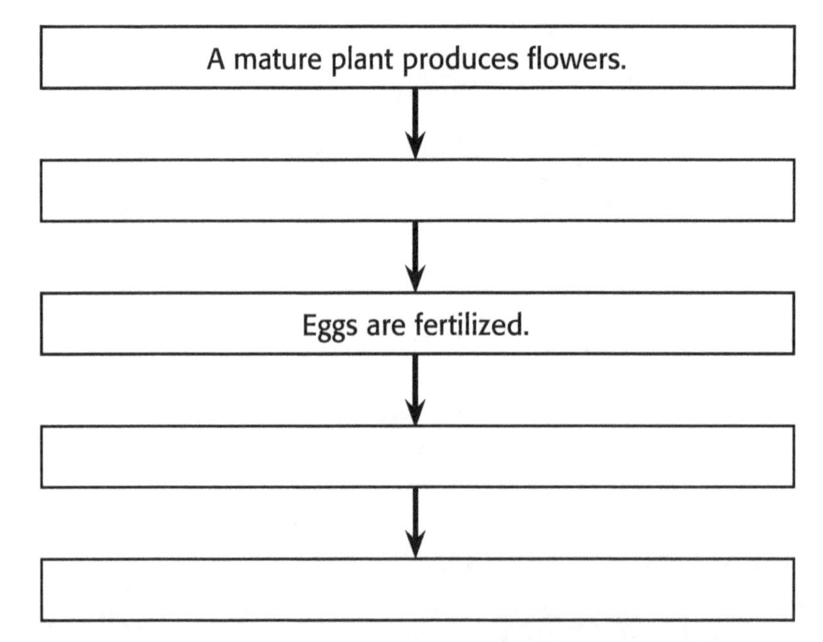

4. Identify Name two environmental conditions that can cause a seed to become dormant.

____

5. List What are three structures a flowering plant can use to reproduce asexually?

____

6. Infer Why do you think roots are the first part of a plant to grow?

____

____

Name ______________________ Class ______________ Date ______________

CHAPTER 13 Plant Processes

SECTION 3

Plant Development and Responses

BEFORE YOU READ

After you read this section, you should be able to answer these questions:

- How do plants respond to the environment?
- Why do some trees lose their leaves?

National Science Education Standards

LS 2b, 2c, 3a, 3c, 3d, 5b

How Do Plants Respond to the Environment?

What happens when you get cold? Do you shiver? Do your teeth chatter? These are your responses to an environmental stimulus such as cold air. A *stimulus* (plural, *stimuli*) is anything that causes a reaction in your body. Plants also respond to environmental stimuli, but not to the same ones we do and not in the same way. Plants respond to stimuli such as light and the pull of gravity. ☑

STUDY TIP

Summarize With a partner, take turns summarizing the text under each header.

READING CHECK

1. Define What is a stimulus?

____

____

____

Some plants respond to a stimulus, such as light, by growing in a particular direction. Growth in response to a stimulus is called a **tropism**. A tropism is either positive or negative. Plant growth toward a stimulus is a positive tropism. Plant growth away from a stimulus is a negative tropism.

PLANT GROWTH IN RESPONSE TO LIGHT

Recall that plants need sunlight in order to make food. What would happen to a plant that could get light from only one direction, such as through a window? To get as much light as possible, it would need to grow toward the light. This growth in response to light is called *phototropism*.

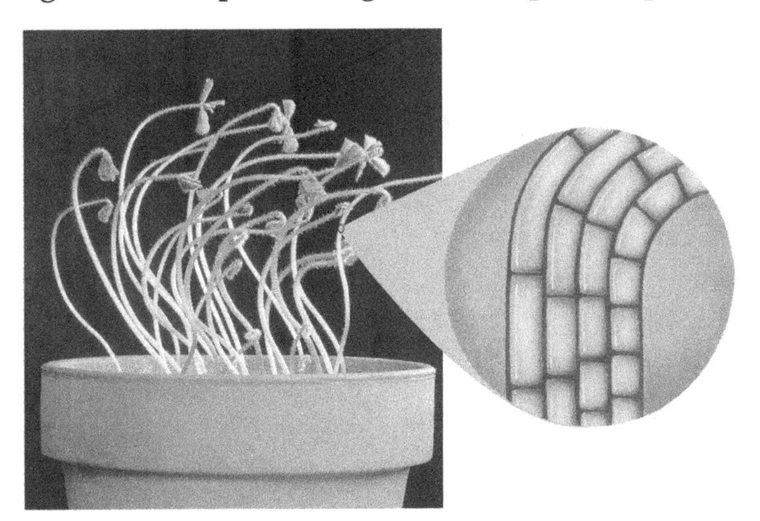

TAKE A LOOK

2. Explain Place an **X** on the picture to show where the light must be coming from. Explain your answer.

____

____

____

PLANT GROWTH IN RESPONSE TO GRAVITY

Gravity can change the direction in which a plant's roots and shoots grow. Most shoot tips grow upward, away from the center of Earth. Most root tips grow downward, toward the center of Earth. If a plant is placed on its side or turned upside down, the roots and shoots will change direction. Shoots will turn to grow away from the Earth. Roots will turn to grow toward the Earth. This response is called *gravitropism.*

Critical Thinking

3. Apply Concepts Look at the plant on the left. Draw an arrow on the flower pot to show the direction the roots are probably growing.

Gravitropism

To grow away from the pull of gravity, this plant has grown upward.

TAKE A LOOK

4. Explain Look at the plant on the right. What do you think made its stem bend?

____

____

____

____

____

What Happens to Plants When Seasons Change?

Have you ever noticed that some plants will drop their leaves in the fall even before the weather turns cool? How do the plants know that fall is coming? We often notice the changing seasons because the temperature changes. Plants, however, respond to change in the length of the day.

Math Focus

5. Calculate It must be dark for 70% of a 24-hour period before a certain plant will bloom. How many hours of daylight does this plant need to bloom?

____

SHORT DAY AND LONG DAY PLANTS

Days are longer in summer and shorter in winter. The change in amount of daylight is a stimulus for many plants. Some plants that bloom in winter, such as poinsettias, need shorter periods of daylight to reproduce. They are called *short-day plants.* Others, such as clover, reproduce in spring or summer. They are called *long-day plants.*

EFFECT OF SEASONS ON LEAF COLOR

The leaves of some trees may change color as seasons change. As the days shorten in fall, the chlorophyll in leaves breaks down. This makes the orange and yellow pigments in the leaves easier to see. During the summer, chlorophyll hides other pigments.

Amount of Leaf Pigment Based on Season

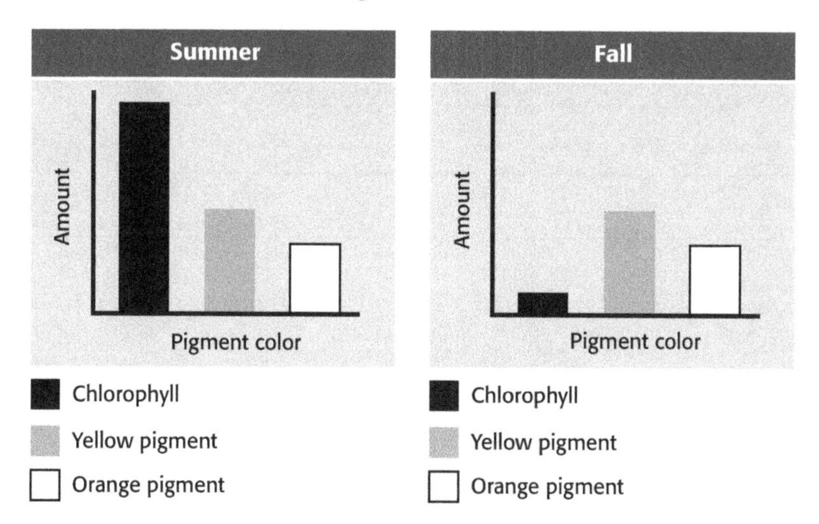

TAKE A LOOK

6. Identify Which pigment's level decreases between summer and fall?

____

7. Identify Which pigments' levels will stay the same between summer and fall?

____

____

LOSS OF LEAVES

Every tree loses leaves throughout its life. Leaves are shed when they become old. For example, pine trees lose some of their leaves, or *needles*, year-round. Because leaves are lost and replaced throughout the year, the tree always has some leaves. These trees are called *evergreen*. A leaf of an evergreen tree is covered with a thick cuticle. The cuticle protects the leaf from cold and dry weather.

Deciduous trees lose all their leaves at about the same time each year. This generally happens as days shorten. The loss of leaves helps these plants survive cold or dry weather. In colder areas, deciduous trees usually lose their leaves before winter begins. In areas that have wet and dry seasons, deciduous trees lose their leaves before the dry season.

Describe What is your favorite kind of tree? Use the Internet or reference books to find out if the tree is evergreen or deciduous. Describe to the class what the tree looks like and where it lives.

Name ______________________ Class ______________ Date ______________

Section 3 Review

NSES LS 2b, 2c, 3a, 3c, 3d, 5b

SECTION VOCABULARY

tropism growth of all or part of an organism in response to an external stimulus, such as light	

1. **Compare** What is the difference between a negative tropism and a positive tropism?

____

____

____

2. **Explain** What happens when a plant gets light from only one direction?

____

____

____

3. **Define** What is gravitropism?

____

____

4. **Identify** What stimulus causes seasonal changes in many plants?

____

5. **Compare** What is the difference between short-day plants and long-day plants?

____

____

____

6. **Explain** Why do leaves look green during the summer even though they have orange and yellow pigments?

____

____

7. **Explain** Many evergreen trees live in areas with long, cold winters. How can they keep their leaves all year?

____

____

Name ______________________ Class ______________ Date ______________

CHAPTER 14 Animals and Behavior

SECTION 1

What Is an Animal?

BEFORE YOU READ

After you read this section, you should be able to answer these questions:

- What do all animals have in common?
- What are vertebrates and invertebrates?

National Science Education Standards

LS 1a, 1b, 1d, 2a, 2b, 4b, 5a

What Are Animals?

What do you think of when you hear the word animal? You may think of a dog or cat. You may think of giraffes or grizzly bears. Would you think of a sponge?

How many kinds of animals are in the picture below? You may be surprised to learn that feather stars and coral are animals. Some animals have four legs and fur. Other animals have wings and feathers. Some animals are so small that they can only be seen with a microscope. Other animals are bigger than a bus. Scientists have named more than 1 million species of animals. However, there are still millions of species that have not been named.

STUDY TIP

Outline As you read, make an outline descrbing the characteristics of animals. Be sure to include examples of each characteristic.

Critical Thinking

1. Infer Why do you think there are many species of animals that have not been named?

____

____

____

____

Although the feather star and coral look like plants, they are actually animals.

CHARACTERISTICS OF ANIMALS

Animals may look very different from one another, but they all have some things in common. While other organisms have some of these characteristics, only animals have all five:

1. They are multicellular.
2. Almost all can reproduce sexually.
3. They have specialized parts.
4. They can move.
5. They eat other organisms.

Say It

Describe Choose your favorite animal. Describe to the class how this animal shows the five characteristics of animals.

MULTICELLULAR

All animals are *multicellular*. This means they are made of more than one cell. Your own body is made of trillions of cells. Animal cells are *eukaryotic*, which means they have a nucleus. Plants are also multicellular. However, plant cells have cell walls and animal cells do not. Animal cells are surrounded only by cell membranes. ☑

READING CHECK

2. Define What does multicellular mean?

____

____

____

REPRODUCTION

Most animals can reproduce sexually. They make sex cells called eggs and sperm. When an egg and sperm join, they form a new cell. This cell grows and divides to form an embryo. An **embryo** is the early stage of development of an organism. A few animals can reproduce asexually. For example, flatworms called *Planaria* break off part of their bodies to form offspring.

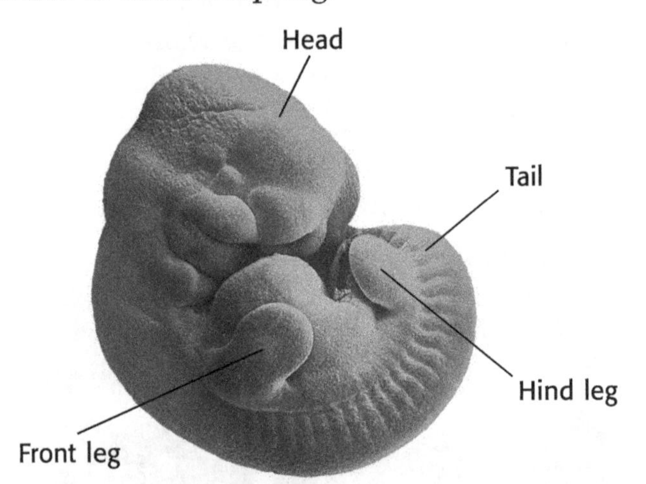

Cells in the mouse embryo will differentiate as the mouse develops. These cells will produce skin, muscles, nerves, and all the other parts of the mouse's body.

TAKE A LOOK

3. Explain What happens to the cells in an embryo as the embryo develops?

____

____

CELL SPECIALIZATION

As a fertilized egg develops, it divides into many cells. As the organism develops, its cells *differentiate*. When cells differentiate, they develop different structures so they can perform different functions. Some cells become skin cells, while others become muscle or bone cells. Groups of similar cells form *tissues*. For example, muscle cells form muscle tissue. ☑

READING CHECK

4. Explain What happens when cells differentiate?

____

____

Tissues can group together to form *organs*. An organ is a group of tissues that carry out a special job in the body. Your heart, lung, and kidneys are all organs. Each organ in an animal's body does different jobs. The figure on the next page shows some of the organs in a shark's body.

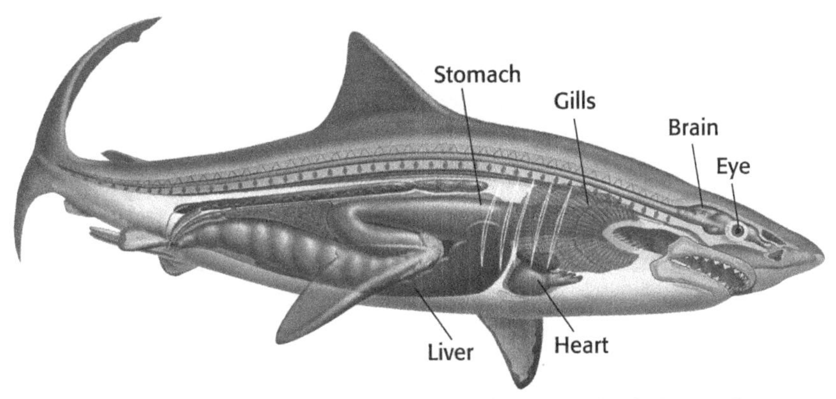

Like most animals, sharks have organs for digestion, circulation, and sensing the environment.

MOVEMENT

Most animals can move from place to place. They can fly, run, swim, or jump. Some animals use movements to find food, shelter, and mates. However, not all animals can move their entire lives. For example, young sea anemones swim to find their food. As adults, sea anemones attach to rocks or the ocean floor and wait for food to drift by.

EATING

Unlike plants, animals cannot make their own food. They are consumers. A **consumer** must eat other organisms to survive. Some animals eat plants or parts of plants. Some animals eat other animals.

What Are the Two Main Groups of Animals?

VERTEBRATES

Most animals don't look like humans. However, humans are part of a group of animals called vertebrates. A *vertebrate* is an animal that has a backbone or spine. Vertebrates include fishes, amphibians, reptiles, birds, and mammals. Humans are a kind of mammal.

INVERTEBRATES

Although you are probably more familiar with vertebrates, most animals are actually invertebrates. *Invertebrates* are animals without a backbone. Insects, snails, jellyfish, and worms are examples of invertebrates. More than 95% of all animals are invertebrates. In fact, beetles make up more then 30% of all animal species! ☑

STANDARDS CHECK

LS 4b Populations of organisms can be categorized by the functions they serve in an ecosystem. Plants and some microorganisms are producers—they make their own food. All animals, including humans, are consumers, which obtain their food by eating other organisms. Decomposers, primarily bacteria and fungi, are consumers that use waste materials and dead organisms for food. Food webs identify the relationship among producers, consumers, and decomposers in an ecosystem.

Word Help: categorize
to put in groups or classes

Word Help: function
use or purpose

5. Explain Are animals producers or consumers? Explain your answer.

____

____

____

____

____

READING CHECK

6. Complete Most animals on Earth are ______________________

______________________.

Name ______________________ Class ______________ Date ______________

Section 1 Review

NSES 1a, 1b, 1d, 2a, 2b, 4b, 5a

SECTION VOCABULARY

consumer an organism that eats other organisms or organic matter	**embryo** a plant or animal at an early stage of development

1. **Identify** What is the largest group of invertebrates on Earth?

____

2. **Summarize** Use the organizer below to show the five characteristics of animals.

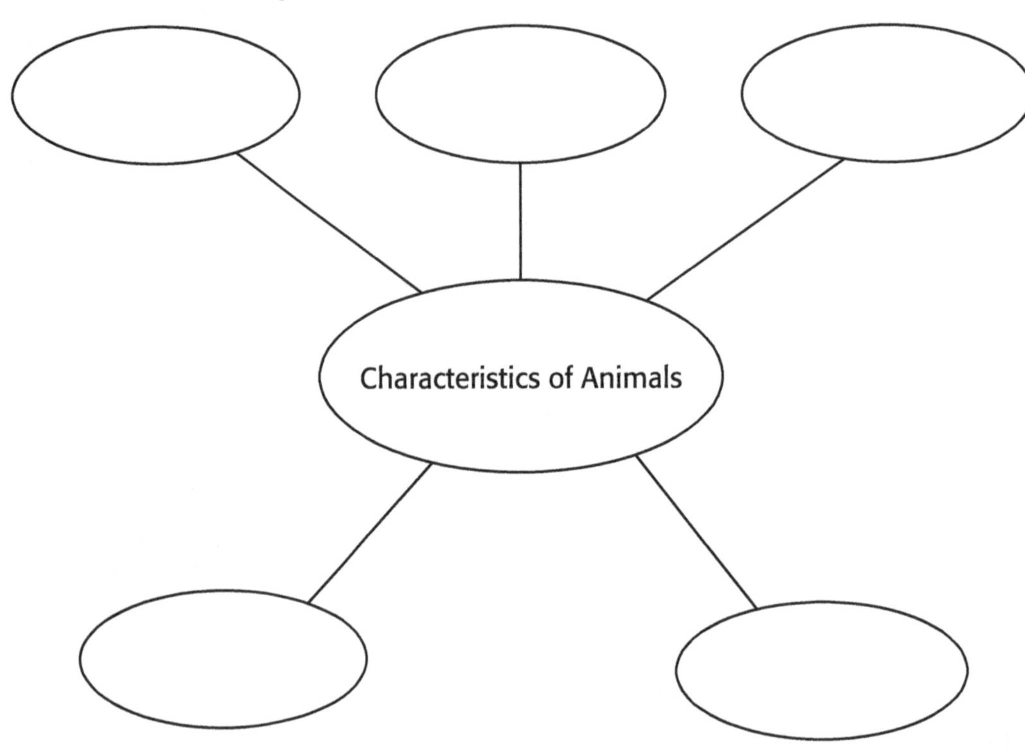

3. **Apply Concepts** Like animals, plants are multicellular and have specialized parts. Why are these considered characteristics of animals?

____

____

____

4. **Explain** An adult sea anemone cannot move. Why is it considered an animal?

____

____

5. **Compare** What is the main difference between vertebrates and invertebrates? Give an example of each kind of animal.

____

____

Name ______________________ Class ______________ Date ______________

CHAPTER 14 Animals and Behavior

SECTION 2

Animal Behavior

BEFORE YOU READ

After you read this section, you should be able to answer these questions:

- How do learned and innate behaviors differ?
- What types of behaviors do animals use to survive?

National Science Education Standards

LS 1a, 3a, 3b, 3c, 3d, 5b

What Is Behavior?

Suppose you look out a window and see a bird flying away from a tree. Is the bird leaving a nest to look for food? Is it escaping from danger?

An animal may run or hide from an enemy. It may also search for food and look for a safe place to build its home. All of these activities are called *behaviors*. Behavior is the way an organism acts in different situations. Animals can be born with some behaviors. They may also learn other behaviors as they grow.

STUDY TIP

Summarize Make combination notes describing different survival behaviors and examples of each.

STANDARDS CHECK

LS 3c Behavior is one kind of response an organism can make to an internal or environmental stimulus. A behavioral response requires coordination and communication at many levels, including cells, organ systems, and whole organisms. Behavioral response is a set of actions determined in part by heredity and in part from experience.

1. Compare How do innate behaviors and learned behaviors differ?

____

____

____

What Are The Two Main Types of Behavior?

Innate behaviors are behaviors that an animal does automatically. Animals are born with innate behaviors. For example, puppies like to chew and bees know how to fly. Some innate behaviors begin when the animal is born. Newborn whales, for example, can swim. Other innate behaviors begin months or years after an animal is born.

Animals can also learn a behavior. A **learned behavior** is a behavior that develops with experience or from watching other animals. For example, humans are born with an ability to speak. However, we must learn a language in order to speak. Humans are not the only animals that can learn.

A male bowerbird collects colorful objects for its nest. These objects attract a female bowerbird.

TAKE A LOOK

2. Identify If this bowerbird collected objects for its nest without ever doing it before or seeing another bird do it, what type of behavior is the bird showing?

____

What Are Some Survival Behaviors?

Animals use their behaviors to survive. To stay alive and pass on its genes, an animal has to do many things. It must find food and a place to live. It also must protect itself against predators. Animals also find mates so that they can have offspring. Some survival behaviors are described below.

FINDING FOOD

Different kinds of animals use different behaviors to find food. Some animals, such as koalas, climb trees to get their food. Other animals, such as tigers, chase prey.

Animals that eat other animals are called *predators*. The animal being eaten is called the *prey*. So, if a frog eats an insect, the frog is the predator and the insect is the prey. If the frog is then eaten by a snake, the snake becomes the predator and the frog is the prey. ☑

READING CHECK

3. Complete An animal eaten by another animal is called _______________.

Chimpanzees make and use tools to get food out of hard-to-reach places.

CLAIMING TERRITORY

Sometimes, members of the same species must compete for the same resources. These include food, mates, and places to live. Since resources are often limited, animals must often fight for them. To avoid competing for resources in one area, some animals claim territories. ☑

READING CHECK

4. List What are three resources animals of the same species may compete for?

____

____

____

A **territory** is an area where one animal or a group of animals live. These animals do not let any other members of their species live there. For example, birds will often sing to warn other birds not to enter their area. Animals use their territories for mating, raising young, and finding food.

PROTECTING THEMSELVES

Animals often have to protect their mates and offspring. They also have to protect their resources, such as food and mates. For example, a dog may growl when another animal enters its territory. Male lions fight to defend mates. Some birds, such as killdeer, will pretend to be hurt. This distracts a predator so that it will not attack the bird's young.

Animals often use their defensive behaviors to protect themselves from being eaten. Rabbits will freeze to blend into the background so that predators do not see them. If a predator does see them, they will run. Some animals, such as bees and wasps, will sting their attackers.

FINDING A MATE

For an individual's genes to survive, the individual must reproduce. For an animal to have offspring, it must first find a mate. Some animals have special behaviors that help them find mates. These behaviors are called *courtship*. Some birds and fish build nests to attract a mate. Other animals use special movements or sounds. ☑

5. Identify What is the name for behaviors that an animal uses to find a mate?

____

RAISING OFFSPRING

Following a courtship behavior and mating, offspring are usually born. Some animals, such as caterpillars, can take care of themselves as soon as they are born. However, many young animals depend on their parents for survival.

Different animals take care of their young for different amounts of time. Some adult birds bring food to their young only until they can fly and get their own food. Other animals, such as killer whales, spend years teaching their young how to hunt for food.

Adult killer whales teach their young how to hunt in the first years of life.

TAKE A LOOK

6. Apply Concepts Is hunting in killer whales a learned or innate behavior?

____

What Do Animals Do As Seasons Change?

When seasons change, humans wear different clothes and do different activities. Animals also act differently during different seasons. During winter in some parts of the world, animals need to protect themselves from the cold. For example, frogs burrow into mud to stay warm. Some animals, such as squirrels, store food to use in winter. Different behaviors in different seasons help animals adjust to the environment. ☑

READING CHECK

7. Explain Why do many animals have different behaviors in different seasons?

____
____
____
____

MIGRATION

Many animals avoid cold weather by traveling to warmer places. When animals travel from one place to another it is called *migration.* Animals migrate to find food, water, and safe places to have offspring. Whales, salmon, bats, and butterflies all migrate. When animals migrate, they use *landmarks* to find their way. Landmarks are fixed objects, such as mountain ranges, rivers, and stars, that animals use to find their way. ☑

READING CHECK

8. List Give three reasons animals migrate.

____
____
____
____

HIBERNATION

During winter, some animals hibernate. During **hibernation**, an animal lowers its body temperature and is inactive. When they are hibernating, animals will not look for food or mates. Hibernating animals get nutrients from stored body fat. Most hibernating animals, such as mice, lower their heart rate and breathing. This lets their bodies use less energy. Their body temperature also lowers to just above freezing.

TAKE A LOOK

9. Explain Do bears hibernate? Explain your answer.

____
____
____
____
____
____

Bears do not actually hibernate. They slow their bodies down for the winter, but their body temperatures do not drop as low as hibernating animals. They also sleep for shorter periods of time.

ESTIVATION

Some animals slow down their bodies when it is hot. Desert squirrels and mice slow down their bodies during the hottest part of summer. This lets them survive when there is little food or water. Reduced activity in the summer or during hot periods is called **estivation**.

Critical Thinking

10. Compare How is hibernation different from estivation?

How Do Animals Know When to Do Certain Behaviors?

Animals need to keep track of time so that they know when to store food or to migrate. An animal's natural cycles are called its *biological clock*. The biological clock is a control inside animals' bodies. Animals often use the length of the day and the temperature to set their biological clocks.

SHORT CYCLES

Some biological clocks keep track of daily cycles. These daily cycles are called **circadian rhythms**. Most animals, including humans, wake up and get sleepy at about the same time each day and night. This is an example of a circadian rhythm.

LONG CYCLES

Biological clocks can also control long cycles. Almost all animals have seasonal cycles, or cycles that change with the seasons. For example, many animals hibernate in one season and have offspring in another season. The start of migration is also controlled by seasonal changes.

Biological clocks can also control changes inside an animal. For example, insects such as treehoppers go through several changes during their lives. They start as eggs, hatch as nymphs, and then develop into adults.

Math Focus

11. Calculate Suppose that an animal's circadian rhythms tell it to eat a meal every 4 hours. How many meals will that animal eat in a day?

The treehopper's biological clock tells the animal when to shed, or get rid of, its skin.

Name ______________________ Class ______________ Date ______________

Section 2 Review

NSES LS 1a, 3a, 3b, 3c, 3d, 5b

SECTION VOCABULARY

circadian rhythm a biological daily cycle

estivation a period of inactivity and lowered body temperature that some animals undergo in summer as a protection against hot weather and lack of food

hibernation a period of inactivity and lowered body temperature that some animals undergo in winter as a protection against cold weather and lack of food

innate behavior an inherited behavior that does not depend on the environment or experience

learned behavior a behavior that has been learned from experience

territory an area that is occupied by one animal or a group of animals that do not allow other members of the species to enter

1. Explain If humans are born with the ability to speak, why isn't talking an innate behavior?

____

____

2. List Name five survival behaviors.

____

____

3. Infer Can an animal be both a predator and prey? Explain your answer.

____

____

4. Explain When animals migrate, how do they find their way?

____

____

5. Apply Concepts People who travel to different time zones often suffer from *jet lag.* Jet lag makes it hard for people to wake up or go to sleep at the right time. Why do you think people get jet lag?

____

____

____

6. Identify Name two clues animals use to set their biological clocks.

____

7. Identify Give an example of a seasonal cycle controlled by a biological clock.

____

Name ______________________ Class ______________ Date ______________

CHAPTER 14 Animals and Behavior

SECTION 3

Social Relationships

BEFORE YOU READ

After you read this section, you should be able to answer these questions:

- How do animals communicate?
- Why do some animals live in groups?

National Science Education Standards

LS 1a, 3c, 3d, 4a

What Is Social Behavior?

Animals often interact with each other—in groups and one on one. They may work together or compete. Maybe you have seen two dogs barking at each other. We may not know exactly why they behave this way, but it is clear that they are interacting. These interactions are called social behavior. **Social behavior** is the interaction among animals that are the same species. Animals use communication for their social interactions. ☑

STUDY TIP

Organize As you read, make a chart of the different ways animals communicate. List an example from the text and add another example of your own.

READING CHECK

1. Define What is social behavior?

____

____

____

____

How Do Animals Communicate?

Imagine what life would be like if people could not read or talk. There would be no telephones, no books, no Internet, and no radios. Language is an important way for people to communicate. In **communication**, a signal is sent from one animal to another animal. The animal that receives the signal then answers back. Humans are the only animals that use complex words and grammar. However, other animals do communicate.

Animals, including humans, can communicate by using sound, touch, smells, and sight. Communication helps animals survive. Many animals, such as wolves, communicate to defend a territory or to identify family members. Animals also communicate to find food, to warn others of danger, to frighten predators, and to find mates.

TAKE A LOOK

2. Apply Concepts Give two possible reasons these wolves could be howling.

____

____

____

SOUND

Many animals communicate by making noises. Wolves howl. Dolphins use whistles and make clicking noises. Male birds often sing to attract mates or defend their territories.

Sound can travel over large distances. Animals such as humpback whales and elephants use sounds to let others know where they are. Elephants make some sounds that are too low for humans to hear. Although humans cannot hear them, elephants that are kilometers away can.

Critical Thinking

3. Infer Why do you think sound is a good way to communicate for animals that live in large areas?

____

____

____

____

____

Elephants communicate using low-pitched sounds. When an elephant makes these sounds, the skin on its forehead flutters.

TOUCH

Animals may also touch to communicate. For example, chimpanzees will groom each other. *Grooming* is when one animal picks dirt out of another animal's fur. Chimpanzees use grooming to calm and comfort each other. With touch, animals can also communicate friendship and support.

SMELL

Some animals, including humans, make chemicals called pheromones. A **pheromone** is a chemical that an animal uses to communicate with other members of its species. Ants and other insects make several pheromones. Some pheromones warn other ants of danger. Other pheromones are used by insects to find mates. Some insects that live in colonies, such as fire ants, use pheromones to control which colony members can reproduce. ☑

READING CHECK

4. Define What are pheromones?

____

____

____

____

SIGHT

When you smile at a friend, you are using *body language*. Body language is the movements made by an animal that communicate an idea to another animal.

When dogs want to play, they drop down on their forelegs.

To scare others, an animal may ruffle its feathers, try to look bigger, or show its teeth. Animals can also use body language to show that they are friendly. The figure above shows how a dog behaves when it wants to play. Visual displays are also used in courtship. For example, fireflies blink signals to attract each other.

Why Do Some Animals Live In Groups?

Some animals live in groups because it can be safer than living alone. Group members can warn each other of danger. These animals work together to defend the entire group. By hunting in groups, animals can kill larger animals than they could by themselves.

However, there are some problems caused by living in a group. All the animals in a group must compete for the same food, shelter, and mates. Sometimes groups of animals must move around to find enough food to feed the group. Large groups of animals usually attract more predators. Sickness and disease can also spread quickly through a group.

Discuss Like many other animals, humans also live in groups. With your classmates, discuss ways that living in groups is helpful to humans. Then, discuss ways that living in groups can cause problems.

A ground squirrel whistles a loud alarm to alert the other ground squirrels that danger is near.

TAKE A LOOK

5. Identify Which advantage of living in a group does this figure show?

____

____

____

Name ______________________ Class ______________ Date ______________

Section 3 Review

NSES LS 1a, 3c, 3d, 4a

SECTION VOCABULARY

communication a transfer of a signal or message from one animal to another that results in some type of response	**pheromone** a substance that is released by the body and that causes another individual of the same species to react in a predictable way **social behavior** the interaction between animals of the same species

1. **Analyze** In parts of Africa, lions and crocodiles may fight over prey. Why is this *not* an example of social behavior?

____

____

____

2. **List** Name six reasons animals communicate.

____

____

____

3. **Summarize** Use the chart below to summarize the advantages and disadvantages of living in a group.

Advantages of living in a group	Disadvantages of living in a group

4. **Analyze** Imagine you are on a safari. You see a group of six lions feeding on a dead zebra. As another animal moves closer, the lions stand up and growl. Name some types of behaviors the lions are showing.

____

____

____

5. **List** What are the four ways animals communicate?

____

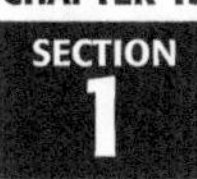

CHAPTER 15 Invertebrates

SECTION 1

Simple Invertebrates

BEFORE YOU READ

After you read this section, you should be able to answer these questions:

- What are the characteristics of invertebrates?
- What are the four groups of simple invertebrates?

National Science Education Standards

LS 1d, 1f, 2a, 3a, 3c, 5a

What Are Invertebrates?

About 96% of all animal species are invertebrates. An **invertebrate** is an animal that does not have a backbone. Worms, jellyfish, octopuses, and butterflies are some examples of invertebrates. More than 1 million invertebrates have already been named. Scientists think there are millions of invertebrates that have not been found or named.

STUDY TIP

Outline As you read, make an outline of this section. Use the header questions to help you make your outline.

BODY PLANS

Invertebrates have three basic body plans: bilateral, radial, or asymmetrical. Most animals have bilateral symmetry. The three basic body plans are decribed below.

Animal Body Plans

Bilateral Symmetry

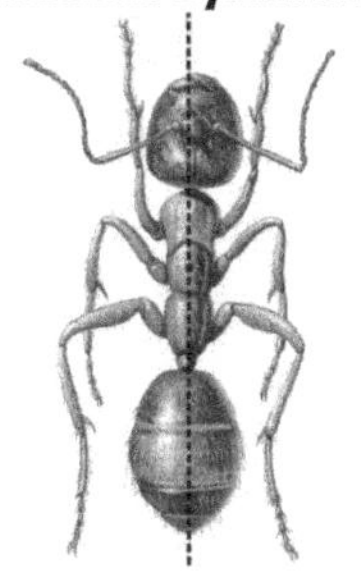

This ant has **bilateral symmetry**. The two sides of its body mirror each other. On each side of its body, the ant has one eye, one antenna, and three legs.

Radial Symmetry

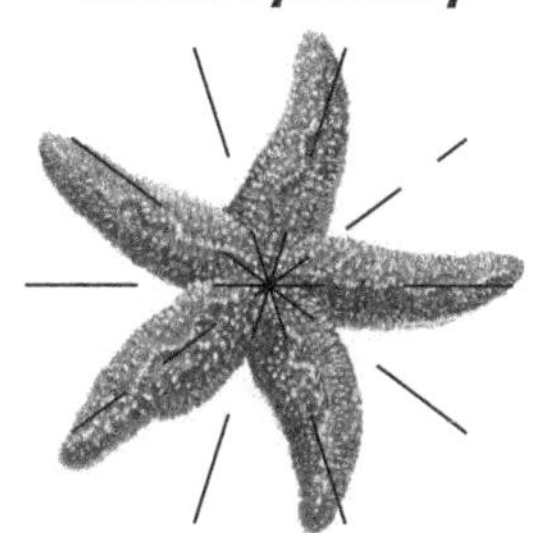

This sea star has **radial symmetry**. Its body is organized around the center, like spokes on a wheel.

Asymmetry

This sponge is **asymmetrical**. You cannot draw a straight line to divide its body into equal parts. Its body is not organized around a center.

TAKE A LOOK

1. Describe What is bilateral symmetry?

____

____

____

2. Apply Concepts What kind of symmetry does a butterfly have?

____

NEURONS AND GANGLIA

All animals, except sponges, have special tissues that allow them to sense their environment. These tissues make fibers called *neurons*. Neurons carry messages around the body to control the animal's actions. Simple invertebrates have nerve cords. *Nerve cords* are bundles of neurons that carry messages along a single path.

Some invertebrates have nerve cells called ganglia (singular, *ganglion*). A **ganglion** is a mass of nerve cells grouped together. Each ganglion controls a different part of the body. The ganglia are connected by nerve cords. In complex invertebrates, a brain controls the ganglia and nerves throughout the body. ☑

READING CHECK

3. Identify In an invertebrate's body, what connects all the ganglia?

____

GUTS

Most animals digest food in a gut. A **gut** is a hollow space lined with cells. The cells make chemicals to break down food into small particles. The cells then absorb the food particles.

In complex animals, the gut is inside a coelom. A **coelom** is a body cavity that surrounds the gut. The coelum keeps other organs, such as the heart, separated from the gut. That way, movements in the gut do not disturb the actions of other organs. In simple invertebrates, the gut is not surrounded by a coelom.

TAKE A LOOK

4. Identify What is the name of the body cavity that surrounds the gut?

____

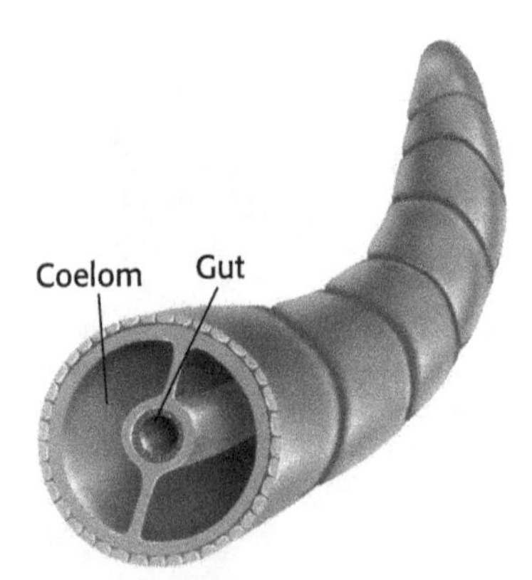

Earthworms have a coelom filled with fluid. The gut is inside the coelom.

What Are Sponges?

Sponges are the simplest invertebrates. They are asymmetrical. They have no tissues, gut, or neurons. Adult sponges may move only a few millimeters each day or not all.

Scientists once thought sponges were plants. However, because sponges cannot make their own food, they are classified as animals.

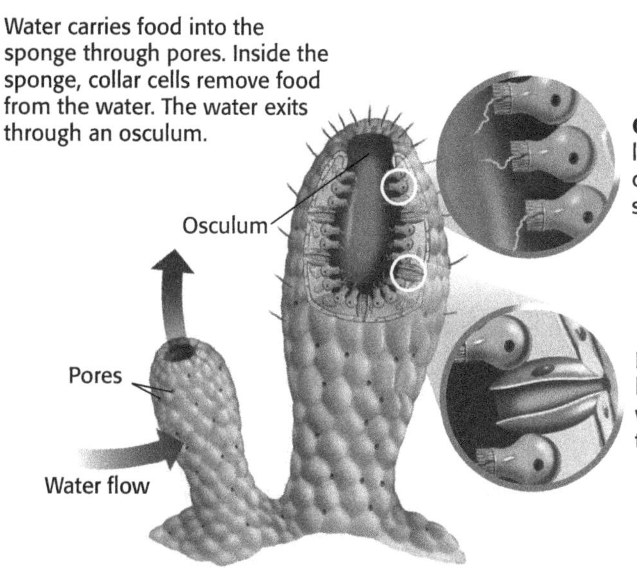

TAKE A LOOK

5. Identify What is the function of collar cells?

____

____

6. Identify Through what structure does water leave a sponge?

____

HOW SPONGES FEED

As shown above, a sponge has a special way of getting food. Water flows into the body of a sponge through pores. *Pores* are holes on the outside of a sponge. The water flows into a cavity in the middle of the sponge. *Collar cells* line this cavity. Collar cells take in tiny plants and animals from the water and digest them. Water leaves through a hole at the top of the sponge. This is called the *osculum.*

REGENERATION

Sponges have some unusual abilities. If a sponge is forced through a strainer, the cells can come back together to re-form a new sponge. If part of a sponge is broken off, the missing part can grow back. This ability is called *regeneration.* A part of the sponge that is broken off can also grow into a new sponge. This means sponges can use regeneration as a way to reproduce.

SUPPORT IN A SPONGE BODY

Most sponges have skeletons that support their bodies and protect them from predators. This skeleton is not made of bones, however. It is made of small, hard fibers called *spicules.* Some spicules are straight, some are curved, and some are star-shaped.

STANDARDS CHECK

LS 2a Reproduction is a characteristic of all living systems; because no living organism lives forever, reproduction is essential to the continuation of every species. Some organisms reproduce asexually. Others reproduce sexually.

7. Apply Concepts When a new sponge grows from a piece of sponge that has broken off, is this asexual or sexual reproduction? Explain your answer.

____

____

____

____

What Are Cnidarians?

If you have ever been stung by a jellyfish, then you have touched a cnidarian. Jellyfish are members of a group of invertebrates that have stinging cells. Animals in this group are called *cnidarians.* ☑

READING CHECK

8. Identify What do all cnidarians have in common?

____

Cnidarians have complex tissues and a gut for digesting food. They also have a system of nerve cells within their bodies. When some cnidarians are broken into pieces, the pieces can become new cnidarians.

TWO BODY FORMS

A cnidarian body can have one of two forms—medusa or polyp. The *medusa* and the *polyp* forms are shown below. Medusas swim through the water. Polyps usually attach to a surface. Some cnidarians change forms during their lives. However, many cnidarians are polyps for their whole lives.

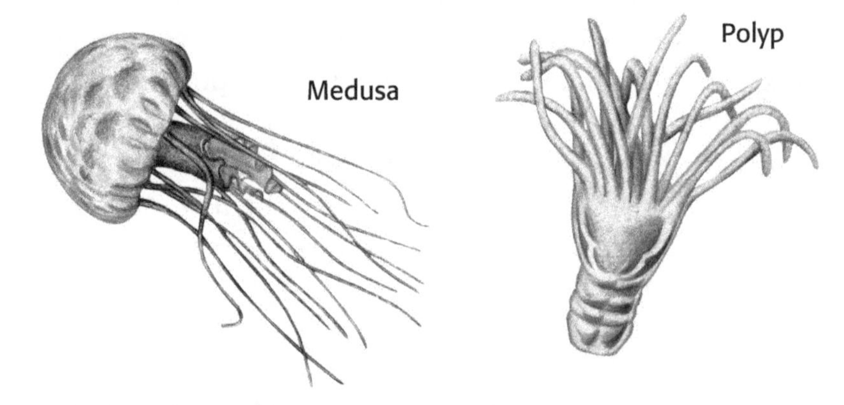

Both the medusa and the polyp forms of a jellyfish have radial symmetry

TAKE A LOOK

9. Identify What kind of symmetry do jellyfish have?

____

STINGING CELLS

All cnidarians have tentacles covered with stinging cells. The touch from another organism activates these stinging cells. Each stinging cell shoots a tiny spear into the organism. Each spear carries poison. Cnidarians use stinging cells to protect themselves and to catch food.

A tiny barbed spear is coiled inside each stinging cell.

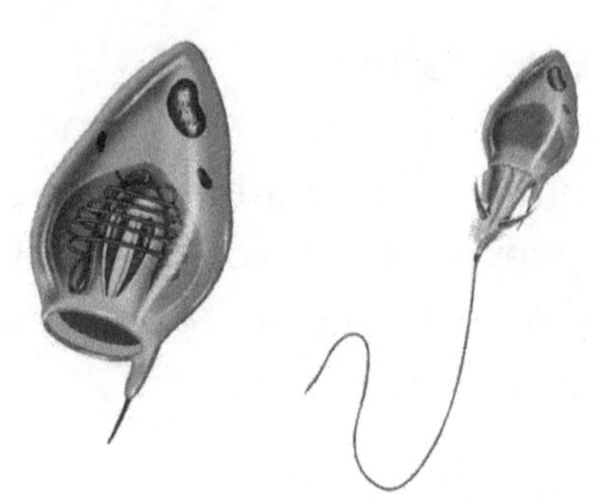

When the tiny spear is fired, the long barbed strand is shot into the prey.

TAKE A LOOK

10. List Name two reasons cnidarians use stinging cells.

____

____

KINDS OF CNIDARIANS

There are three classes of cnidarians: hydrozoans, jellyfish, and sea anemones and corals. The figure below shows each type of cnidarian.

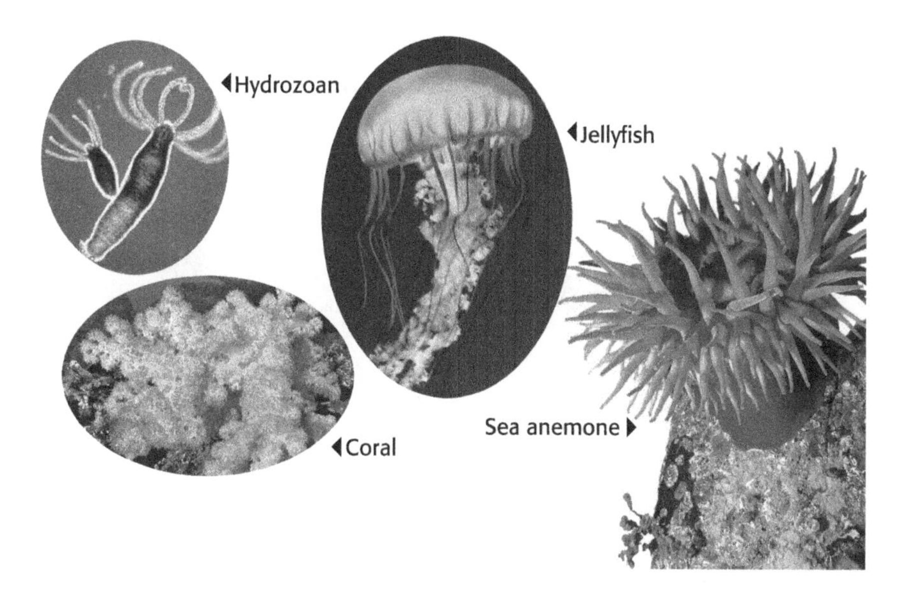

TAKE A LOOK

11. Apply Concepts Which of the cnidarians in this figure have the polyp body form?

____

____

____

Hydrozoans live in both freshwater and saltwater. Most hydrozoans live as polyps their entire lives. Jellyfish use their tentacles to catch fish and other invertebrates. They spend most of their lives as medusas. Sea anemones and corals spend their lives as polyps. Corals can build large and colorful reefs.

What Are Flatworms?

When you think of worms, you probably think of earthworms. However, there are many other kinds of worms. Many of them are too small to see without a microscope. The simplest worms are the flatworms. ☑

Flatworms are divided into three classes: planarians, flukes, and tapeworms. All flatworms have bilateral symmetry. Most flatworms have a head and two eyespots. Some flatworms have a bump on each side of their heads. These are called sensory lobes. The flatworm uses sensory lobes to find food.

READING CHECK

12. Identify What are the simplest worms?

____

____

PLANARIANS

Planarians are flatworms that live in fresh water or damp places. As shown below, they have a head, eyespots, and sensory lobes. Most planarians are predators. They eat other animals or parts of animals and digest the food in a gut. Planarians use their sensory lobes to find food. Planarians also have a nervous system with a brain. ☑

READING CHECK

13. Explain What do planarians use their sensory lobe for?

____

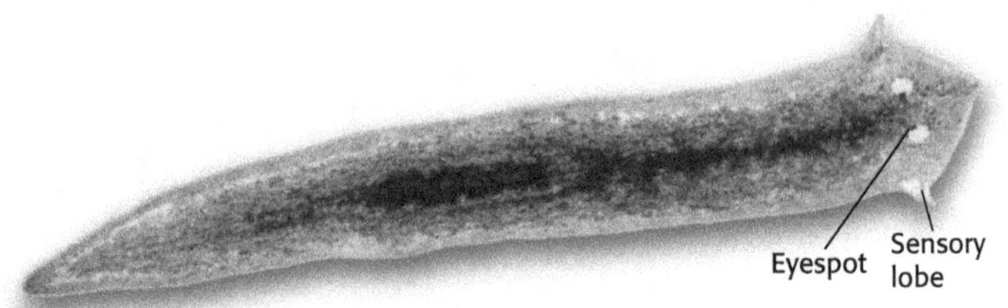

Planarians are often about 15 mm long.

FLUKES

Flukes are parasites. A *parasite* is an organism that feeds on another living thing. This organism is called the *host.*

Most flukes live and reproduce inside a host animal. The fertilized eggs of a fluke pass out of the animal's body with its waste products. If these eggs infect water or food, animals may eat them. These eggs would then grow into new flukes inside these animals. Flukes have no eyespots or sensory lobes. They use suckers to attach to their hosts.

Flukes use suckers to attach to their hosts. Most flukes are just a few millimeters long.

TAKE A LOOK

14. Identify What structures do flukes use to attach to a host?

____

TAPEWORMS

Tapeworms are also parasites. Like flukes, tapeworms live, feed, and reproduce inside a host. They have no eyespots, sensory lobes, or gut. Tapeworms simply attach to the intestines of their hosts and absorb nutrients. Nutrients move directly through the tapeworm's tissues. Some tapeworms can live in humans.

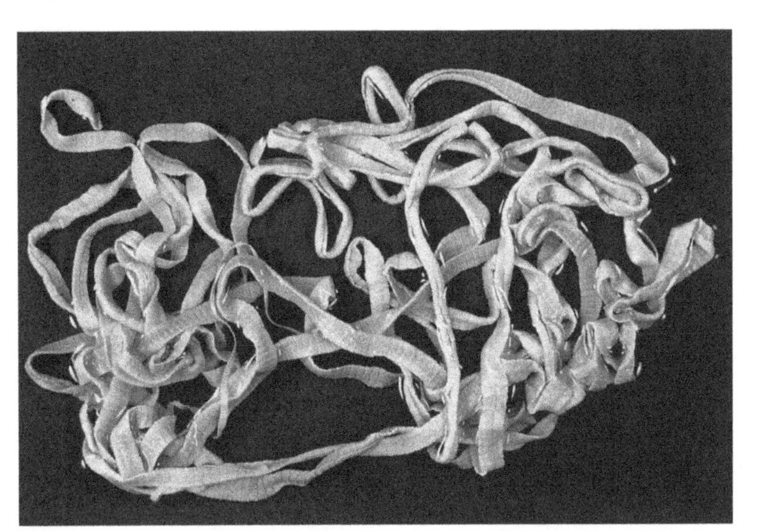

Tapeworms can grow very long. Some are longer than a bus!

Critical Thinking

15. Compare How does the size of a tapeworm compare to the size of a planarian?

____

____

____

____

____

What Are Roundworms?

Roundworms have bodies that are long, slim, and round, like short pieces of spaghetti. Like other worms, they have bilateral symmetry. Roundworms have a simple nervous system and brain. ☑

READING CHECK

16. Describe Describe the body of a roundworm.

____

____

____

Most species of roundworms are very small. A single rotten apple could contain 100,000 roundworms! These tiny worms get nutrients by breaking down the tissues of dead plants and animals. This process helps put some nutrients back into the soil.

Some roundworms are parasites. Several roundworms, such as pinworms and hookworms, can infect humans. Humans can become infected with roundworms by eating meat that has not been fully cooked. In humans, roundworms can cause fever, tiredness, and digestive problems.

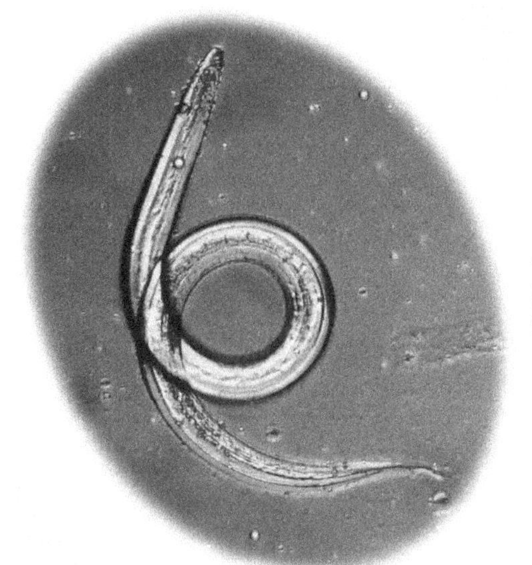

This hookworm is a tiny larva. Even as an adult, it will be less than 15 mm.

Name ________________________ Class ______________ Date ______________

Section 1 Review

NSES LS 1d, 1f, 2a, 3a, 3c, 5a

SECTION VOCABULARY

coelom a body cavity that contains the internal organs

ganglion a mass of nerve cells

gut the digestive tract

invertebrate an animal that does not have a backbone

1. **Identify** What do all invertebrates have in common?

____

2. **List** What are the four groups of simple invertebrates?

____

3. **Explain** How does the gut of a complex animal differ from that of a simple animal?

____

4. **List** Give two ways sponges use regeneration.

____

5. **Identify** What are the two cnidarian body forms?

____

6. **Infer** Why do you think it would be important to a parasite that its host survive?

____

____

____

7. **Compare** Complete the chart below to compare the different types of flatworms.

Type of flatworm	Parasitic or nonparasitic?	Features
Planarians		head, eyespots, sensory lobes, brain

8. **Explain** Are all roundworms parasites? Explain your answer.

____

____

Name ______________________ Class ______________ Date ______________

CHAPTER 15 Invertebrates

SECTION 2

Mollusks and Annelid Worms

BEFORE YOU READ

After you read this section, you should be able to answer these questions:

- What are the four features of mollusks?
- What are the three kinds of annelid worms?

National Science Education Standards

LS 1a, 1d, 3a, 5a

What Are Mollusks and Annelid Worms?

Have you ever eaten clams? Have you ever seen earthworms on the sidewalk after it rains? If so, then you have already seen mollusks and annelid worms. These invertebrates are more complex than the simple invertebrates. For example, mollusks and annelid worms have circulatory systems that carry materials throughout their bodies.

Underline As you read, use colored pencils to underline the characteristics of mollusks and annelid worms. Use one color for mollusks and another for annelid worms.

What Are Mollusks?

Snails, slugs, clams, oysters, squids, and octopuses are all mollusks. Most mollusks live in the ocean. However, some live in fresh water and some live on land.

There are three classes of mollusks: gastropods, bivalves, and cephalopods. The *gastropods* include snails and slugs. The *bivalves* include shellfish with two shells, such as clams and oysters. *Cephalopods* include squids and octopuses. ☑

READING CHECK

1. Identify Which class of mollusks do snails belong to?

____

HOW MOLLUSKS EAT

Each kind of mollusk has its own way of eating. As shown below, snails and slugs eat with a special organ called the *radula*. The radula is a tongue covered with curved teeth. It lets the animal scrape food off rocks, seaweed, or plants. Clams and oysters use gills to filter tiny organisms from the water. Squids and octopuses use tentacles to grab their food and place it in their jaws.

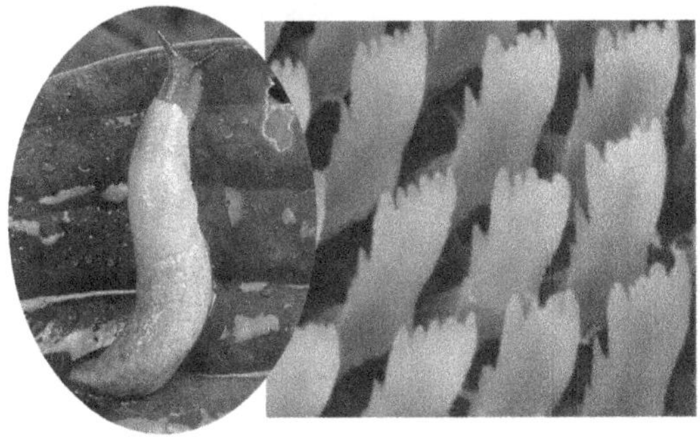

The rows of teeth on a slug's radula help scrape food from surfaces. This radula has been magnified 2,000 times.

TAKE A LOOK

2. Identify What is the function of the radula?

____

____

GANGLIA AND BRAINS

All mollusks have complex ganglia. They have special ganglia to control breathing, movement, and digestion. Cephalopods have the most advanced nervous system of all invertebrates. An octopus, for example, has a large brain that connects all of its ganglia. Cephalopods are thought to be the smartest invertebrates. ☑

READING CHECK

3. Identify Which group of invertebrates has the most advanced nervous systems?

____

PUMPING BLOOD

Unlike simple invertebrates, mollusks have a circulatory system. The circulatory system moves materials through the body in blood. Most mollusks have open circulatory systems. In an **open circulatory system**, a heart pumps blood through blood vessels that empty into spaces in the animal's body. These spaces are called *sinuses*.

Squids and octopuses have closed circulatory systems. In a **closed circulatory system**, a heart pumps blood through a network of blood vessels that form a loop.

Critical Thinking

4. Infer What kind of circulatory system do you have?

____

MOLLUSK BODIES

Although mollusks can look quite different from one another, their bodies have similar parts. The body parts of mollusks are described below.

Body Parts of Mollusks

Mollusks have a broad, muscular **foot**. The foot helps the animals move. In gastropods, the foot makes mucus that the animal slides along.

The gills, gut, and other organs form the **visceral mass**. It lies in the center of the mollusk's body.

A layer of tissue called the **mantle** covers the visceral mass. The mantle protects the bodies of mollusks that do not have a shell.

In most mollusks, the outside of the mantle makes a **shell**. The shell protects the mollusks from predators. It also keeps land mollusks from drying out.

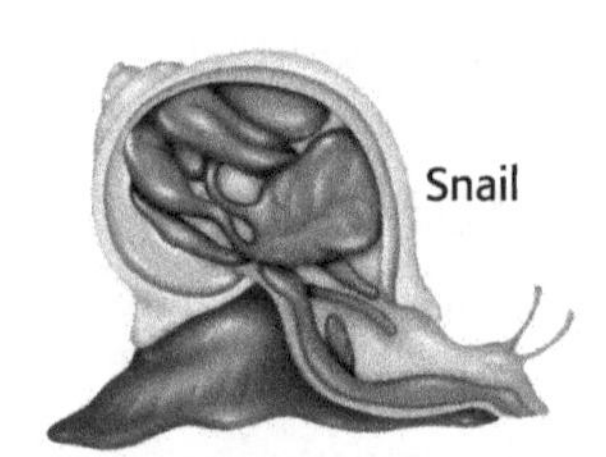

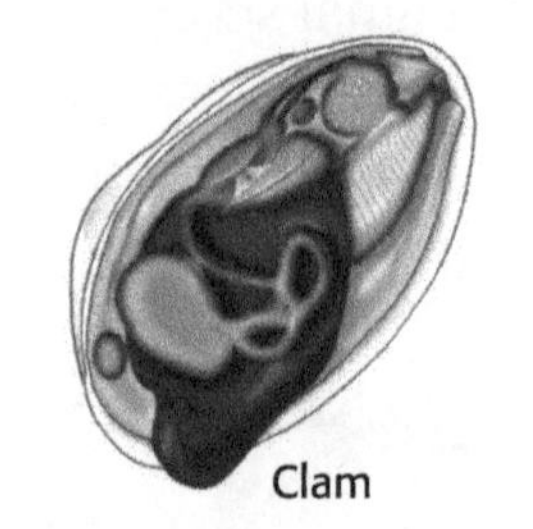

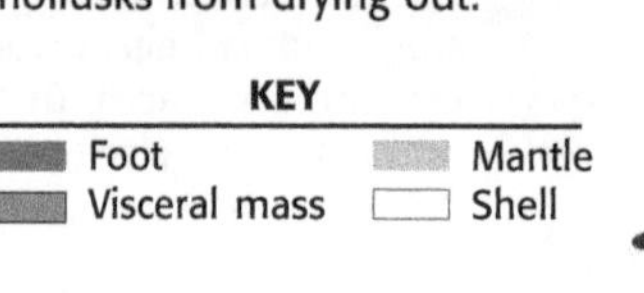

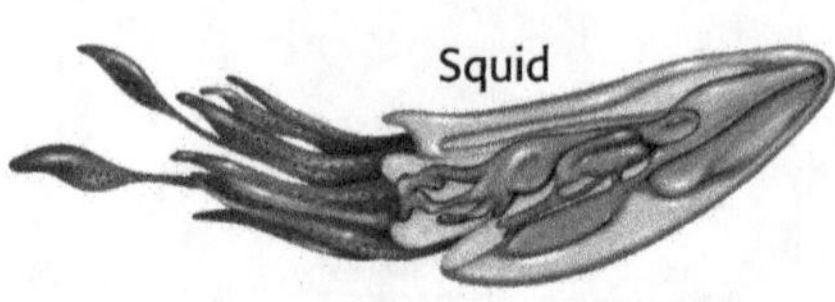

TAKE A LOOK

5. Identify Which mollusk does not have an obvious shell?

____

6. List What are two functions of a mollusk's shell?

____

____

____

What Are Annelid Worms?

The bodies of annelid worms are segmented. Annelid worms are sometimes called *segmented worms*. A **segment** is an identical, or almost identical, repeating body part.

Annelid worms are more complex than other worms. They have a closed circulatory system. They also have a complex nervous system with a brain. A nerve cord connects the brain to a ganglion in each segment.

Annelid worms live in salt water, fresh water, or on land. There are three kinds of annelid worms: earthworms, marine worms, and leeches. ☑

READING CHECK

7. List What are the three kinds of annelid worms?

____

____

____

EARTHWORMS

Earthworms are the most familiar annelid worms. An earthworm has between 100 and 175 segments. Most segments look the same. However, some segments have special jobs, such as eating and reproduction.

As they feed, earthworms break down plant and animal matter in soil. They leave behind wastes called *castings*. Castings make soil richer by adding nutrients near the surface. Earthworms also improve soil by digging tunnels. These tunnels let air and water reach down into the soil. ☑

READING CHECK

8. Define What are castings?

____

____

____

____

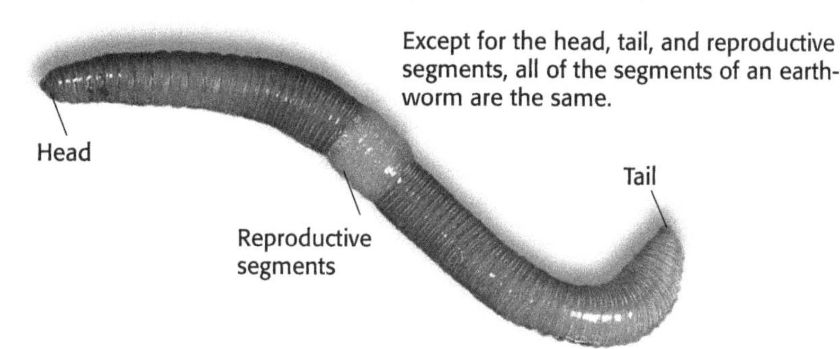

Except for the head, tail, and reproductive segments, all of the segments of an earthworm are the same.

MARINE WORMS

Some worms live in the ocean. These marine worms are called *polychaetes*. They are covered with hairlike bristles and come in many bright colors. Marine worms eat small animals and mollusks. Some also filter food from the water.

LEECHES

Many people think of leeches as parasites that suck other animals' blood. This is true of some leeches. However, not all leeches are parasites. Some feed on dead animals. Others eat insects, slugs, and snails.

Name ______________________ Class ____________ Date ____________

Section 2 Review

NSES LS 1a, 1d, 3a, 5a

SECTION VOCABULARY

closed circulatory system a circulatory system in which the heart circulates blood through a network of blood vessels that form a closed loop

open circulatory system a circulatory system in which the circulatory fluid is not contained entirely within vessels

segment any part of a larger structure, such as the body of an organism, that is set off by natural or arbitrary boundaries

1. Compare What is the difference between on open circulatory system and a closed circulatory system?

____

____

____

2. List What are the four body parts that all mollusks have?

____

3. Summarize Complete the organizer below to show the classes of mollusks.

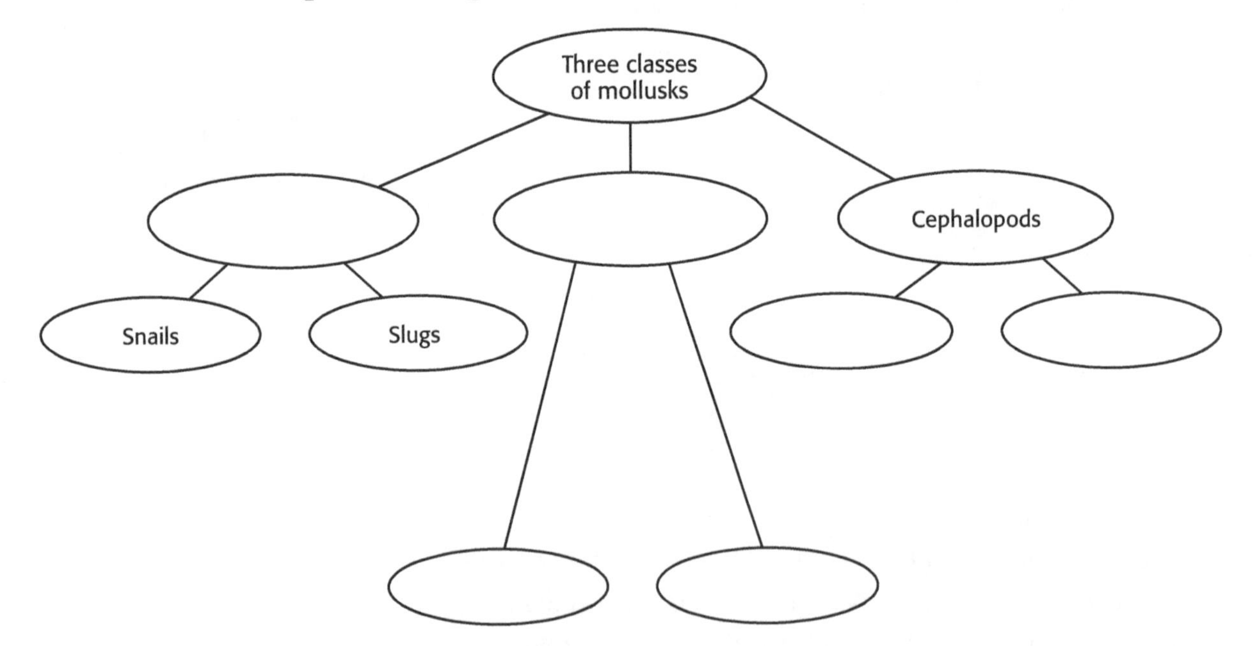

4. Describe Describe two ways earthworms improve soil.

____

____

7. Explain Are all leeches parasites? Explain your answer.

____

____

Name ______________________ Class ______________ Date ______________

CHAPTER 15 Invertebrates

SECTION 3

Arthropods

BEFORE YOU READ

After you read this section, you should be able to answer these questions:

- What are the four main characteristics of arthropods?
- What are two types of metamorphosis in insects?

National Science Education Standards

LS 1a, 1d, 1f, 5a

What Are the Characteristics of Arthropods?

How many animals do you think can live in one acre of land? If you could find all the arthropods in that acre, you would count more than a million animals!

Arthropods are the largest group of animals on Earth. Around 75% of all animal species are arthropods. Insects, spiders, crabs, and centipedes are all arthropods. ☑

Arthropods share four characteristics:

1. a segmented body with specialized parts
2. jointed limbs
3. an exoskeleton
4. a well-developed nervous system.

STUDY TIP

Organize As you read, make a chart listing the different kinds of arthropods, their characteristics, and examples of each.

1. Identify What percentage of all animals are arthropods?

____

SEGMENTED AND SPECIALIZED PARTS

Like annelid worms, arthropods are segmented. Some arthropods, such as centipedes, have segments that are mostly the same. Only the head and tail segments are different. However, most arthropods have segments with specialized structures. These specialized structures can be wings, antennae, gills, pinchers, and claws. As an arthropod develops, some of its segments grow together. This forms three main body parts: head, thorax, and abdomen.

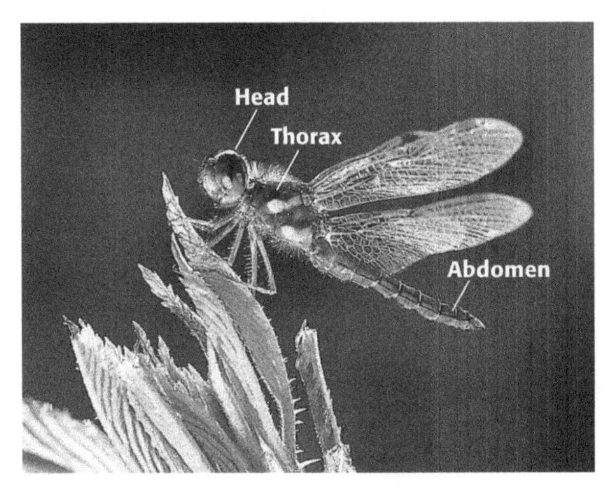

Like most arthropods, this dragonfly has a head, a thorax, and an abdomen.

TAKE A LOOK

2. Identify What are the three main body parts of an arthropod?

____

____

JOINTED LIMBS

Jointed limbs give arthropods their name. In Greek, *arthro* means "joint" and *pod* means "foot." Jointed limbs are legs or other body parts that bend at the joints. Having jointed limbs makes it easier for arthropods to move.

AN EXTERNAL SKELETON

Arthropods have an exoskeleton. An **exoskeleton** is a hard outer layer that covers the body. It is made of protein and a substance called *chitin*. An exoskeleton does some of the same things that an internal skeleton does. It supports the body and allows the animal to move. ☑

3. Identify What is an exoskeleton made of?

____

____

An exoskeleton also does things that an internal skeleton cannot do. It can protect the animal's body like a suit of armor. By keeping water inside the body, the exoskeleton helps keep the animal from drying out.

The jointed limbs of a ghost crab let it move quickly across the sand. The exoskeleton protects the crab's body from drying out on land.

SENSING SURROUNDINGS

All arthropods have a head and a well-developed brain and nerve cord. The nervous system receives information from sense organs such as eyes and bristles. For example, tarantulas use bristles to detect movement and vibration. Some arthropods have simple eyes that only allow them to see light. Most arthropods have compound eyes. A **compound eye** is an eye made of many identical light sensors. ☑

4. Identify What kind of eyes do most arthropods have?

____

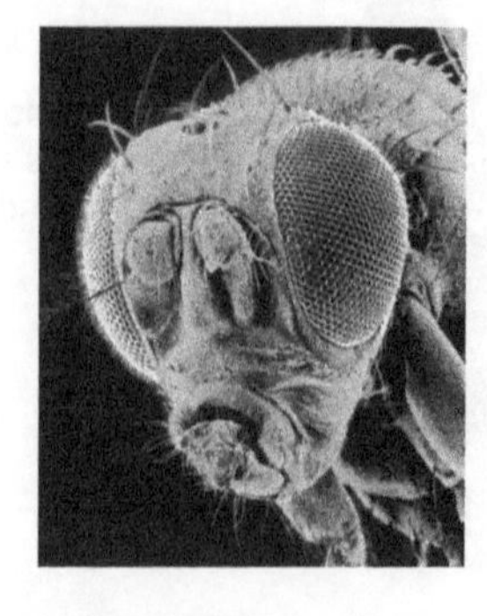

Compound eyes are made of many light-sensitive units that work together.

What Are the Different Kinds of Arthropods?

Arthropods are grouped by the kinds of body parts they have. You can tell the difference between kinds of arthropods by looking at how many legs, eyes, and antennae they have. Arthropods include centipedes and millipedes, crustaceans, arachnids, and insects.

CENTIPEDES AND MILLIPEDES

Centipedes and millipedes have many segments. Centipedes have two legs on each segment and millipedes have four.

Centipedes and millipedes also have hard heads with one pair of antennae. An **antenna** is a feeler that senses touch, taste, or smell. Centipedes and millipedes also have one pair of mandibles. *Mandibles* are mouthparts that can stab and chew food.

Centipedes eat other animals. Millipedes eat plants.

CRUSTACEANS

Shrimp, barnacles, crabs, and lobsters are *crustaceans*. Most crustaceans live in water. They have gills for breathing, mandibles for eating, and two compound eyes. Each eye is at the end of an eyestalk. Unlike other arthropods, crustaceans have two pairs of antennae. ☑

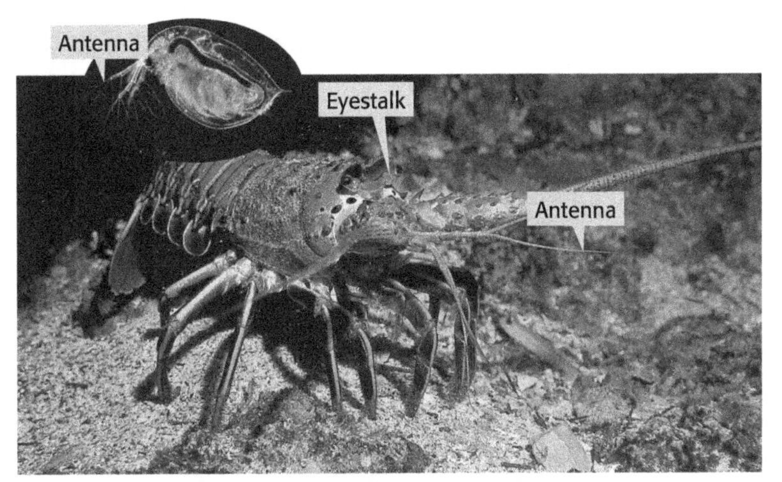

Water fleas and lobsters are two kinds of crustaceans.

Math Focus

5. Calculate If a millipede has 752 legs, how many segments does it have? If a centipede has 354 legs, how many segments does it have?

____

____

READING CHECK

6. Identify What is one difference between crustaceans and other arthropods?

____

____

ARACHNIDS

Spiders, scorpions, mites, and ticks are arachnids. *Arachnids* have two main body parts: the cephalothorax and the abdomen. The *cephalothorax* is made of a head and thorax. Most arachnids have eight legs and two simple eyes. They also have claw-like mouthparts called *chelicerae*. Spiders use their chelicerae to catch insects. ☑

READING CHECK

7. Identify What are the two main body parts of an arachnid?

____

____

Spiders can be helpful to humans becasue they kill many insect pests. Though some people are afraid of spiders, most spiders are not harmful. The mouthparts of most spiders are too small to bite humans. However, some spider bites do need medical treatment.

Ticks are parasites that feed on a host's blood. These animals are just a few millimeters long. They live in forests, brushy areas, and even lawns. Some ticks carry diseases, such as Lyme disease.

TAKE A LOOK

8. Identify What are the mouthparts of an archnid called?

____

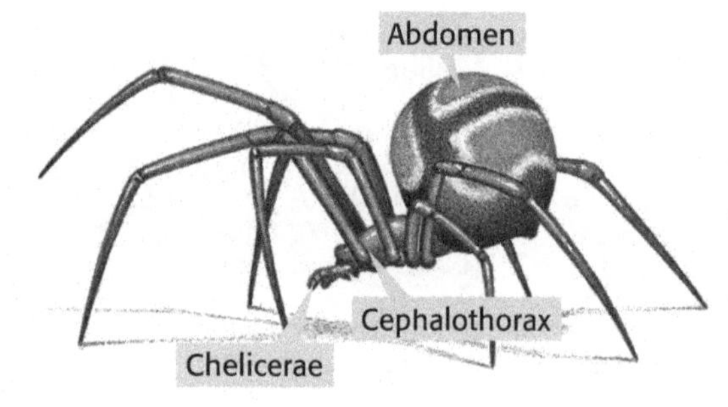

Arachnids have two main body parts. They have special mouthparts called chelicerae.

INSECTS

Insects make up the largest group of arthropods. The only place on Earth where insects do not live is in the oceans. Although some insects are pests, many insects are helpful. Many flowering plants need insects to carry pollen. Farmers depend on insects to pollinate fruit crops.

All insects have three main body parts, six legs, and two antennae. On their heads they have antennae and a pair of compound eyes. They use mandibles for eating. Some insects have wings on their thorax.

Investigate Choose an insect to research. Find out where it lives, what it eats, and whether it is considered helpful or harmful to humans. Present your findings to the class.

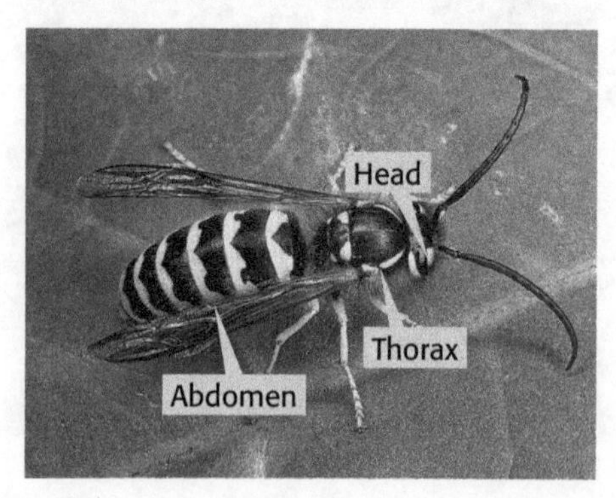

METAMORPHOSIS

As an insect develops, it changes form. This process is called **metamorphosis**. Most insects, including butterflies, beetles, flies, wasps, and ants, go through complete metamorphosis. Complete metamorphosis has four stages, as shown below.

Stages of Complete Metamorphosis

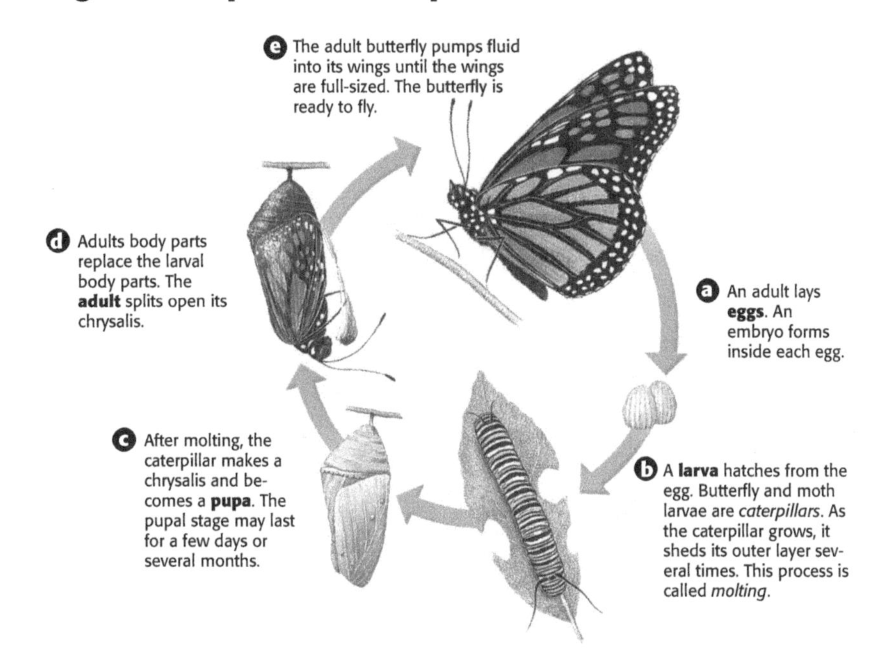

TAKE A LOOK

9. List What are the four stages of complete metamorphosis?

____

____

Some insects, such as grasshoppers and cockroaches, go through incomplete metamorphosis. *Incomplete metamorphosis* has three stages: egg, nymph, and adult. Nymphs look like small adults. However, nymphs do not have wings.

Critical Thinking

10. Compare How is incomplete metamorphosis different from complete metamorphosis?

____

____

____

____

____

____

____

____

Stages of Incomplete Metamorphosis

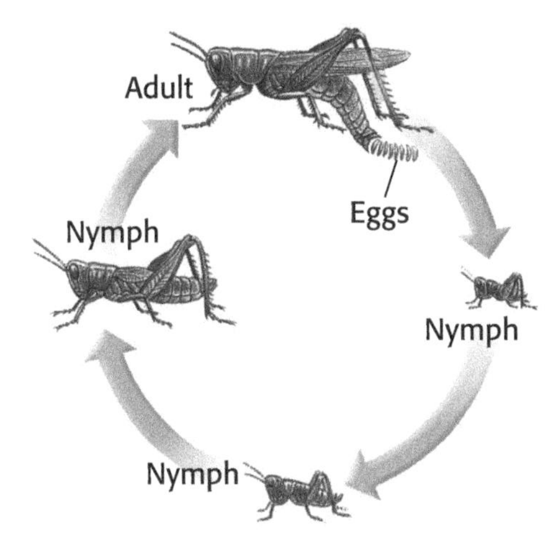

Name ______________________ Class ______________ Date ______________

Section 3 Review

NSES LS 1a, 1d, 1f, 5a

SECTION VOCABULARY

antenna a feeler that is on the head of an invertebrate, such as a crustacean or an insect, that senses touch, taste, or smell

compound eye an eye composed of many light detectors

exoskeleton a hard, external, supporting structure

metamorphosis a phase in the life cycle of many animals during which a rapid change from the immature form of an organism to the adult form takes place

1. Summarize Complete the organizer to name the four characteristics of arthropods.

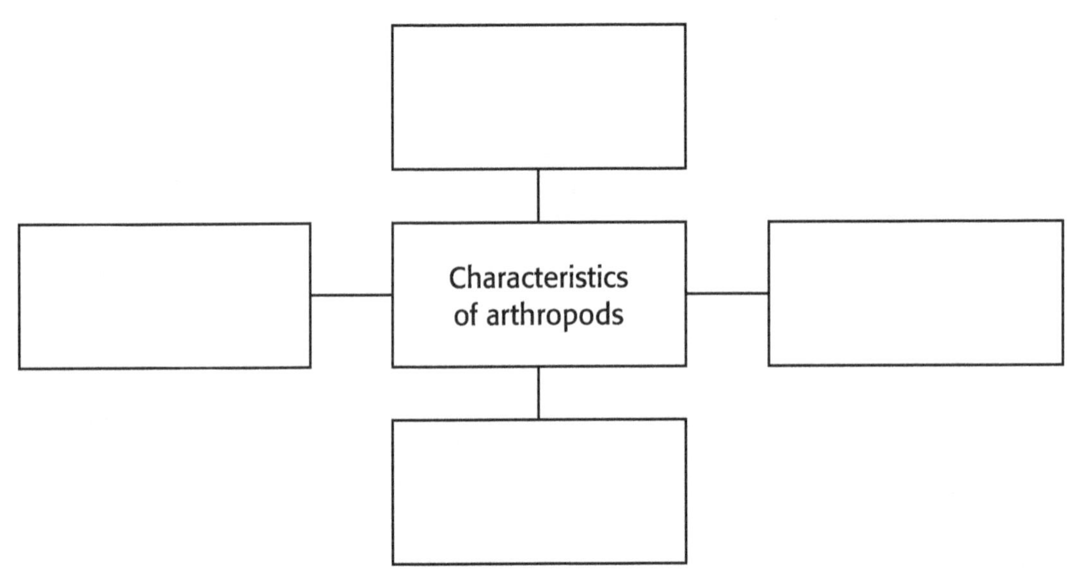

2. Compare How is the body of an arachnid different from that of other arthropods?

____

____

____

3. Apply Concepts Scientists have discovered a new organism. They know it is an arthropod because it has jointed legs and an exoskeleton. The organism also has mandibles and two pairs of antennae. Should they classify this organism as an insect or a crustacean? Explain your answer.

____

____

____

4. Explain How are spiders helpful to humans?

____

5. Identify What is the largest group of arthropods?

____

Name ______________________ Class ______________ Date ______________

CHAPTER 15 Invertebrates

SECTION 4

Echinoderms

BEFORE YOU READ

After you read this section, you should be able to answer these questions:

- What are the characteristics of echinoderms?
- What are the five classes of echinoderms?

National Science Education Standards

LS 1a, 1d, 3a, 5a

What Are Echinoderms?

Sea stars, sea urchins, and sand dollars are all echinoderms. Echinoderms are marine animals. They live on the sea floor in all parts of the world's oceans. Some echinoderms eat shellfish, some eat dead plants and animals, and others eat algae.

STUDY TIP

Organize As you read this section, make a Spider Map that shows the different kinds of echinoderms.

SPINY SKINNED

Echinoderms have spines. In Greek, the name echinoderm means "spiny skinned." The spines of an echinoderm are found on its skeleton. Echinoderms have an **endoskeleton**, which is found inside their bodies. The spines grow from the endoskeleton. The animal's skin then covers the spines.

SYMMETRY

As echinoderms grow, their body symmetry changes. When they are larvae, they have bilateral symmetry. However, as adults, they have radial symmetry. ☑

READING CHECK

1. Identify What kind of body symmetry does an adult echinoderm have?

____

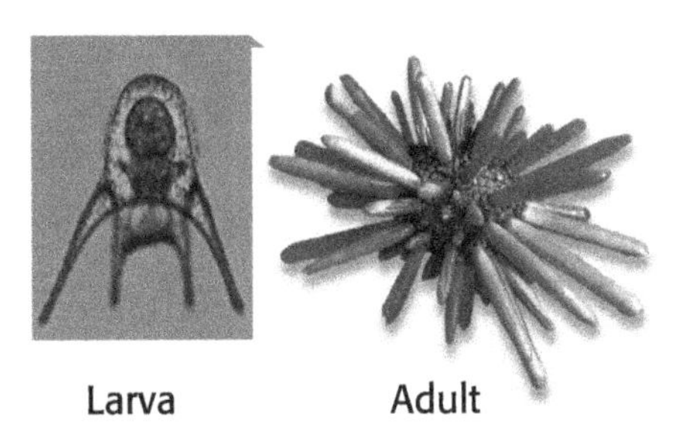

The sea urchin larva has bilateral symmetry. The adult sea urchin has radial symmetry.

NERVOUS SYSTEM

Echinoderms have a simple nervous system. Around the mouth is a circle of nerve fibers called the *nerve ring*. In sea stars, a *radial nerve* runs along each arm to the nerve ring. The radial nerves let a sea star move its arms. Sea stars have a simple eye at the tip of each arm. These eyes sense light. Other cells that cover the sea star's body sense touch and chemical signals in the water. ☑

READING CHECK

2. Explain What do the radial nerves allow a sea star to do?

____

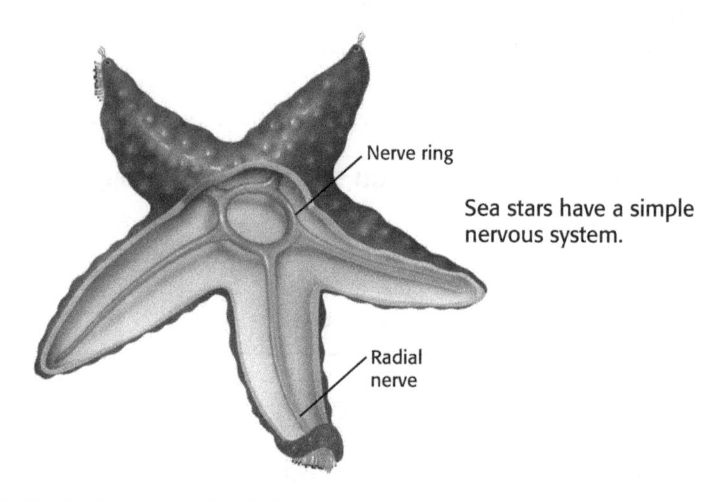

Sea stars have a simple nervous system.

TAKE A LOOK

3. Circle On the diagram, circle the areas where the eyespots of a sea star are found.

WATER VASCULAR SYSTEM

Only echinoderms have a water vascular system. A **water vascular system** is a system of tubes filled with fluid. Echinoderms use the water pressure in these tubes to move, eat, breathe, and sense their environment.

Water Vascular System

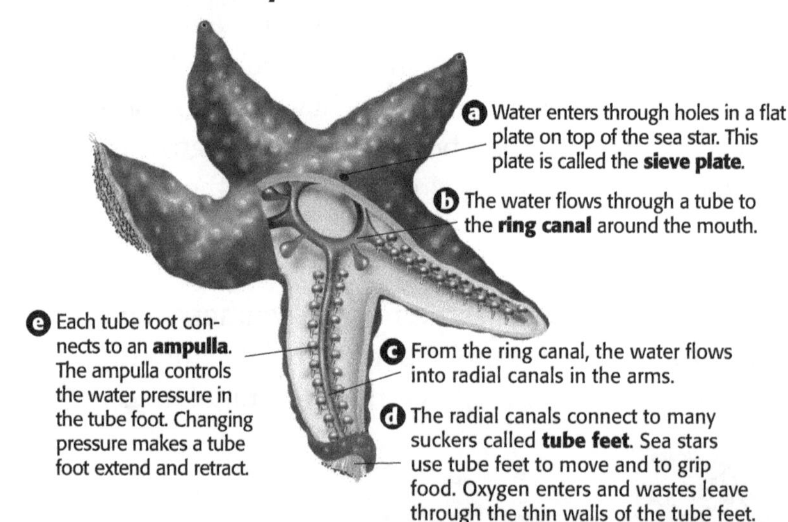

TAKE A LOOK

4. Identify Through what structure does water enter the water vascular system?

____

5. Explain What does a sea star use its tube feet for?

____

____

What Are the Different Kinds of Echinoderms?

There are five main classes of echinoderms. Sea stars are probably the most familiar class. The other classes are brittle stars and basket stars, sea urchins and sand dollars, sea lilies and feather stars, and sea cucumbers. Sea stars, brittle stars, basket stars, and feather stars can all grow back lost arms.

BRITTLE STARS AND BASKET STARS

Brittle stars and basket stars look like their close relatives, sea stars. However, their arms are longer, thinner, and smaller than those of sea stars.

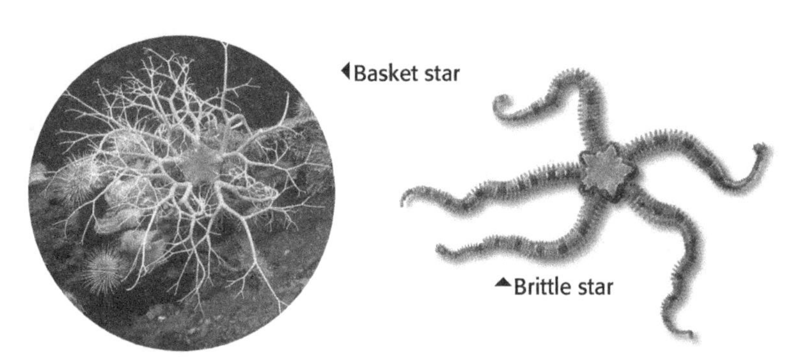

Basket stars and brittle stars move around more than other echinoderms do.

Critical Thinking

6. Infer Suppose you found a brittle star that had four long arms and one short one. How could you explain this?

SEA URCHINS AND SAND DOLLARS

Sea urchins and sand dollars are round animals and have no arms. Like sea stars, they use tube feet to move. Sea urchins and sand dollars have shell-like endoskeletons. Sea urchins feed on algae they scrape from rocks. Sand dollars eat tiny particles they find in sand.

Sea urchins and sand dollars use their spines to protect themselves and to move.

TAKE A LOOK

7. Identify What are the functions of spines on sea urchins and sand dollars?

SEA LILIES AND FEATHER STARS

Sea lilies and feather stars have 5 to 200 feathery arms. They use these arms to catch pieces of food in the water. A sea lily's arms sit on top of a stalk, making the animal look like a flower.

SEA CUCUMBERS

Unlike other echinoderms, sea cucumbers are long and shaped like fat worms. They have soft, leathery bodies and no arms. Like other echinoderms, sea cucumbers move with tube feet.

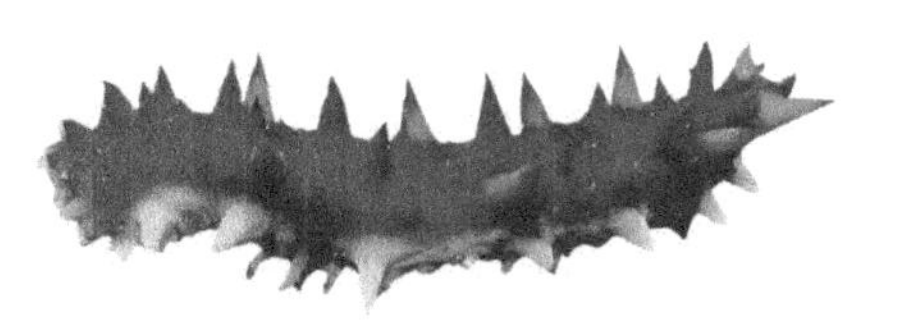

Sea Cucumber

TAKE A LOOK

8. Compare How are sea cucumbers, sand dollars, and sea urchins similar?

Name ______________________ Class ______________ Date ______________

Section 4 Review

NSES LS 1a, 1d, 3a, 5a

SECTION VOCABULARY

endoskeleton an internal skeleton made of bone or cartilage	**water vascular system** a system of canals filled with a watery fluid that circulates throughout the body of an echinoderm

1. **Describe** How does the body symmetry of an echinoderm change as the larva develops into an adult?

____

____

2. **List** What are the four main features of a sea star's nervous system?

____

____

3. **List** What are the five classes of echinoderms?

____

____

4. **Compare** How are sea cucumbers different from other echinoderms?

____

5. **Explain** How does a sea star get oxygen from the water?

____

6. **Summarize** Complete the Process Chart below to show how water moves through the water vascular system.

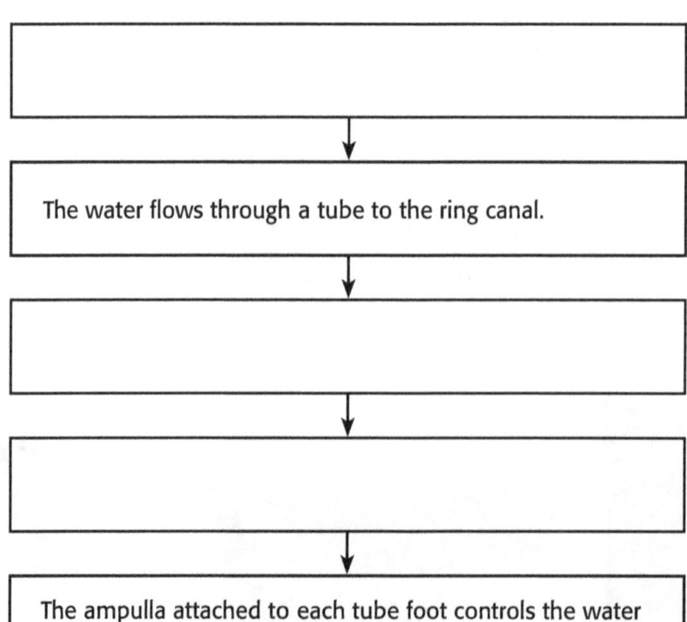

Name ______________________ Class ______________ Date ______________

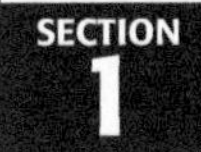

CHAPTER 16 Fishes, Amphibians, and Reptiles

SECTION 1

Fishes: The First Vertebrates

BEFORE YOU READ

After you read this section, you should be able to answer these questions:

- What are the two main characteristics of vertebrates?
- What is the difference between an ectotherm and an endotherm?
- What are the three classes of living fishes?

National Science Education Standards

LS 1a, 1d, 2a, 3a, 3b, 3c, 5a, 5c

What Are Chordates?

Many of the animals you are most familiar with, including fishes, reptiles, and mammals, belong to phylum Chordata. Members of this phylum are called *chordates*. Some chordates may be less familiar to you. Lancelets and tunicates are simple chordates.

STUDY TIP

Compare After you read this section, make a Concept Map using the vocabulary words from this section.

Tunicates

Lancelet

All chordates show four characteristics at some time in their lives. Sometimes the characteristic is found only in embryos.

- a tail
- a stiff flexible rod called the *notocord* that supports the body
- a hollow nerve cord, called a *spinal cord* in vertebrates
- structures called *pharyngeal pouches*, found in embryos that later become gills or other body parts ☑

READING CHECK

1. List What are the four characteristics shared by chordates?

____

____

____

____

Chordate Body Parts

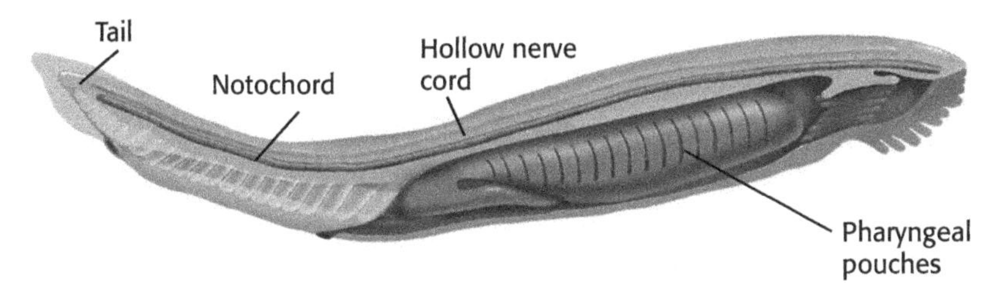

What Are Vertebrates?

The largest group of chordates is the vertebrates. **Vertebrates** are animals that have backbones. Fishes, amphibians, reptiles, birds, and mammals are vertebrates.

Vertebrates have two characteristics that lancelets and tunicates do not have—backbones and heads protected by skulls. A backbone is a strong, flexible column of bones called *vertebrae* (singular *vertebra*). It protects the spinal cord and helps support the rest of the body. The backbone replaces the notocord in most vertebrates.

In vertebrate embryos, the backbone, the skull, and the rest of the skeleton are made of cartilage. *Cartilage* is the tough material that makes up the flexible parts of our ears and nose. As most vertebrates grow, the cartilage is replaced by bone. Bone is much harder than cartilage. ☑

2. Identify What material makes up the skeleton of vertebrate embryos?

____

Because bone is so hard, it can be fossilized easily. Scientists have discovered many fossils of vertebrates. The fossils give scientists clues about how organisms are related to one another and when some organisms evolved.

STANDARDS CHECK

LS 3b Regulation of an organism's internal environment involves sensing the internal environment and changing physiological activities to keep conditions within the range required to survive.

Word Help: environment the surrounding natural conditions that affect an organism

3. Infer If an ectotherm's body temperature became too high, what would it need to do? Explain your answer.

____

____

____

____

____

____

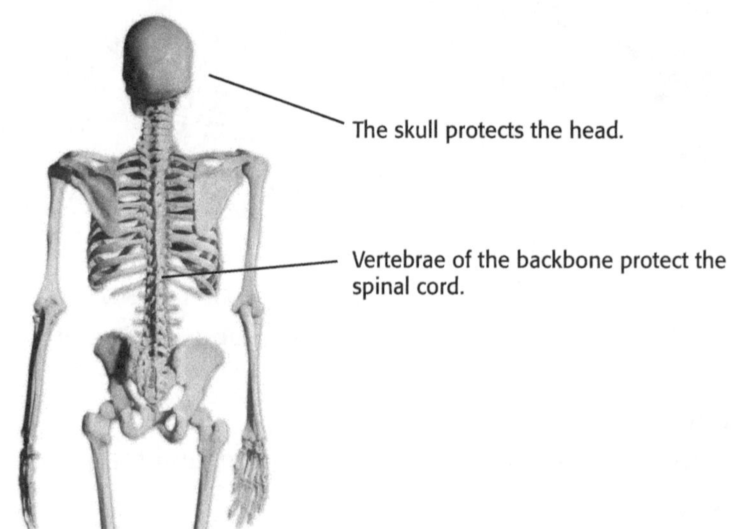

How Do Vertebrates Keep Their Bodies at the Right Temperature?

If an animal's body is too hot or too cold, its cells cannot work properly. Vertebrates control their body temperatures in two ways.

Ectotherms depend on their surroundings to stay warm. Their body temperatures change as the temperature of the environment changes. Most fish, and almost all amphibians and reptiles, are ectotherms. They are sometimes called *cold-blooded animals*.

Most fishes, including this leafy sea dragon, are ectotherms.

Endotherms generally have stable body temperatures, even when the temperature of the environment changes. They heat their bodies with energy from the chemical reactions in their cells. Birds and mammals are endotherms. They are sometimes called *warm-blooded animals*. ☑

READING CHECK

4. Explain How do endotherms heat their bodies?

____

____

____

What Are the Characteristics of Fishes?

Fossil evidence shows that fish evolved about 500 million years ago. They were the first vertebrates on Earth.

There are more than 25,000 species of fishes, and they come in many shapes, sizes, and colors. All fishes share several characteristics.

- a strong body for swimming, with muscles attached to the backbone
- fan-shaped structures called *fins* that help fishes swim, steer, stop, and balance
- scales to cover and protect the body
- gills
- senses of vision, hearing, and smell, including a lateral line system

The **lateral line** is a row of tiny sense organs along each side of the body. It detects water vibrations, such as those caused by other fish swimming by. ☑

READING CHECK

5. Identify What is the function of the lateral line system?

____

____

Fish use **gills** to breathe. Water passes over the thin gill membranes. Oxygen moves from the water to the blood. Carbon dioxide moves from the blood to the water.

Fishes are many shapes and sizes, but all fish have gills, fins, and a tail.

TAKE A LOOK

6. Complete Fill in the missing labels on the diagram.

How Do Fishes Reproduce?

Most fishes reproduce by *external fertilization*. The female lays unfertilized eggs in the water, and the male drops sperm on them.

Some fishes use *internal fertilization*. The male deposits sperm inside the female. Usually, the female lays fertilized eggs that have embryos inside. In some species, the embryos develop inside the female.

What Are the Three Classes of Living Fishes?

There were once five classes of fishes, but two of them are now extinct. The three classes of fishes living today are *jawless fishes*, *cartilaginous fishes*, and *bony fishes*.

JAWLESS FISHES

The first fishes were probably jawless fishes. Today, the two kinds of jawless fishes are hagfish and lampreys. Hagfish and lampreys look like eels and have round mouths without jaws. They do not need jaws because they do not bite or chew. Instead, they attach to other animals and suck their blood and flesh. ☑

READING CHECK

7. Explain How do jawless fishes eat?

____

____

____

Jawless fishes share the following characteristics:

- no jaws
- a notocord
- a skeleton made of cartilage

Jawless Fishes

Hagfish can tie their flexible bodies into knots. They slide the knot from their tails to their heads to remove slime from their skin or to escape predators.

Lampreys can live in salt water or fresh water, but they must reproduce in fresh water.

TAKE A LOOK

8. Identify Give two reasons hagfish tie themselves in knots.

____

____

____

CARTILAGINOUS FISHES

In most vertebrates, soft cartilage in the embryo is slowly replaced by bone. The skeletons of sharks, skates, and rays, however, are cartilage, and never change to bone. They are called *cartilaginous fishes.*

Cartilaginous fishes share the following characteristics:

- jaws
- a backbone
- a skeleton made of cartilage

Cartilaginous fishes store a lot of oil in their livers. Oil is less dense than water and this helps cartilaginous fishes float. However, they will sink if they stop swimming. Some cartilaginous fishes must also keep swimming to move water over their gills. Others can lie on the ocean floor and pump water across their gills. ☑

READING CHECK

9. Explain Do all cartilaginous fishes need to keep swimming in order to breathe? Explain your answer.

____

____

____

____

Cartilaginous Fish

Skates lay eggs. They move their fins up and down to swim.

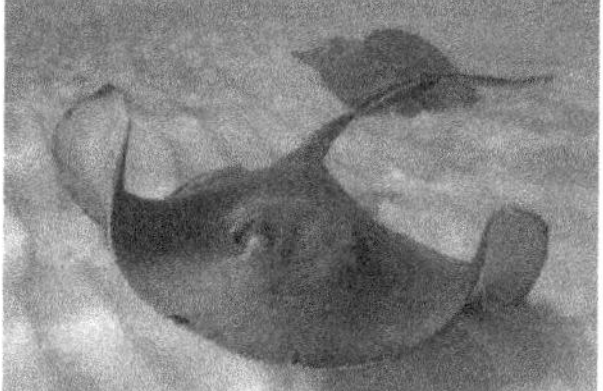

Rays give birth to live young. Rays also use their fins to swim They feed on organisms from the sea floor, such as crabs and worms.

Sharks, such as this hammerhead shark, rarely prey on humans. They usually eat other fish.

Critical Thinking

10. Compare Give two differences between cartilaginous fishes and bony fishes.

____

____

____

____

____

____

____

____

BONY FISHES

Ninety-five percent of all fishes are bony fishes. This group includes goldfish, tuna, trout, and catfish. Bony fishes have the following characteristics:

- jaws
- a backbone
- a skeleton made of bone
- a swim bladder

A **swim bladder** is a balloon-like organ filled with gases. It is also called a *gas bladder*. The swim bladder keeps bony fish from sinking. This lets them rest in one place without swimming.

What Are the Two Main Groups of Bony Fishes?

There are two main groups of bony fishes: *ray-finned fishes* and *lobe-finned fishes*.

Almost all bony fishes are ray-finned fishes. Their fins are supported by thin rays of bone. Eels, herrings, trout, minnows, and perch are all ray-finned fishes.

Lobe-finned fishes have thick, muscular fins. There are only seven species of lobe-finned fishes. Six of these species are lungfishes. Lungfishes are lobe-finned fishes with air sacs that can fill with air, like lungs. Scientists think that one group of lobe-finned fishes were the ancestors of amphibians. ☑

READING CHECK

11. Identify What group of fishes do scientists think were the ancestors of amphibians?

____

____

Bony Fishes

Lungfishes live in shallow waters that often dry up in the summer.

Masked butterfly fish live in warm waters around coral reefs.

Pikes are fast predators. They move in quick bursts of speed to catch fish and invertebrates.

How Are Fishes Related to Other Vertebrates?

The Concept Map below shows how fishes are related to each other and to other vertebrates and chordates. Vertebrates are the largest group of chordates. Bony fishes are the largest group of fishes. Ray-finned fishes are the largest group of bony fishes. ☑

READING CHECK

12. Identify What is the largest group of fishes?

____

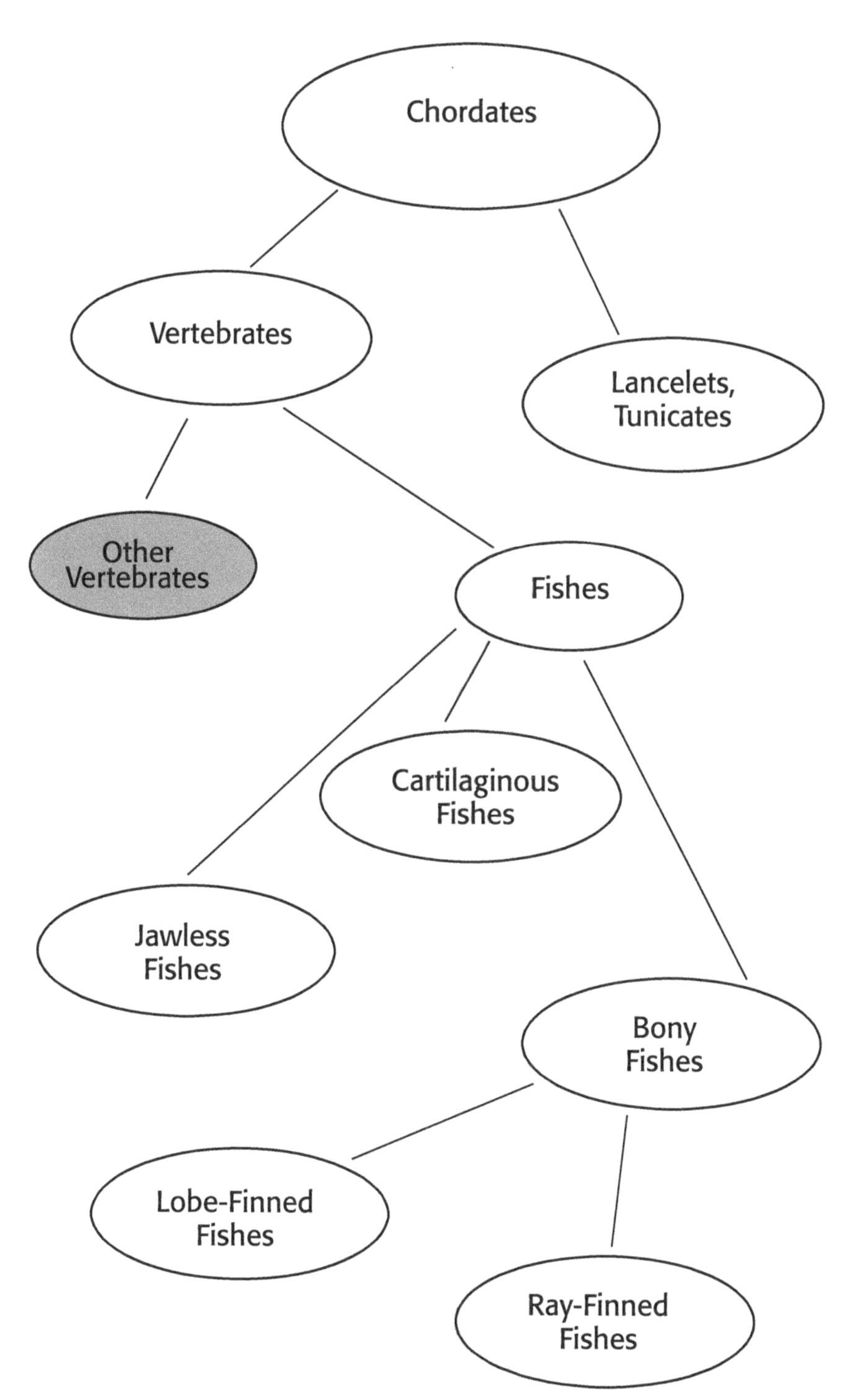

TAKE A LOOK

13. Identify Relationships Are lobe-finned fishes more closely related to jawless fishes or lancelets?

____

Name ______________________ Class ______________ Date ______________

Section 1 Review

NSES LS 1a, 1d, 2a, 3a, 3b, 3c, 5a, 5c

SECTION VOCABULARY

ectotherm an organism that needs sources of heat outside of itself

endotherm an animal that can use body heat from chemical reactions in the body's cells to maintain a constant body temperature

gill a respiratory organ in which oxygen from the water is exchanged with carbon dioxide from the blood

lateral line a faint line visible on both sides of a fish's body that runs the length of the body and marks the location of sense organs that detect vibrations in water

swim bladder in bony fishes, a gas-filled sac that is used to control buoyancy; also known as a gas bladder

vertebrate an animal that has a backbone

1. **Identify** What are two characteristics that only vertebrates have?

____

____

2. **List** What are three structures shared by all fishes?

____

____

____

3. **Explain** What is the difference between external fertilization and internal fertilization?

____

____

____

4. **Compare** Complete the table below to compare the three classes of living fishes.

Characteristic	Jawless fishes	Cartilaginous fishes	Bony fishes
Jaws (yes or no?)			
Structure for support (notocord or backbone?)			
Skeleton (cartilage or bone?)			

5. **Identify** In general, which groups of animals are ectotherms? Which groups are endotherms?

____

____

Name ______________________ Class ______________ Date ____________

SECTION 2

Amphibians

BEFORE YOU READ

After you read this section, you should be able to answer these questions:

- How do amphibians breathe?
- What are the three groups of amphibians?
- What is metamorphosis ?

National Science Education Standards

LS 1a, 2a, 3a, 3c, 3d, 5a, 5b

What Are Amphibians?

Amphibians are animals that have lungs and legs and can live in water. A **lung** is a saclike organ that delivers oxygen from the air to the blood.

Scientists think amphibians evolved from ancient lobe-finned fishes. These fishes developed lungs that could take oxygen from the air. They also had strong fins that could have evolved into legs. Fossils show that the first amphibians looked like a cross between a fish and a salamander. ☑

STUDY TIP

Organize As you read this section, make combination notes to describe the different groups of amphibians.

READING CHECK

1. Describe What type of animals do scientists think amphibians evolved from?

____

____

What Are the Characteristics of Amphibians?

Most amphibians must spend part of their lives in or near water. Most amphibians live in the water after hatching and later develop into adults that can live on land.

THIN SKIN

Amphibians have thin, smooth, moist skin. They can absorb water through their skin instead of drinking. However, they can also lose water through their skin. Because of this, they must live in water or in wet habitats. Some amphibians can breathe through their skin. The skin of some amphibians contains poison glands to protect the animals from predators.

Critical Thinking

2. Analyze Ideas The word *amphibian* means "double life." Explain why this is a good description for amphibians.

____

____

____

____

METAMORPHOSIS

Most amphibians don't just get bigger as they grow into adults. They change form, or shape, as they grow. **Metamorphosis** is the change from an immature form to a different adult form. For example, a **tadpole** is an immature frog or toad. As it grows, it changes form. It loses its gills and develops structures such as lungs and legs that the adult will need to live on land.

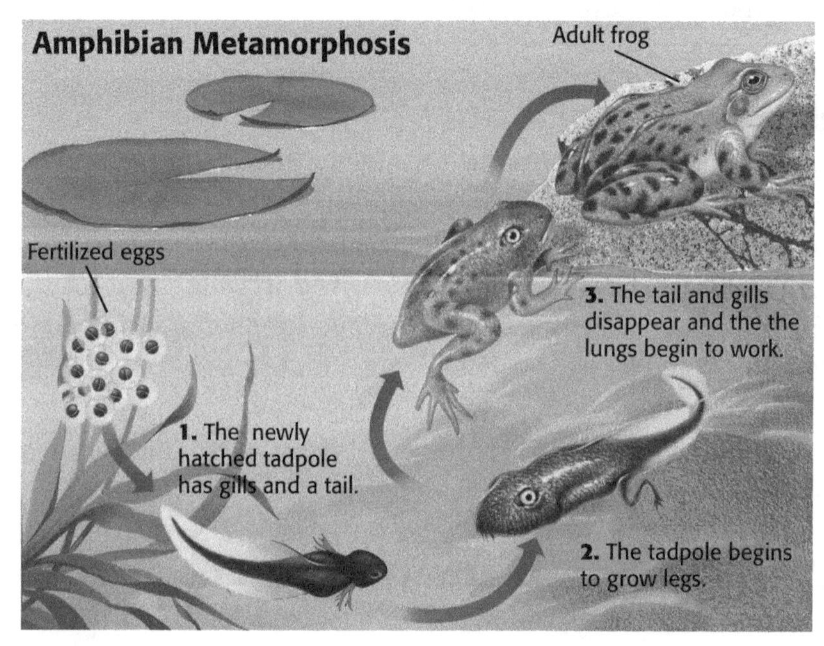

TAKE A LOOK

3. Compare How does breathing differ in tadpoles and adult frogs?

____

____

____

A few types of amphibians do not go through full metamorphosis. Their young look like tiny adults when they hatch, but they have gills.

EGGS WITHOUT SHELLS

Most of the eggs you are probably familiar with are bird eggs, such as chicken, or duck. Unlike bird eggs, however, amphibian eggs do not have shells or membranes. Without shells, amphibian eggs dry out very easily. Because of this, amphibian embryos must develop in wet environments. ☑

4. Explain Why do amphibian embryos need to develop in wet environments?

____

____

____

BODY TEMPERATURE

Amphibians are ectotherms. Their body temperature depends on the temperature of the environment.

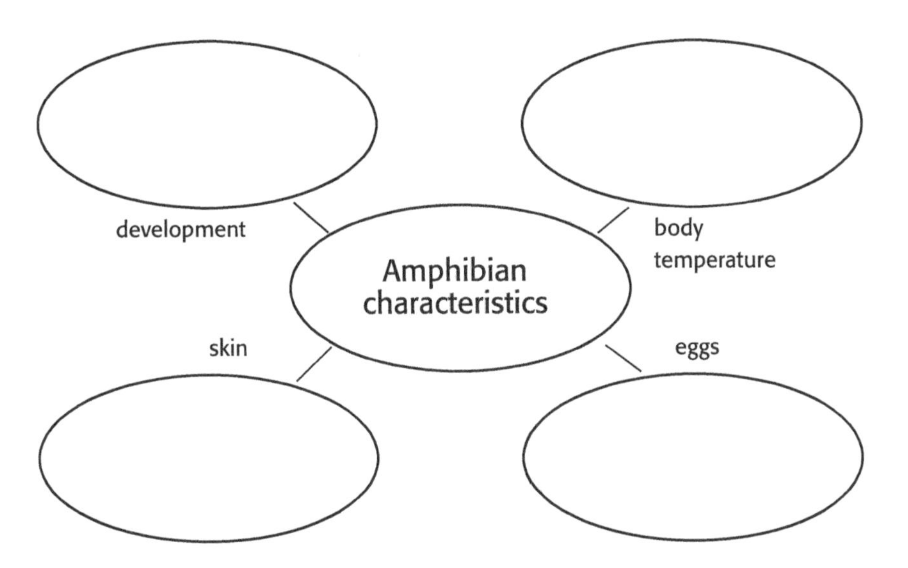

TAKE A LOOK

5. Describe Complete the Spider Map to describe the characteristics of amphibians.

What Are the Three Groups of Amphibians?

More than 5,400 species of amphibians are alive today. They belong to three groups: caecilians, salamanders, and frogs and toads.

CAECILIANS

Most people are not familiar with caecilians. However, scientists have discovered more than 160 species. Caecilians live in warm climates in Asia, Africa, and South America. They have thin, moist skin like other amphibians. However, unlike other amphibians, some have bony scales in their skin. Caecilians also have no legs. They look like earthworms or snakes. ☑

READING CHECK

6. Identify Name two ways caecilians are different from most amphibians.

____

____

____

Caecilian

SALAMANDERS

There are about 500 known species of salamanders. Most adult salamanders live under stones and logs in the woods of North America. Salamanders have long tails and four strong legs. Some look like tadpoles when they hatch, others look like small adults. Most salamanders lose their gills and grow lungs as they develop. Some never lose their gills.

TAKE A LOOK

7. Compare What is one major difference between these two salamander species?

____

____

____

The **marbled salamander** lives in damp places, such as under rocks or logs or among leaves.

This **axolotl** is an unusual salamander. It keeps its gills and never leaves the water.

FROGS AND TOADS

About 90% of all amphibians are frogs or toads. Frogs and toads are very similar. In fact, toads are a type of frog. Frogs tend to have smooth, moist skin. Toads spend less time in the water and have drier, bumpier skin. Frogs and toads share the following characteristics:

- strong leg muscles for jumping
- well-developed ears for hearing
- a long sticky tongue for catching insects
- vocal cords

Math Focus

8. Calculate Percentages A certain toad species spends two months of its life as a tadpole and three years as an adult. What percentage of its life is spent in the water?

____

Frogs, such as this **bullfrog**, have smooth, moist skin.

Toads, such as this **Fowler's toad**, spend less time in water than frogs do. Their skin is drier and bumpier.

How Do Frogs Sing?

Frogs are well known for singing, or croaking, at night, However, many frogs sing during the daytime, too. Like we do, frogs make sounds by forcing air across their vocal cords. They have a sac of skin, called the vocal sac around their vocal cords. When a frog sings, its *vocal sac*, fills with air. The sac makes the song louder. Frogs sing to attract mates and to mark territory.

Critical Thinking

9. Infer Why do you think a frog needs a vocal sac to make its song louder?

____

____

____

Most frogs that sing are males.

What Can Amphibians Tell Us About the Environment?

Unhealthy amphibians can be an early sign of changes in an ecosystem. Because of this, they are called *ecological indicators*. Amphibians are very sensitive to changes in the environment, such as chemicals in the air or water. Climate change may also affect amphibians. When large numbers of amphibians grow abnormally or die, there may be a problem with the environment. ☑

READING CHECK

10. Define What is an ecological indicator?

____

____

____

____

How Are Amphibians Related to Other Vertebrates?

The Concept Map below shows how amphibians are related to each other and to other vertebrates and chordates. Vertebrates are the largest group of chordates. Frogs and toads are the largest group of amphibians.

Chordates

Vertebrates

Other chordates

Other Vertebrates

Amphibians

Fishes

Caecilians

Salamanders

Frogs and Toads

Name ______________________ Class ______________ Date ______________

Section 2 Review

NSES LS 1a, 2a, 3a, 3c, 3d, 5a, 5b

SECTION VOCABULARY

lung a respiratory organ in which oxygen from the air is exchanged with carbon dioxide from the blood	**metamorphosis** a phase in the life cycle of many animals during which a rapid change from the immature form of an organism to the adult form takes place **tadpole** the aquatic, fish-shaped larva of a frog or toad

1. **Explain** Why do amphibians need to live in moist environments?

____

____

____

2. **Explain** Why do frogs sing?

____

____

3. **Identify** What body parts do tadpoles lose when they go through metamorphosis? What body parts do they gain?

____

____

4. **List** What are three organs amphibians can use to breathe?

____

5. **Compare** Fill in the Venn Diagram to compare salamanders and frogs.

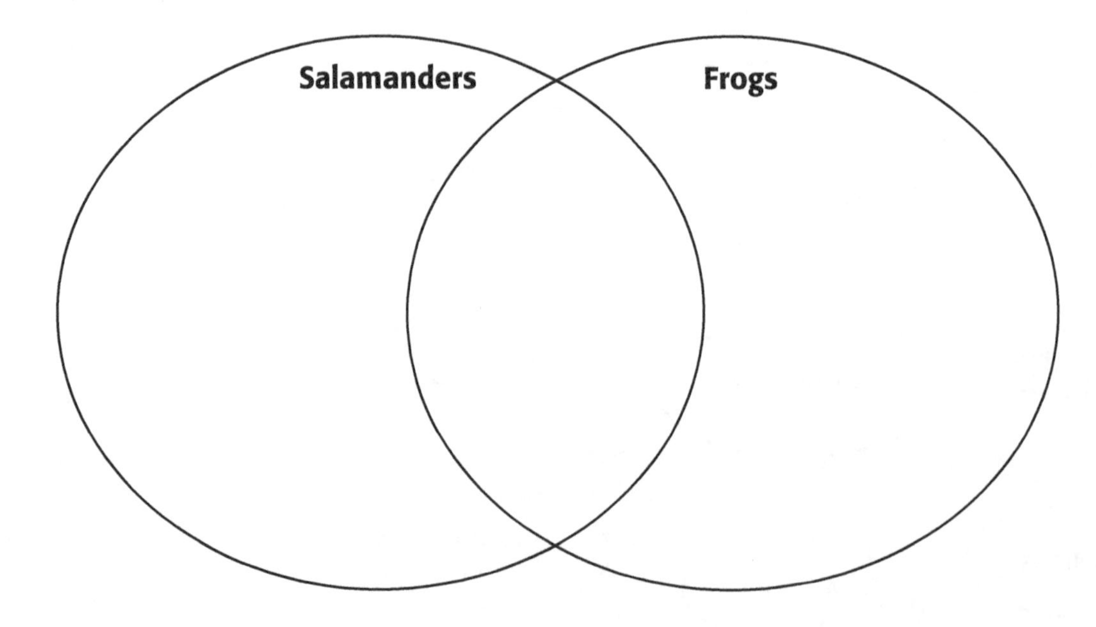

Name ______________________ Class ______________ Date ______________

CHAPTER 16 Fishes, Amphibians, and Reptiles

SECTION 3

Reptiles

BEFORE YOU READ

After you read this section, you should be able to answer these questions:

- What characteristics of reptiles allow them to live on land?
- What is an amniotic egg?
- What are the four groups of living reptiles?

National Science Education Standards

LS 1a, 1d, 2a, 2b, 3a, 3c, 3d, 5a, 5b, 5c

How Did Reptiles Evolve?

After amphibians moved onto land, some of them began to change. They grew thick, dry skin. Their legs changed and grew stronger. They also laid eggs that did not dry out. They had become reptiles, the first animals to live out of the water their whole lives.

Many reptiles are now extinct. Dinosaurs that lived on land are the most well-known prehistoric reptiles. However, there were many other ancient reptiles. Some could swim, some could fly, and many were similar to reptiles alive today.

Organize As you read, make a chart comparing the four different groups of reptiles.

Investigate Use your school's media center to learn about a prehistoric reptile that was not a dinosaur. Describe this reptile to the class.

What Are the Characteristics of Reptiles?

Reptiles are well-adapted for life on land. Unlike amphibians, most reptiles can spend their whole lives on land.

LUNGS

All reptiles—even reptiles that live in water—have lungs to breathe air. Most reptiles cannot breathe through their skin and get oxygen only from their lungs.

THICK SKIN

Thick, dry skin is a very important adaptation for life on land. This skin forms a watertight layer that keeps cells from losing too much water by evaporation. ☑

READING CHECK

1. Explain How is thick skin important for reptiles?

____

____

____

Reptiles, such as this crocodile, have thick skin.

STANDARDS CHECK

LS 5b Biological evolution accounts for the diversity of species developed through gradual processes over many generations. Species acquire many of their unique characteristics through biological adaptation, which involves the selection of naturally occurring variations in populations. Biological adaptations include changes in structures, behaviors, or physiology that enhanace survival and reproductive success in a particular environment.

Word Help: diversity
variety

Word Help: process
a set of steps, events, or changes

Word Help: selection
the process of choosing

Word Help: structure
a whole that is built or put together from parts

2. Infer How did the amniotic egg allow reptiles to live on land?

____

____

____

____

BODY TEMPERATURE

Nearly all reptiles are ectotherms. They cannot keep their bodies at a stable temperature. They must use their surroundings to control their temperature. They are active when it is warm outside, and they slow down when it is cool. A few reptiles can get some heat from their own body cells.

AMNIOTIC EGG

The most important adaptation to life on land is the amniotic egg. An **amniotic egg** has membranes and a shell that protect the embryo. A reptile's amniotic egg can be laid under rocks, in the ground, or even in the desert.

What Are the Parts of the Amniotic Egg?

The parts of the amniotic egg protect the embryo from predators, infections, and water loss.

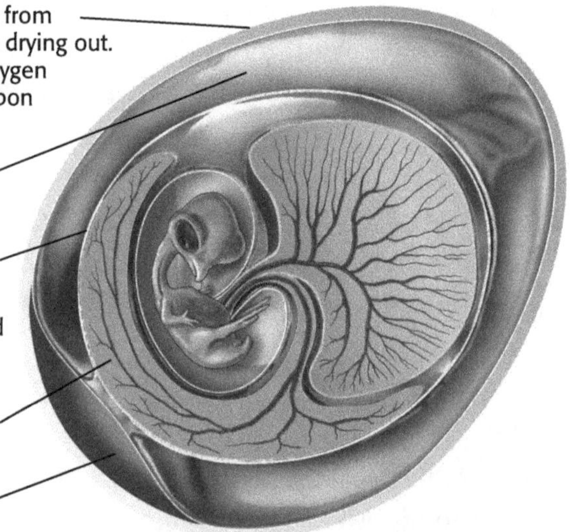

TAKE A LOOK

3. Compare Name one thing you can see in this picture that reptile eggs have but the amphibian eggs do not.

____

Amphibian eggs

Reptile eggs

The reptile eggs are amniotic but the amphibian eggs are not.

How Do Reptiles Reproduce?

Reptiles usually reproduce by internal fertilization. The egg is fertilized inside the female. Then a shell forms, and the female lays the egg. A few reptiles do not lay eggs. In this case, the embryos develop inside the mother and the young are born live. ☑

Unlike amphibians, reptiles do not go through metamorphosis. The embryo develops into a young reptile that looks like a tiny adult.

4. Identify Which type of fertilization do reptiles use?

____

What Are the Four Kinds of Living Reptiles?

Many reptiles, including dinosaurs, are now extinct. Today, about 8,000 species of reptiles are known to exist. These species make up four main groups of reptiles.

TURTLES AND TORTOISES

In general, tortoises live on land and turtles spend all or much of their lives in water. Even if they live in the sea, turtles leave the water to lay their eggs on land.

Turtles and tortoises are unique because of their shells. Shells make turtles and tortoises slow, so they cannot run away from predators. However, many of these reptiles can pull their heads and limbs into their shells. The shells protect them like suits of armor.

Green sea turtle

Texas tortoise

Math Focus

5. Calculate Suppose a sea turtle laid 104 eggs. After the young hatch, 50% reach the ocean alive. Of those survivors, 25% grow to be adults. How many adults resulted from the eggs?

____

CROCODILES AND ALLIGATORS

Crocodiles and alligators spend most of their time in the water. Their eyes and nostrils are on the top of their flat heads. This lets them hide under water and watch their surroundings. They eat invertebrates, fishes, turtles, birds, and mammals. The figure on the next page shows some differences between crocodiles and alligators.

TAKE A LOOK

6. Describe How can you tell the difference between an alligator and a crocodile?

____

____

____

____

____

A crocodile, such as this **American crocodile**, has a narrow head and pointed snout.

An alligator, such as this **American alligator**, has a broad head and a rounded snout.

Critical Thinking

7. Analyze Ideas Rattlesnakes do not see very well. However, they can sense changes in temperature as small as 3/1000 of a degree Celsius. How could this ability be useful to snakes?

____

____

____

SNAKES AND LIZARDS

Snakes and lizards are the most common reptiles alive today. Snakes are *carnivores*, or meat-eaters. They use a special organ in their mouths to smell prey. When a snake flicks out its tongue, molecules in the air stick to it and are brought back to the organ in the mouth. Snakes can open their mouths very wide. This lets them swallow animals and eggs whole. Some snakes squeeze prey to suffocate it. Others use fangs to inject venom into prey to kill it.

Cape Cobra

Frilled lizard

Most lizards eat insects and worms, but some eat plants. One giant lizard—the komodo dragon—eats deer, pigs, and goats! Lizards have loosely connected lower jaws. This lets them open their jaws wide. However, lizards do not swallow large prey whole. Many lizards can escape predators by breaking off their tails. They can grow back new tails.

TUATARAS

Tuataras live only on a few islands off the coast of New Zealand. Tuataras look like lizards but are different from lizards in the following ways. Tuataras do not have visible ear openings on the outside of their bodies. Tuataras are most active when the temperature is low. They rest during the day and search for food at night. ☑

8. Describe When are tuataras most active?

____

____

____

Tuataras have survived without changing for about 150 million years.

How Are Reptiles Related to Other Vertebrates?

Vertebrates are the largest group of chordates. Snakes and lizards are the most common reptiles alive today.

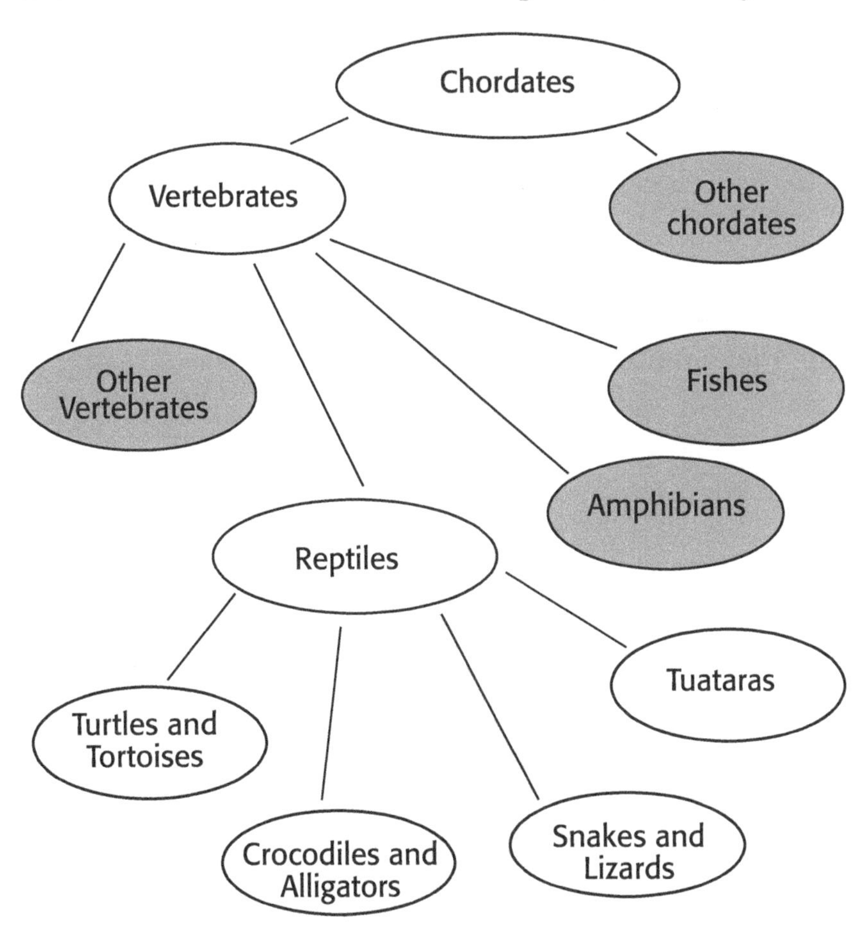

Name ______________________ Class ______________ Date ______________

Section 3 Review

NSES LS 1a, 1d, 2a, 2b, 3a, 3c, 3d, 5a, 5b, 5c

SECTION VOCABULARY

amniotic egg a type of egg that is surrounded by a membrane, the amnion, and that in reptiles, birds, and egg-laying mammals contains a large amount of yolk and is surrounded by a shell.	

1. **List** What are the four main groups of reptiles alive today?

____

____

2. **Explain** What are the functions of the shell of an amniotic egg?

____

____

3. **Explain** How do reptiles reproduce?

____

____

4. **Explain** How does a reptile embryo get oxygen?

____

5. **Compare** Give two ways tuataras differ from lizards.

____

____

____

6. **Describe** Describe the adaptations reptiles have that allow them to live on land.

____

____

____

7. **Apply Concepts** Mammals give birth to live young. The embryo develops and is nourished inside the female's body. Which parts of a reptilian amniotic egg could a mammal do without? Explain your answer.

____

____

____

Name ______________________ Class ______________ Date ______________

CHAPTER 17 Birds and Mammals

SECTION 1

Characteristics of Birds

BEFORE YOU READ

After you read this section, you should be able to answer these questions:

- What are the characteristics of birds?
- How do birds fly?

National Science Education Standards
LS 1a, 1d, 2a, 3a, 3b, 3c, 5a, 5b

What Are the Characteristics of Birds?

What do a powerful eagle, a fast-swimming penguin, and a tiny hummingbird have in common? They all have feathers, wings, and beaks. These features make them all birds.

STUDY TIP

Outline As you read this section, make an outline describing the characteristics of birds.

HOW BIRDS AND REPTILES ARE ALIKE

Birds share many characteristics with reptiles. Both birds and reptiles are vertebrates. They both have feet and legs covered with scales. They also have eggs with shells and amniotic sacs.

HOW BIRDS AND REPTILES DIFFER

Birds also have some characteristics that reptiles do not have. Birds have feathers, wings, and beaks. Unlike reptiles, birds are endotherms. They keep their bodies warm from heat produced by chemical reactions in their body cells. Their body temperatures are generally stable. ☑

READING CHECK

1. Explain How do birds keep their body temperatures stable?

____

____

____

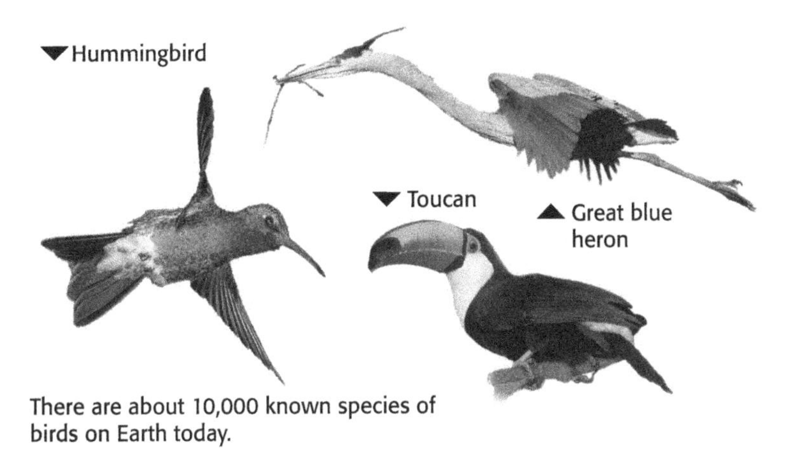

There are about 10,000 known species of birds on Earth today.

TAKE A LOOK

2. List Name three things that all the animals in this picture have in common.

____

____

What Are the Functions of Feathers?

Feathers are one of the most familiar features of birds. They help birds stay dry and warm, attract mates, and fly. Therefore, birds take good care of their feathers.

PREENING AND MOLTING

A bird uses its beak to spread a special oil on its feathers. It gets the oil from a gland near it tail. This process is called **preening**. Preening oil helps waterproof the feathers and keeps them clean. ☑

READING CHECK

3. Identify What are two functions of preening oil?

____

____

When feathers wear out, birds shed them and grow new ones. This process is called **molting**. Most birds molt at least once a year.

KINDS OF FEATHERS

Birds have two main kinds of feathers—down feathers and contour feathers. **Down feathers** are fluffy and lie close to a bird's body. These feathers help birds stay warm. When a bird fluffs its down feathers, air gets trapped next to its body. The trapped air keeps heat in, keeping the bird warm.

Contour feathers are stiff and cover a bird's body and wings. These feathers have a stiff shaft with many side branches, called *barbs*. Barbs link together to form a smooth surface. This streamlined surface helps birds fly. Some birds use colorful contour feathers to attract mates.

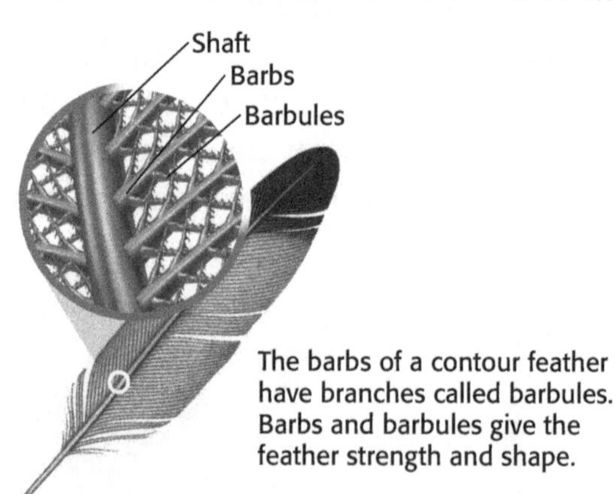

The barbs of a contour feather have branches called barbules. Barbs and barbules give the feather strength and shape.

TAKE A LOOK

4. Complete Feather barbs have branches called

_______________.

How Do Birds Get Energy?

Birds need a lot of energy to fly. To get this energy, birds eat a lot of food. Most birds eat insects, nuts, seeds, or meat. A few, such as geese, eat grass, leaves, and other plants.

Birds do not have teeth, so they cannot chew their food. Instead, food moves from the mouth to the crop. The *crop* stores the food until it moves into the gizzard. The *gizzard* has small stones in it. These stones grind up the food. This grinding action is similar to what happens when we chew our food. The food then moves to the intestines where it is digested.

Critical Thinking

5. Compare How is a bird's gizzard like our teeth?

____

____

____

____

____

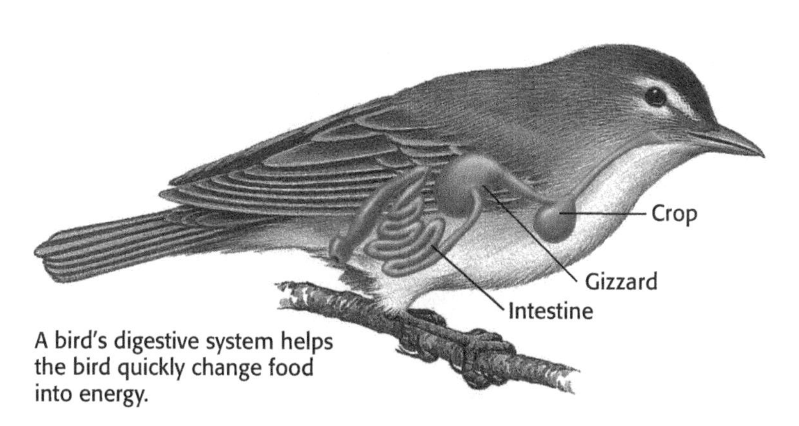

A bird's digestive system helps the bird quickly change food into energy.

TAKE A LOOK
6. Identify Circle the organ where food is stored.

A bird's body also needs to break down its food quickly to provide energy. This process creates a lot of heat. Birds cannot sweat to cool off as humans do. Instead, they lay their wings flat and pant like a dog.

How Do Birds Fly?

Most birds can fly. Even flightless birds, such as ostriches, had ancestors that could fly. Birds have many adaptations that make flight possible. They have wings, lightweight bodies, and strong flight muscles. Their hearts beat very fast to bring large amounts of oxygen to the flight muscles. The figure below describes how parts of a bird's body help it fly.

Math Focus
7. Calculate When a bird flies long distances, such as during migration, it uses a lot of energy. This causes it to lose weight. If a bird that weighs 325 g loses 40% of its body weight during migration, how much does it weigh?

____

Flight Adaptations of Birds

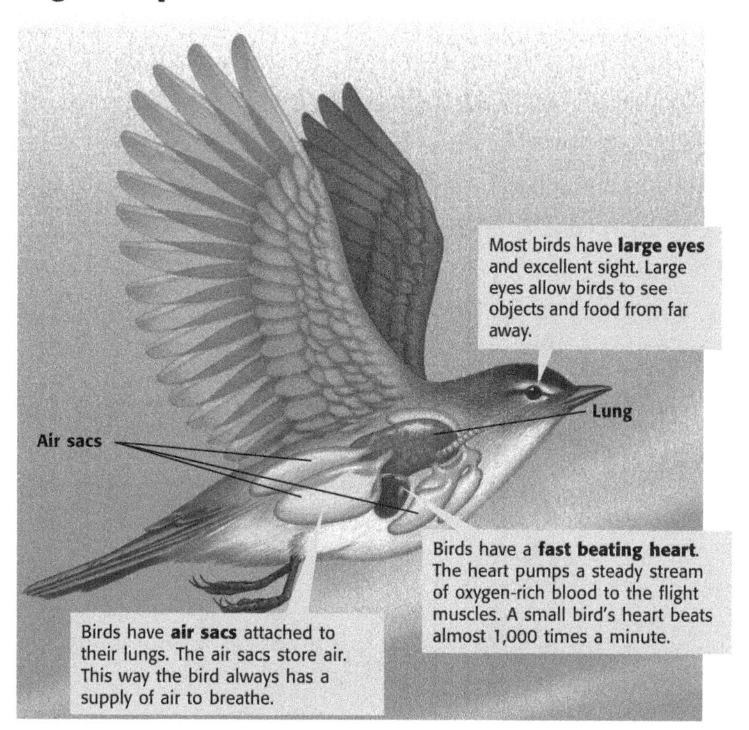

TAKE A LOOK
8. Identify What is the function of air sacs?

____

Critical Thinking

9. Infer Some birds do not fly. These birds generally do not have large keels. Why are large keels not necessary for flightless birds?

____

____

____

____

____

GETTING OFF THE GROUND

In order to fly, birds have to overcome the force of gravity that pulls them to the ground. Birds flap their wings to get into the air. They are able to stay in the air because their wings cause lift. **Lift** is an upward force that pushes on a bird's wings. ☑

The shape of a bird's wing helps create lift. A bird's wing is curved on top and flatter underneath. The air that flows over the top of the wing moves faster than the air underneath. Slower air under the wing pushes upward.

READING CHECK

10. Identify What force must birds overcome in order to get off the ground?

____

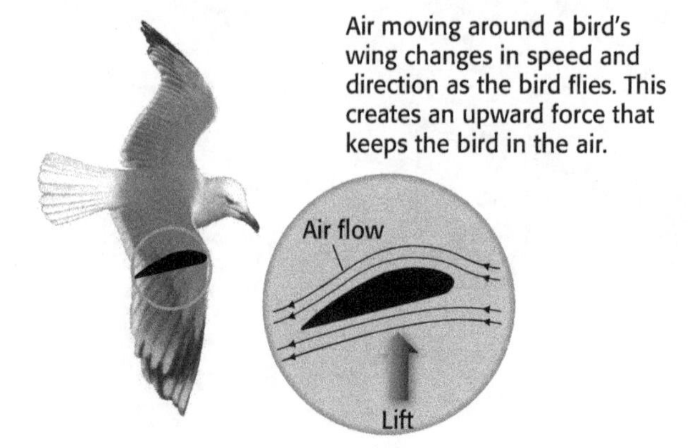

Air moving around a bird's wing changes in speed and direction as the bird flies. This creates an upward force that keeps the bird in the air.

TAKE A LOOK

11. Identify What upward force helps a bird overcome the force of gravity?

____

How Do Birds Reproduce?

Like reptiles, birds reproduce sexually by internal fertilization. They lay amniotic eggs in which embryos grow. However, unlike reptiles, birds must keep their eggs warm so that the embryos can survive. ☑

READING CHECK

12. Explain What kind of reproduction do birds use?

____

____

BROODING

Most birds lay their eggs in nests. They sit on their eggs and use their body heat to keep the eggs warm. This is called **brooding**. Birds sit on the eggs until they hatch. In most species, the female sits on the eggs. In a few species, the males brood the eggs.

This robin's nest is only one example of a bird's nest. Birds build nests of many different shapes and sizes.

PRECOCIAL AND ALTRICIAL

Some chicks, such as chickens and ducks, are active soon after they hatch. These active chicks are called *precocial.* Precocial chicks are covered with downy feathers. They only depend on their parents for warmth and protection. They can walk, swim, and feed themselves.

Other chicks, such as hawks and songbirds, are weak and helpless after hatching. These chicks are called *altricial.* Altricial chicks have no feathers, and their eyes are closed. They cannot walk or fly. Their parents must keep them warm and feed them for several weeks. ☑

READING CHECK

13. Describe What are the characteristics of newly hatched altricial chicks?

____

____

____

____

____

Parents of altricial chicks bring food to the nest.

Name ______________________ Class ______________ Date ______________

Section 1 Review

NSES LS 1a, 1d, 2a, 3a, 3b, 3c, 5a, 5b

SECTION VOCABULARY

brooding to sit on and cover eggs to keep them warm until they hatch; to incubate **contour feather** one of the most external feathers that cover a bird and that help determine its shape **down feather** a soft feather that covers the body of young birds and provides insulation to adult birds	**lift** an upward force on an object that moves in a fluid **molting** the shedding of an exoskeleton, skin, feathers, or hair to be replaced by new parts **preening** in birds, the act of grooming and maintaining their feathers

1. Explain How do birds get the energy they need to fly?

____

____

2. Compare Complete the Venn Diagram to compare reptiles and birds.

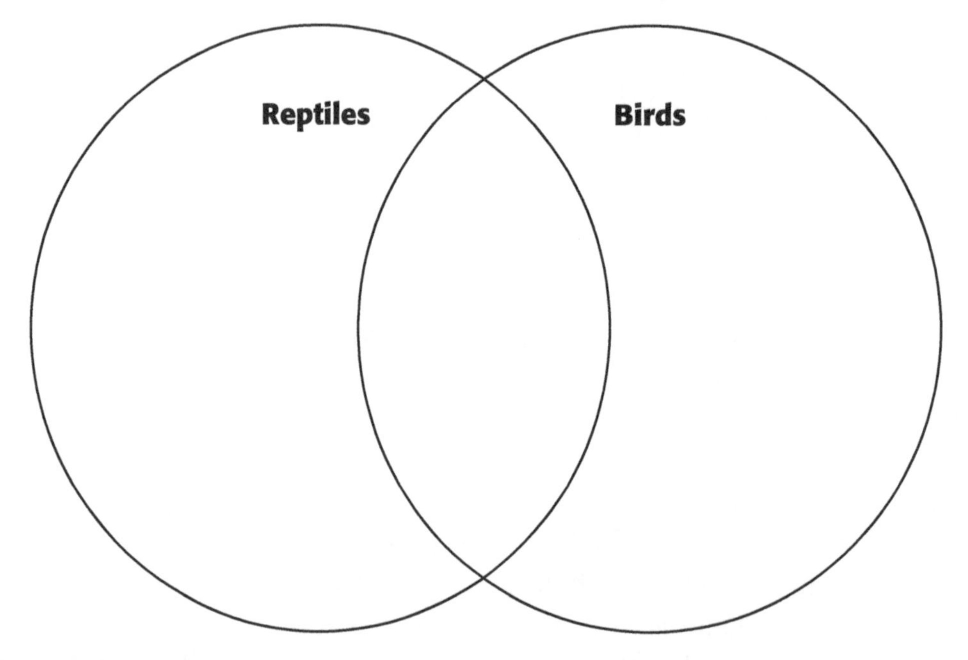

3. Explain How does a fast-beating heart help a bird fly?

____

____

4. Compare How do precocial chicks and altricial chicks differ?

____

____

____

SECTION 2

Kinds of Birds

BEFORE YOU READ

After you read this section, you should be able to answer these questions:

- What are the four main groups of birds?
- What adaptations do birds in each group have?

National Science Education Standards
LS 1a, 3a, 5a, 5b

What Are the Different Groups of Birds?

There are about 10,000 species of birds on Earth. Birds come in many different colors, shapes, and sizes. Scientists group living bird species into 28 different orders. However, birds can also be grouped into four nonscientific categories. These categories are flightless birds, water birds, perching birds, and birds of prey. ☑

STUDY TIP

Describe Make a list of birds you are familiar with. Decide which of the four groups each bird on your list belongs to.

READING CHECK

1. List What are the four main groups of birds?

____

____

____

____

FLIGHTLESS BIRDS

Although all birds have wings, not all birds can fly. Birds that cannot fly are called *flightless*. Most flightless birds do not have a large keel to anchor strong flight muscles. Even though they cannot fly, flightless birds have evolved other ways to move around. Some have powerful legs for running, while others are excellent swimmers. The figure below shows some flightless birds.

Flightless Birds

◀ **Ostriches** are the largest birds. They can reach a height of 2.5 m and a mass of 125 kg. These birds can run up to 60 km/h.

▼ Unlike other flightless birds, **penguins** have large keels and strong flight muscles. They use their wings as flippers to "fly" through the water.

Kiwis are about the size of chickens. At night, they hunt for worms, caterpillars, and berries. ▶

Say It

Describe In small groups, describe the kinds of birds you have seen around your home or school. What do they look like? What time of year do you see them? Do you know what they eat?

TAKE A LOOK

2. Apply Concepts What characteristic of penguins is generally not seen in other flightless birds?

____

WATER BIRDS

Some birds that fly can also swim in water. Cranes, ducks, geese, swans, pelicans, and loons are all water birds. Water birds usually have webbed feet for swimming. Some have long legs for wading.

Water birds find food both in the water and on land. Many of them eat plants, invertebrates, or fish. Some water birds have rounded, flat beaks for eating plants and small invertebrates. Others have sharp, long beaks for catching fish.

STANDARDS CHECK

LS 5b Biological evolution accounts for the diversity of species developed through gradual processes over many generations. Species acquire many of their unique characteristics through biological adaptation, which involves the selection of naturally occurring variations in populations. Biological adaptations include changes in structures, behaviors, or physiology that enhance survival and reproductive success in a particular environment.

Word Help: diversity
variety

Word Help: process
a set of steps, events or changes

Word Help: selection
the process of choosing

Word Help: variation
a difference in form or function

3. Describe Describe two adaptations that can help a fish-eating water bird survive in its environment.

____
____
____
____
____

Water Birds

▼ The **blue-footed booby** is a tropical water bird. These birds attract mates with a courtship dance that includes raising one foot at a time.

▼ Male **wood ducks** have colorful feathers to attract females. Like all ducks, they are strong swimmers and flyers.

◀ The **common loon** can make deep dives when searching for food. It can stay underwater for several minutes.

PERCHING BIRDS

Perching birds can sit on branches. Their legs and feet are adapted to let them hold on to branches. Robins, warblers, and sparrows are all perching birds. When a perching bird lands in a tree, its feet close around a branch. The bird's feet will still stay wrapped around the branch, even if the bird falls asleep.

Critical Thinking

4. Infer Why don't most ducks sit in trees?

____

____

____

____

____

Perching Birds

Parrots have special feet for perching and climbing.

Chickadees are lively little birds. They can hang under a branch by their feet while looking for insects, seeds, or fruits.

BIRDS OF PREY

Birds that hunt and eat other vertebrates are called *birds of prey*. This group includes owls, eagles, falcons, ospreys, vultures, and hawks. These birds may eat insects, mammals, fish, reptiles, and even other birds. Birds of prey have special adaptations that help them hunt. They have sharp claws and beaks, to help them catch and kill prey. They also have good vision for finding prey. Most birds of prey hunt during the day. However, most owls hunt at night. ☑

READING CHECK

5. List Name three adaptations birds of prey have for catching and killing prey.

____

____

____

Birds of Prey

Owls, such as this **northern spotted owl**, hunt for food at night. They have a strong sense of hearing to help them find prey in the dark.

Ospreys fly over water and catch fish with their claws.

Name ______________________ Class ______________ Date ______________

Section 2 Review

NSES LS 1a, 3a, 5a, 5b

1. Organize Fill in the chart below to describe the four main groups of birds.

Group of birds	Characteristics	Examples
Flightless birds		
Water birds		
Perching birds		
Birds of prey		

2. Apply Concepts Why would a perching bird *not* be well adapted to live in the water?

____

____

____

3. Explain Into which group would you place a bird that had a small keel and strong legs?

____

4. Apply Concepts In what type of habitat would you expect to find a bird with long, thin legs? Explain your answer.

____

____

5. Infer Penguins have webbed feet and spend much of their time in the water. Why do you think they are not grouped with the water birds?

____

____

____

Name ______________________ Class ______________ Date ______________

CHAPTER 17 Birds and Mammals

SECTION 3

Characteristics of Mammals

BEFORE YOU READ

After you read this section, you should be able to answer these questions:

- How did mammals evolve?
- What are the seven characteristics of mammals?

National Science Education Standards

LS 1a, 1d, 2a, 3a, 3b, 3c, 5a

What Are the Characteristics of Mammals?

Like a donkey, a bat, a giraffe, and a whale, you are a mammal. Mammals live in almost every environment on Earth. Though there are almost 5,000 different species of mammals, they all share seven common characteristics.

Outline **As** you read this section, make an outline describing the characteristics of mammals.

▲ Mandrill baboon

▲ Rhinoceros

◀ Beluga whale

Although they look very different, all of these animals are mammals.

Describe Before you read the entire section, discuss with your classmates some of the animals you know that are mammals. Make a list of the characteristics you think all mammals share. After you read the section, dicuss whether or not your list of characteristics was correct.

SEXUAL REPRODUCTION

All mammals reproduce sexually. Sperm fertilize eggs inside the female's body. Most mammals give birth to live young instead of laying eggs. At least one parent cares for and protects the young until they are grown.

MAKE MILK

Mammary glands are structures that make milk. While all mammals have mammary glands, only mature females that have had offspring can produce milk. Milk is the main source of food for most young mammals.

All milk is made of water, proteins, fats, and sugars. However, different species have different amounts of each nutrient in their milk. For example, cow milk has twice as much fat as human milk. ☑

READING CHECK

1. List What is milk made of?

____

____

Like all mammals, this calf drinks its mother's milk for its first meals.

ENDOTHERMY

As a mammal's cells use oxygen to break down food, energy is released. Mammals use this energy to keep their bodies warm. Like birds, mammals are *endotherms*. They can keep their body temperatures constant even when outside temperatures change. The ability to stay warm helps many mammals survive in cold areas and stay active in cool weather. ☑

READING CHECK

2. Identify What are two advantages of being endothermic?

____

____

____

____

HAIR

Mammals are the only animals that have hair. All mammals—even whales—have hair at some point in their lives. Mammals that live in in cold climates, such as polar bears, usually have thick coats of hair. This is called *fur*. Mammals in warmer climates, such as elephants, do not need warm fur. However, they do have some hair.

Most mammals also have a layer of fat underneath their skin. Fat traps heat in the body. Mammals that live in cold oceans have a very thick layer of fat. This is called *blubber*.

SPECIALIZED TEETH

Unlike most other animals, mammals have teeth with different shapes and sizes for different jobs. For example, humans have three types of teeth. The front of the mouth has cutting teeth called *incisors*. Next to them are stabbing teeth called *canines*. These help grab and hold onto food. In the back are flat teeth that grind called *molars*. ☑

3. List What are the three types of teeth that humans have?

____

____

____

Each kind of tooth helps mammals eat a certain kind of food. Meat-eating mammals have large canines to help them tear prey. Plant-eating mammals have larger incisors and molars to help them bite off plants and chew them up.

Mountain lions have sharp canine teeth for grabbing their prey.

Horses have sharp incisors in front for cutting plants. They have flat molars in back for grinding plants.

TAKE A LOOK

4. Describe How does a horse's molars help it eat plants?

____

____

____

BREATHE AIR

All animals need oxygen to help their cells break down food. Like birds and reptiles, mammals use lungs to get oxygen from the air. Mammals also have a large muscle called the **diaphragm** that helps them bring air into the lungs. As the diaphragm moves down, air rushes into the lungs. As it moves up, air is pushed out.

LARGE BRAINS

In general, a mammal has a larger brain than a non-mammal of the same size. A large brain lets mammals learn and think quickly. Like many other animals, mammals use vision, hearing, smell, touch, and taste to learn about the world around them. Different senses may be more important to mammals in different environments. For example, mammals that are active at night often depend on their hearing more than their vision.

Critical Thinking

5. Infer Why do you think the first mammals were active at night but the dinosaurs were not?

____

____

____

____

____

____

____

____

How Did the First Mammals Evolve?

Scientists have found fossils of mammals from over 225 million years ago. These early mammals were about the size of mice. They were *endotherms*, which means they did not have to depend on their surroundings to stay warm. This allowed them to look for food at night when temperatures were cooler. Looking for food at night helped them avoid being eaten by dinosaurs.

When the dinosaurs died out, there was more food and land for mammals. The mammals were able to spread out to new environments.

Name ______________________ Class ______________ Date ______________

Section 3 Review

NSES LS 1a, 1d, 2a, 3a, 3b, 3c, 5a

SECTION VOCABULARY

diaphragm a dome-shaped muscle that is attached to the lower ribs and that functions as the main muscle in respiration	**mammary gland** in a female mammal, a gland that secretes milk

1. Summarize Complete the organizer below to identify the seven characteristics shared by all mammals.

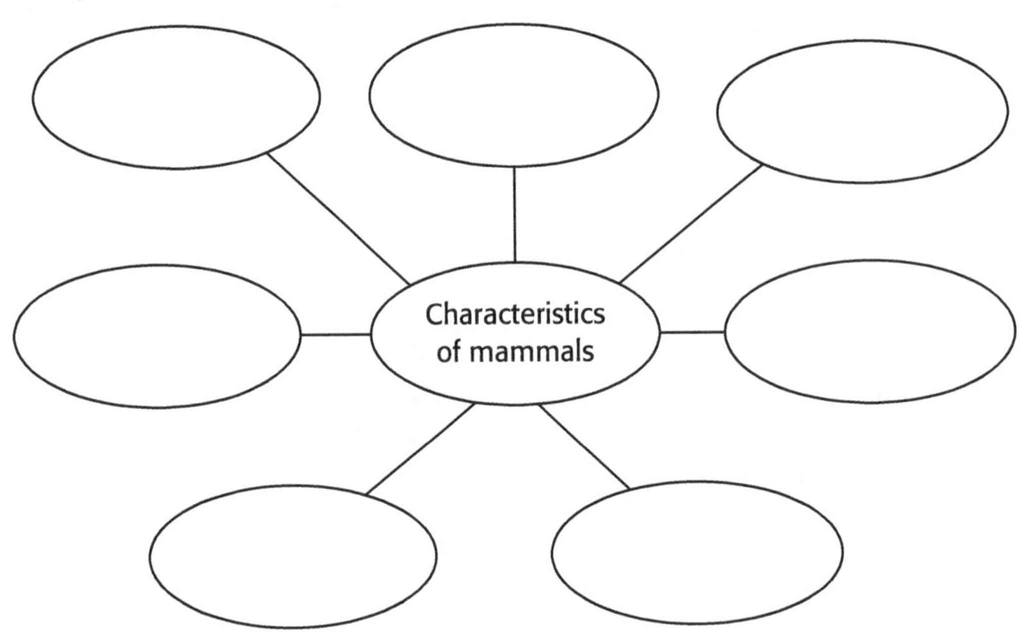

2. Explain How is the milk of different mammal species the same and how is it different?

____

____

3. List Name three characteristics of mammals that help them stay warm.

____

____

4. Apply Concepts A particular mammal skull has large, sharp canine teeth. What does this tell you about the food the mammal ate?

____

____

5. Infer A nurse shark and a dolphin are about the same size. Which one would you expect to have a larger brain? Explain your answer.

____

____

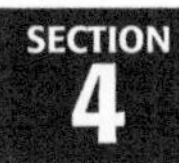

CHAPTER 17 Birds and Mammals

SECTION 4

Placental Mammals

BEFORE YOU READ

After you read this section, you should be able to answer these questions:

- What is a placental mammal?
- What are some examples of placental mammals?

National Science Education Standards
LS 1a, 1d, 2a, 3a, 3c, 5a

How Do Most Mammals Develop?

Mammals are classified by how they develop. There are three main groups: placental mammals, monotremes, and marsupials. A **placental mammal** is a mammal whose embryos develop inside the mother's body. The embryo grows in an organ called the *uterus*. Another organ, called the *placenta*, connects the embryo to the uterus. The placenta brings food and oxygen to the embryo. It also carries wastes away from the embryo. ☑

The amount of time that an embryo grows before it is born is called a **gestation period**. Different placental mammals have different gestation periods. For example, the gestation period for humans is 9 months. For elephants, the gestation period is 23 months.

Summarize As you read, make a chart that lists the different groups of placental mammals described in this section. In your chart, include the characteristics of each group.

READING CHECK

1. Identify What organ connects the embryo to the uterus?

____

What Are Some Kinds of Placental Mammals?

Placental mammals are divided into groups based on how they evolved and how closely they are related.

ANTEATERS, ARMADILLOS, AND SLOTHS

Anteaters, armadillos, and sloths have backbones like all mammals. However, mammals in this group have extra connections between their vertebrae. These connections strengthen the animals' lower backs and hips. Some members of this group have very small teeth and some have no teeth at all. Many mammals in this group catch insects using long, sticky tongues. ☑

2. Explain How do members of this group differ from other mammals?

____

____

____

____

____

▲ **Armadillos** eat insects, frogs, mushrooms, and roots.

◀ **Giant anteaters** eat only a few insects from each nest they find.

MOLES, SHREWS, AND HEDGEHOGS

Most members of this group are small animals with small brains and simple teeth. Many animals in this group rely more on hearing, smell, and touch than on vision. Their long, pointed noses help them smell their food. Members of this group feed mostly on invertebrates, such as insects and worms. Some also eat fish, frogs, lizards, and other small mammals. ☑

READING CHECK

3. Identify What do most members of the moles, shrews, and hedgehogs group eat?

____

____

▲ The **star-nosed mole** has sensitive feelers on its nose. These help the mole find earthworms to eat. Moles have very poor vision.

▲ **Hedgehogs** have spines that keep them safe from most predators.

RODENTS

More than one third of all mammal species are rodents. This group includes squirrels, mice, rats, guinea pigs, and porcupines. Rodents all have one set of incisors in their upper jaws. They chew and gnaw so much that their teeth wear down. However, their incisors grow continuously, so they never wear out completely. Rodents also have sensitive whiskers on their faces. ☑

READING CHECK

4. List What are three characteristics of rodents?

____

____

____

____

▲ Like all rodents, a **porcupine** has incisors that grow continuously.

▼ The **capybara** is the largest rodent in the world. Female capybaras can weigh as much as a grown man.

RABBITS, HARES, AND PIKAS

Rabbits, hares, and pikas belong to another group of mammals. Like rodents, these mammals have sharp, gnawing teeth. However, unlike rodents, these mammals have two sets of incisors in their upper jaws. Their tails are also shorter than rodents' tails. The figure below shows some members of this group. ☑

READING CHECK

5. Compare How do the teeth of rodents differ from those of rabbits, hares, and pikas?

____

____

____

The **black-tailed jack** rabbit uses its large eyes, ears, and its sensitive nose to detect predators.

Pikas are small animals that live high in the mountains.

FLYING MAMMALS

Bats are the only mammals that can fly. Bats are active at night and sleep during the day. Most bats eat insects or other small animals. Some eat fruit or nectar. A few bats, called *vampire bats*, drink the blood of birds and mammals.

Bats make clicking noises as they fly. The clicks echo off trees, rocks, and insects. The size, shape, and distance of an object affect the echo produced. Bats know what is around them by hearing these echoes. Using echoes to find things is called *echolocation*.

Critical Thinking

6. Infer How do you think bats use echolocation?

____

____

____

Fruit bats, also called *flying foxes*, live in tropical regions. They pollinate plants as they go from plant to plant to find fruit.

The **spotted bat** is found in parts of the southwestern United States. Like most bats, it eats flying insects. It uses its large ears during echolocation.

MAMMALS WITH TEETH FOR SHEARING

Mammals in this group evolved from a species of *carnivore*, or an animal that eats meat. Members of this group include dogs, cats, otters, raccoons, and bears. The teeth of these animals are adapted for shearing, or tearing meat off bone. Although the animals in this group evolved from carnivores, some members, such as pandas, are herbivores. Herbivores are plant eaters. Others, such as black bears, are omnivores. They eat both plants and animals.

Some members of this group live in the oceans. They are called *pinnipeds*. Pinnipeds include seals, sea lions, and walruses.

STANDARDS CHECK

LS 5a Millions of species of animals, plants, and microorganisms are alive today. Although different species might look dissimilar, the unity among organisms becomes apparent from an analysis of internal structures, the similarity of chemical processes, and the evidence of common ancestry.

Word Help: evidence information showing whether an idea is true or valid

7. Explain Why do some members of this group have shearing teeth, even though they are herbivores?

____

____

____

____

Coyotes are members of the dog family. They live throughout North America and parts of Central America.

Like all pinnipeds, **walruses** eat in the ocean, and sleep and mate on land. They use their huge canine teeth for defense, to attract mates, and to climb onto ice.

TRUNK-NOSED MAMMALS

Elephants are the only living mammals that have trunks. A trunk is a combination of an upper lip and a nose. An elephant uses its trunk to put food in its mouth. It also uses its trunk to spray water on its back to cool off.

TAKE A LOOK

8. Compare How can you tell the difference between an African elephant and an Indian elephant?

____

____

____

Elephants are social. The females live in herds made up of mothers, daughters, and sisters. These elephants are **African elephants**.

These **Indian elephants** have smaller ears and tusks than African elephants do.

HOOFED MAMMALS

A *hoof* is a thick, hard pad that covers a mammal's toe. A hoof is similar to a toenail or claw, but it covers the entire toe. Horses, pigs, deer, and rhinoceroses all have hoofs. Most hoofed mammals are plant eaters. They have flat molars to help them grind the plants they eat.

Hoofed mammals can have an odd or even number of toes. Odd-toed hoofed mammals have one or three toes. Horses and zebras both have one large hoofed toe on each foot. Rhinoceroses have three toes. Even-toed hoofed mammals have two or four toes on each foot. Pigs, cattle, deer, and giraffes are even-toed.

Tapirs are large mammals with an odd number of toes. They live in Central America, South America, and Southeast Asia.

Camels are even-toed mammals. A camel's hump is a large lump of fat. It provides the camel with energy when there is not much food around. Camels can live without drinking water for a long time. This lets them live in dry places.

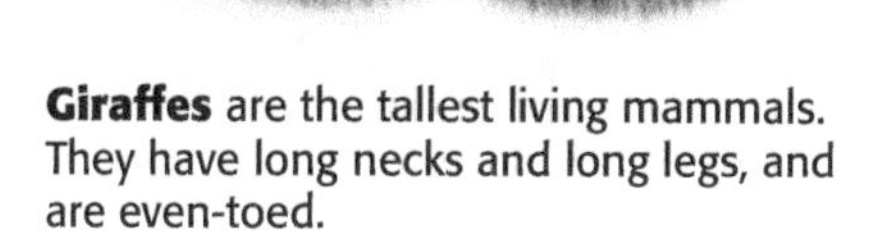

Giraffes are the tallest living mammals. They have long necks and long legs, and are even-toed.

Critical Thinking

9. Infer Which animals do you think are more closely related to giraffes—deer or rhinoceroses? Explain your answer.

____

____

____

____

Say It

Investigate Choose one of the animals from this figure, or one from the text above. Use your school media center to research this animal. Describe this animal to your class. Talk about what it eats, where it lives, and any other interesting facts.

CETACEANS

Cetaceans are a group of mammals that include whales, dolphins, and porpoises. All cetaceans live in the water. Cetaceans may look more like fish than mammals. However, unlike fish, cetaceans have lungs and nurse their young. ☑

Most cetaceans, such as dolphins and porpoises, have teeth. However, some, such as a humpback whale, have no teeth. Instead, they strain water for tiny organisms.

READING CHECK

10. Explain Why are cetaceans mammals, even though they look more like fish?

Spinner dolphins spin like footballs when they leap out of the water. Like all dolphins, they are intelligent and highly social.

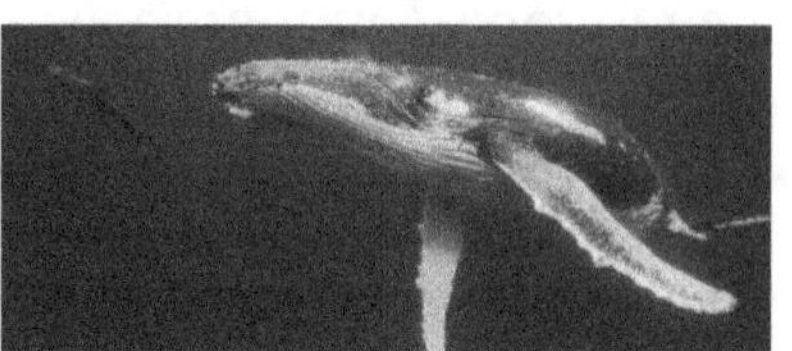

Humpback whales do not have teeth. Like all toothless whales, they strain sea water through large, flat plates in their mouths. The plates trap tiny organisms for the whales to eat.

MANATEES AND DUGONGS

Manatees and dugongs are another group of mammals that live in the water. They swim slowly, using their flippers and their tail. They live in the ocean and in rivers. Manatees and dugongs eat mostly seaweed and water plants. To breathe, these mammals lift their noses out of the water.

Critical Thinking

11. Infer Manatees are often killed or seriously hurt by boats. What characteristics of these mammals make injuries likely?

Manatees are also called sea cows.

PRIMATES

Primates include monkeys, apes, and humans. Members of this group have five fingers on each hand and five toes on each foot. Primates have larger brains than most other mammals. Their eyes face forward. They also have opposable thumbs, which allow them to hold objects. ☑

Many primates live in trees. They climb with grasping hands and feet. They have flexible shoulders that let them swing between branches. Most primates eat leaves and fruit. Some hunt for meat.

12. Identify What adaptation have primates evolved that allows them to hold onto objects?

____

Orangutans and other apes often walk upright. Apes usually have larger brains and bodies than monkeys do.

The **proboscis monkey** has a huge nose! Males have larger noses than females. Some scientists think the male's nose is used to attract females.

Like many monkeys, **spider monkeys** have grasping tails. Their long arms, legs, and tails help them move among the trees.

TAKE A LOOK

13. Identify Which group of primates represented in this figure has the largest brains?

____

14. Identify What adaptation do many monkeys have for swinging through trees?

____

Name ______________________ Class ______________ Date ______________

Section 4 Review

NSES LS 1a, 1d, 2a, 3a, 3c, 5a

SECTION VOCABULARY

gestation period in mammals, the length of time between fertilization and birth	**placental mammal** a mammal that nourishes its unborn offspring through a placenta inside its uterus

1. Describe What is the function of the placenta?

____

____

2. Explain How do scientists organize placental mammals into groups?

____

____

3. Apply Concepts A manatee may look like a pinniped, but they are not classified in the same group. Why do you think scientists put them in different groups?

____

____

____

Use the figure below to answer the questions that follow.

4. Explain How do you know that this animal is not a primate?

____

____

5. Apply Concepts With which group of placental mammals would you classify this organism? Explain why it would not belong in any other group described in this section.

____

____

____

Name ______________________ Class ______________ Date ____________

CHAPTER 17 Birds and Mammals

SECTION 5

Monotremes and Marsupials

BEFORE YOU READ

After you read this section, you should be able to answer these questions:

- What are monotremes?
- What are marsupials?

National Science Education Standards

LS 1a, 1d, 2a, 3a, 3c, 5a, 5c

Do All Mammals Develop Inside Their Mother?

Most mammals are placental. However, not all mammals develop inside their mothers. Some mammals hatch from eggs. These mammals are called monotremes. Other mammals spend the first months of their lives living in their mother's pouch. These mammals are called marsupials.

STUDY TIP

List As you read this section, make lists of the characteristics of monotremes and the characteristics of marsupials. Remember to include the characteristics that all mammals share.

What Are Monotremes?

A **monotreme** is a mammal that lays eggs. Monotremes, like all mammals, have mammary glands, diaphragms, and hair. They also keep their body temperatures constant. A female monotreme lays eggs and keeps them warm with her body. When the eggs hatch, she nurses her offspring. Monotremes do not have nipples as other mammals do. Baby monotremes lick milk from the skin around the mother's mammary glands. ☑

READING CHECK

1. Identify Name two characteristics that are unique to monotreme mammals.

____

____

ECHIDNA

There are only three living species of monotremes. Two of these species are echidnas. *Echidnas* are about the size of a house cat. They have large claws and long snouts. These structures help them dig for ants and termites to eat.

Echidnas

The **long-beaked echidna** lives in New Guinea.

The **short-beaked echidna** lives in Australia and New Guinea.

Math Focus

2. Calculate Percentages Only three of the 5,000 known species of mammals are monotremes. What percentage of mammals are monotremes?

____

____

STANDARDS CHECK

LS 5a Millions of species of animals, plants, and microorganisms are alive today. Although different species might look dissimilar, the unity among organisms becomes apparent from an analysis of internal structures, the similarity of their chemical processes, and the evidence of common ancestry.

Word Help: evidence information showing whether an idea is true or valid

3. List Name three characteristics that monotremes share with other mammals.

____

____

____

PLATYPUS

The other type of monotreme is called a *platypus*. Platypuses live in Australia. They look very different from other mammals. They have webbed feet and flat tails to help them swim. They use their flat bills to search for food. Platypuses have claws on their feet, which they use to dig tunnels in riverbanks. They then lay their eggs in these tunnels.

When underwater, the **duckbill platypus** closes its eyes and ears. It uses its bill to find food.

What Are Marsupials?

Some mammals, such as kangaroos, carry their young in pouches. Mammals with pouches are called **marsupials**. Like all mammals, marsupials have mammary glands, hair, and specialized teeth. Marsupials also give birth to live young. However, the young continue to develop inside their mother's pouch. The young will stay in the pouch for several months. ☑

4. Identify What is one feature that only marsupials have?

____

____

THE POUCH

Marsupials are not very developed when they are born. For example, a newborn kangaroo is only the size of a bumblebee. Newborn marsupials have no hair and can only use their front limbs. They use these limbs to pull themselves into their mother's pouch. In the pouch, the newborn starts drinking milk from its mother's nipples.

Young kangaroos are called *joeys*. Joeys stay in their mother's pouch for several months. When they first leave the pouch, they do so for only short periods of time.

KINDS OF MARSUPIALS

You may be familiar with some well known marsupials such as kangaroos. Have you heard of wallabies, bettongs, and numbats? There are over 280 species of marsupials. Most marsupials live in Australia, New Guinea, and South America. The only marsupial that is native to North America is the opossum. ☑

READING CHECK

5. Identify Where are most marsupials found?

____

____

____

Kinds of Marsupials

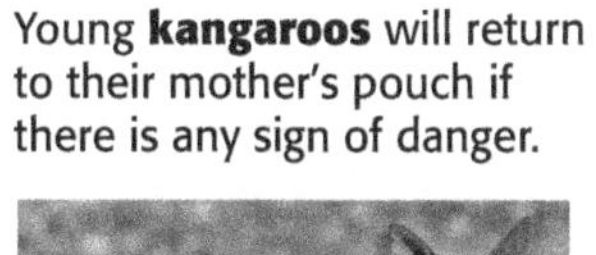

Young **kangaroos** will return to their mother's pouch if there is any sign of danger.

Koalas sleep for about 18 hours each day. They eat eucalyptus leaves.

When in danger, an **opossum** will lie perfectly still. It "plays dead" so predators will leave it alone.

Why Are Many Marsupials Endangered?

The number of living marsupial species is decreasing. At least 22 of Australia's native mammal species have become extinct in the last 400 years. Many more are currently in danger.

Many marsupials have gone extinct or become endangered because of exotic species. Exotic species are organisms that are brought into an area by humans. When Europeans came to Australia in the 1700s and 1800s, they brought animals such as cats, rabbits, and foxes. Many of these animals escaped into the wild. Some of these exotic species now compete with marsupials for food. Others kill marsupials for food.

Exotic species are not the only threat to marsupials. Destruction of habitats has left many maruspials with no place to live and no food. Today, Australia is trying to protect its marsupials.

Critical Thinking

6. Infer Why were cats and rabbits considered exotic species in Australia?

____

____

____

____

Name ______________________ Class ______________ Date ______________

Section 5 Review

NSES LS 1a, 1d, 2a, 3a, 3c, 5a, 5c

SECTION VOCABULARY

marsupial a mammal that carries and nourishes its young in a pouch	**monotreme** a mammal that lays eggs

1. Identify What are the two kinds of monotremes?

____

2. Compare Fill in the Venn Diagram to compare monotremes and marsupials.

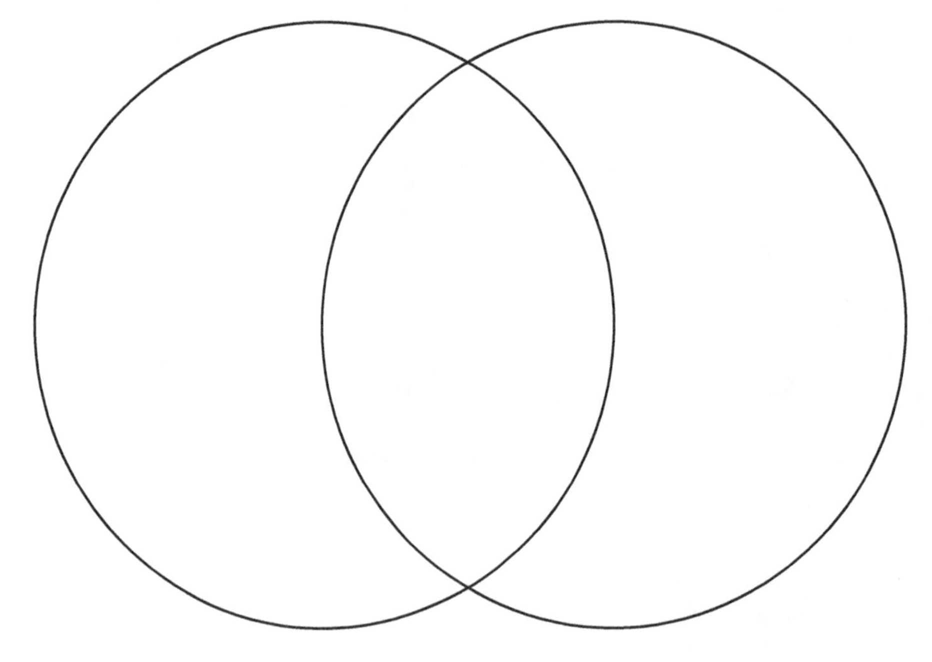

3. Explain How does a platypus use its bill?

____

4. List Give three examples of marsupials.

____

____

5. Describe Describe a newborn marsupial.

____

____

6. Explain Why have many of Australia's marsupials become endangered or gone extinct?

____

____

____

Name ______________________ Class ______________ Date ______________

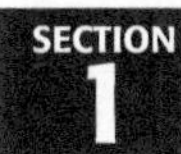

CHAPTER 18 Interactions of Living Things

SECTION 1

Everything Is Connected

BEFORE YOU READ

After you read this section, you should be able to answer these questions:

- What do organisms in an ecosystem depend on for survival?
- What are biotic and abiotic factors?
- What are the levels of organization in the environment?

National Science Education Standards
LS 4a, 4b, 4c, 4d

What Is the Web of Life?

All organisms, or living things, are linked together in the web of life. In this web, energy and resources pass between organisms and their surroundings. The study of how different organisms interact with one another and their environment is **ecology**.

An alligator may hunt along the edge of a river. It may catch a fish, such as a gar, that swims by too closely. As it hunts, the alligator is interacting with its environment. Its environment includes other organisms living in the area. The alligator depends on other organisms to survive, and other organisms depend on the alligator.

However, one organism eating another is not the only way living things interact. For example, when it gets too hot, the alligator may dig a hole in the mud under water. When the alligator no longer uses the hole, fish and other organisms can use it. They may live in the hole when the water level in the rest of the river is low.

Underline As you read, underline any new science terms. Find their definitions in the section review or a dictionary. Make sure you learn what each term means before you move to the next section.

Living things in an environment interact.

Discuss With a partner, talk about the organisms in this picture. How do you think each type of organism interacts with the others? What kind of things do you think each of these organisms needs to survive?

What Are the Two Parts of an Environment?

An organism's environment is made up of biotic and abiotic parts. **Biotic** describes the living parts of the environment, such as fish. **Abiotic** describes the nonliving parts of the environment, such as rivers. Organisms need both biotic and abiotic parts of the environment to live. ☑

READING CHECK

1. Compare What is the difference between biotic and abiotic factors?

____

____

____

How Is the Environment Organized?

The environment can be organized into five levels. Individual organisms are at the first level. The higher levels include more and more parts of the environment. The highest level is called the biosphere. It is the largest level, and includes all the other levels.

1. An *individual* is a single organism.

Organism

Critical Thinking

2. Identify Relationships How are the first two levels of organization related?

____

____

____

____

2. A **population** is a group of individuals of the same species in the same area. For example, all the alligators in the same river make a population. The whole population uses the same area for food and shelter.

Population

3. A **community** is made up of all the different populations that live and interact in the same area. The different populations in a community depend on each other. For example, alligators eat other animals, including fish. Alligators create water-filled holes where fish and other organisms in the river can live during dry seasons.

Critical Thinking

3. Infer Could a community be made up of only one population of organisms? Explain.

____

____

____

____

Community

4. An **ecosystem** is made up of a community and its abiotic environment. The abiotic factors provide resources for all the organisms and energy for some. A river, for example, can provide water for river plants and many animals, and shelter for water insects. It can provide nutrients for plants, as well as food for fish and alligators.

Ecosystem

TAKE A LOOK

4. Identify Use colored pencils to make circles on the picture.

Circle an individual in red.

Circle a population in blue.

Circle a community in brown.

Circle an ecosystem in green.

5. The **biosphere** is the part of Earth where life exists. The biosphere is the largest environmental level. It reaches from the bottom of the ocean and the Earth's crust to high in the sky. Scientists study the biosphere to learn how organisms interact with abiotic parts of the environment. These abiotic parts include Earth's atmosphere, water, soil, and rock.

Biosphere

Math Focus

5. Calculate From sea level, the biosphere goes up about 9 km and down about 19 km. What is the thickness of the biosphere in meters?

____

Name ______________________ Class ______________ Date ______________

Section 1 Review

NSES LS 4a, 4b, 4c, 4d

SECTION VOCABULARY

abiotic describes the nonliving part of the environment, including water, rocks, light, and temperature

biosphere the part of Earth where life exists

biotic describes living factors in the environment

community all of the populations of species that live in the same habitat and interact with each other

ecology the study of the interactions of living organisms with one another and with their environment

ecosystem a community of organisms and their abiotic, or nonliving, environment

population a group of organisms of the same species that live in a specific geographical area

1. Compare What is the difference between a community and an ecosystem?

____

____

____

2. Organize Complete the chart below to describe the five levels of the environment, from smallest to largest.

Level	Description
	a single organism
Population	
	all of the populations of species that live in the same habitat and interact with one another
Ecosystem	
Biosphere	

3. Identify What two kinds of factors does an organism depend on for survival?

____

____

4. Infer Would all the birds in an area make up a population? Explain your answer.

____

____

____

Name ______________________ Class ______________ Date ______________

CHAPTER 18 Interactions of Living Things

SECTION 2

Living Things Need Energy

BEFORE YOU READ

After you read this section, you should be able to answer these questions:

- How do producers, consumers, and decomposers get energy?
- What is a food web?

National Science Education Standards

LS 4a, 4b, 4c, 4d

How Do Organisms Get Energy?

Eating gives organisms two things they cannot live without—energy and nutrients. Prairie dogs, for example, eat grasses and seeds to get their energy and nutrients. Like all organisms, prairie dogs need energy to live.

Organisms in any community can be separated into three groups based on how they get energy: producers, consumers, and decomposers.

STUDY TIP

Circle Choose different colored pencils for producers, primary consumers, secondary consumers, and decomposers. As you read, circle these terms in the text with the colors you chose. Use the same colors to circle animals in any figures that are examples of each group.

PRODUCERS

Producers are organisms that use the energy from sunlight to make their own food. This process is called *photosynthesis*. Most producers are green plants, such as grasses on the prairie and trees in a forest. Some bacteria and algae also photosynthesize to make food. ☑

READING CHECK

1. Explain Why is sunlight important to producers?

____

____

____

CONSUMERS

Consumers cannot make their own food. They need to eat other organisms to obtain energy and nutrients. Consumers can be put into four groups based on how they get energy: herbivores, carnivores, omnivores, and scavengers.

Critical Thinking

2. Apply Concepts What types of consumers are the following organisms?

tigers ______________

deer ______________

humans ______________

Herbivore

Carnivore

Omnivore

Scavenger

STANDARDS CHECK

LS 4b Populations of organisms can be categorized by the functions they serve in an ecosystem. Plants and some microorganisms are producers—they make their own food. All animals, including humans, are consumers, which obtain their food by eating other organisms. Decomposers, primarily bacteria and fungi, are consumers that use waste materials and dead organisms for food. Food webs identify the relationship among producers, consumers, and decomposers in an ecosystem.

3. Define What is the role of decomposers in an ecosystem?

____

____

____

____

An **herbivore** is a consumer that eats only plants. Prairie dogs and bison are herbivores. A **carnivore** is a consumer that eats other animals. Eagles and cougars are carnivores. An **omnivore** is a consumer that eats both plants and animals. Bears and raccoons are omnivores.

A *scavenger* is a consumer that eats dead plants and animals. Turkey vultures are scavengers. They will eat animals and plants that have been dead for days. They will also eat what is left over after a carnivore has had a meal.

DECOMPOSERS

Decomposers recycle nature's resources. They get energy by breaking down dead organisms into simple materials. These materials, such as carbon dioxide and water, can then be used by other organisms. Many bacteria and fungi are decomposers.

What Is a Food Chain?

When an organism eats, it gets energy from its food. If that organism is then eaten, the energy stored in its body is passed to the organism eating it. A **food chain** is the path energy takes from one organism to another. Producers form the beginning of the food chain. Energy passes through the rest of the chain as one organism eats another.

A Prairie Ecosystem Food Chain

TAKE A LOOK

4. Identify Label the food chain diagram with the following terms: energy, producer, primary consumer, secondary consumer, tertiary consumer, decomposer.

In a food chain:

- Producers are eaten by *primary consumers.*
- Primary consumers are eaten by *secondary consumers.*
- Secondary consumers are eaten by *tertiary consumers.*

In the food chain above, the grasses are the producers. The grasses are eaten by prairie dogs, which are the primary consumers. The prairie dogs are eaten by coyotes, which are the secondary consumers. When coyotes die, they are eaten by turkey vultures, which are the tertiary consumers. The tertiary consumer is usually the end of the food chain.

What Is a Food Web?

In most ecosystems, organisms eat more than one thing. Feeding relationships in an ecosystem are shown more completely by a food web. A **food web** is a system of many connected food chains in an ecosystem. Organisms in different food chains may feed upon one another. ☑

As in a food chain, in a food web, energy moves from one organism to the next in one direction. The energy in an organism that is eaten goes into the body of the organism that eats it.

READING CHECK

5. Explain Why does a food web show feeding relationships better than a food chain?

____

____

____

Critical Thinking

6. Predict What do you think would happen if all of the plants were taken out of this food web?

____

____

____

____

____

Simple Food Web

MANY AT THE BASE

An organism uses much of the energy from its food for life processes such as growing or reproducing. When this organism is eaten, only a small amount of energy passes to the next consumer in the chain. Because of this, many more organisms have to be at the base, or bottom, of the food chain than at the top. For example, in a prairie community, there is more grass than prairie dogs. There are more prairie dogs than coyotes.

What Is an Energy Pyramid?

Energy is lost as it passes through a food chain. An **energy pyramid** is a diagram that shows this energy loss. Each level of the pyramid represents a link in the food chain. The bottom of the pyramid is larger than the top. There is less energy for use at the top of the pyramid than at the bottom. This is because most of the energy is used up at the lower levels. Only about 10% of the energy at each level of the energy pyramid passes on to the next level.

Math Focus

7. Calculate How much energy is lost at each level of the energy pyramid?

____

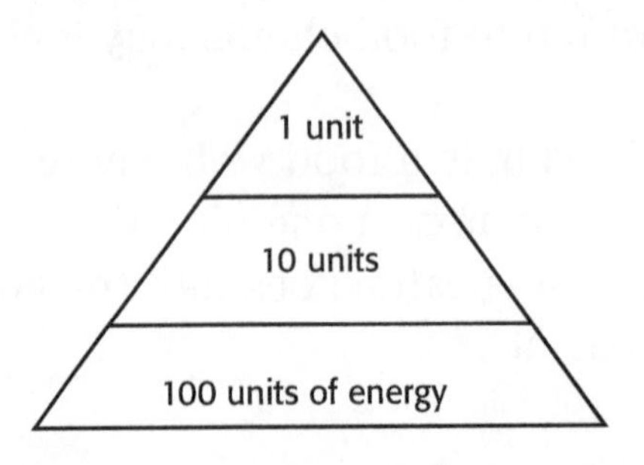

Energy Pyramid

TAKE A LOOK

8. Explain In which level of this energy pyramid do you think deer would belong? Explain your answer.

____

____

____

EFFECT OF ONE SPECIES

A single species can change the flow of energy in an ecosystem. For example, gray wolves are at the top of their food chains. They eat a lot of different organisms but are usually not eaten by any other animal. By eating other organisms, wolves help control the size of those populations.

At one time, wolves were found across the United States. As settlers moved west, many wolves were killed. With few wolves left to feed on the primary consumers, such as elk, those populations began to grow. The elk ate all the grass, and there was none left for the smaller herbivores, such as hares. As these small herbivores died, there was less food for the secondary consumers. When wolves were removed from the food web, the whole ecosystem was affected. ☑

READING CHECK

9. Summarize Why did a change in the wolf population affect the other organisms in the community?

____

____

____

____

____

When wolves were removed from the ecosystem, other organisms were affected.

Name ______________________ Class ______________ Date ______________

Section 2 Review

NSES LS 4a, 4b, 4c, 4d

SECTION VOCABULARY

carnivore an organism that eats animals

energy pyramid a triangular diagram that shows an ecosystem's loss of energy, which results as energy passes through the ecosystem's food chain

food chain the pathway of energy transfer through various stages as a result of the feeding patterns of a series of organisms

food web a diagram that shows the feeding relationships between organisms in an ecosystem

herbivore an organism that eats only plants

omnivore an organism that eats both plants and animals

1. Explain Why are producers important in an ecosystem?

____

____

____

2. Connect Make a food chain using the following organisms: mouse, snake, grass, hawk. Draw arrows showing how energy flows through the chain. Identify each organism as a producer, primary consumer, secondary consumer, or tertiary consumer.

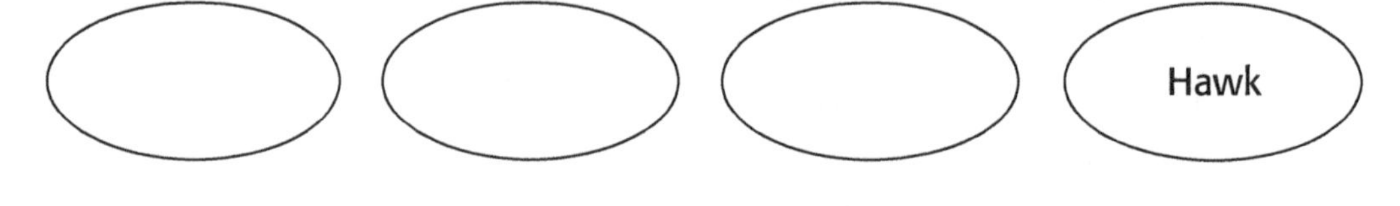

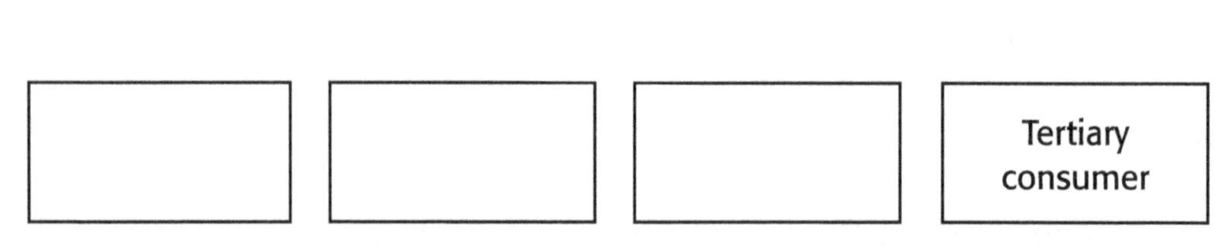

3. Apply Concepts Organisms can be part of more than one food chain. Make a food chain that includes one of the organisms above.

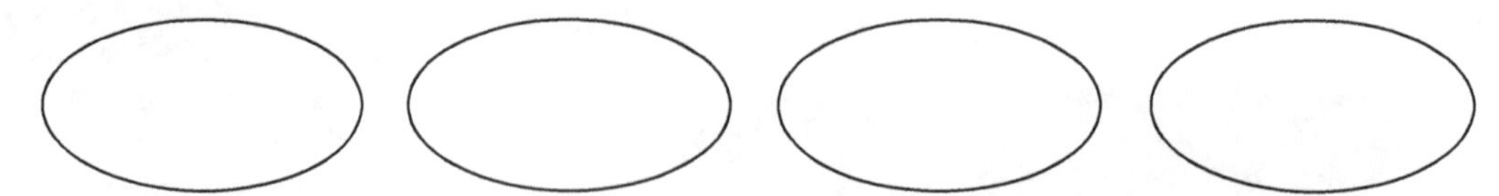

4. Infer Do you think you could find a food chain that had 10 organisms? Explain.

____

____

____

SECTION 3

Types of Interactions

BEFORE YOU READ

After you read this section, you should be able to answer these questions:

- What determines an area's carrying capacity?
- Why does competition occur?
- How do organisms avoid being eaten?
- What are three kinds of symbiotic relationships?

National Science Education Standards
LS 3a, 3c, 4b, 4d

How Does the Environment Control Population Sizes?

Most living things have more offspring than will survive. A female frog, for example, may lay hundreds of eggs in a small pond. If all of the eggs became frogs, the pond would soon become very crowded. There would not be enough food for the frogs or other organisms in the pond. But in nature, this usually does not happen. The biotic and abiotic factors in the pond control the frog population so that it does not get too large.

Populations cannot grow without stopping because the environment has only a certain amount of food, water, space, and other resources. A resource that keeps a population from growing forever is called a *limiting factor.* Food is often a limiting factor in an ecosystem.

All plants need sunlight. In this forest, sunlight may be a limiting factor. Not all plants can get the same amount of light.

STUDY TIP

Make a List As you read this section, write down any questions you may have. Work with a partner to find the answers to your questions.

STANDARDS CHECK

LS 4d The number of organisms an ecosystem can support depends on the resources available and abiotic factors, such as quantity of light and water, range of temperatures, and soil composition. Given adequate biotic and abiotic resources and no disease or predators, populations (including humans) increase at rapid rates. Lack of resources and other factors, such as predation and climate, limit the growth of populations in specific niches in the ecosystem.

Word Help: resource
anything that can be used to take care of a need

1. Define What is a limiting factor?

____

____

____

What Is Carrying Capacity?

The largest number of organisms that can live in an environment is called the **carrying capacity**. When a population grows beyond the carrying capacity, limiting factors will cause some individuals to leave the area or to die. As individuals die or leave, the population decreases.

The carrying capacity of an area can change if the amount of the limiting factor changes. For example, the carrying capacity of an area will be higher in seasons when more food is available. ☑

READING CHECK

2. Explain Why can the carrying capacity of an area change?

____

____

____

How Do Organisms Interact in an Ecosystem?

Populations are made of individuals of the same species. Communities are made of different populations that interact. There are four main ways that individuals and populations affect one another in an ecosystem: in competition, as predator and prey, through symbiosis, and coevolution. ☑

READING CHECK

3. List What are four ways that organisms in an ecosystem interact?

____

____

____

____

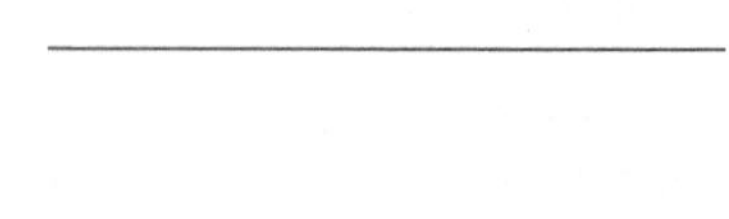

Why Do Organisms Compete?

Competition happens when more than one individual or population tries to use the same resource. There may not be enough resources, such as food, water, shelter, or sunlight, for all the organisms in an environment. When one individual or population uses a resource, there is less for others to use.

Competition can happen between organisms in the same population. For example, in Yellowstone National Park, elk compete with one another for the same plants. In the winter, when there are not many plants, competition is much higher. Some elk will die because there is not enough food. In spring, when many plants grow, there is more food for the elk, and competition is lower.

Competition can also happen between populations. In a forest, different types of trees compete to grow in the same area. All of the plant populations must compete for the same resources: sunlight, space, water, and nutrients.

Critical Thinking

4. Predict In a prairie ecosystem, which two of the following organisms most likely compete for the same food source: elk, coyotes, prairie dogs, vultures?

____

____

How Do Predators and Prey Interact?

Another way organisms interact is when one organism eats another to get energy. The organism that is eaten is called the **prey**. The organism that eats the prey is called the **predator**. When a bird eats a worm, for example, the bird is the predator, and the worm is the prey.

PREDATORS

Predators have traits or skills that help them catch and kill their prey. Different types of predators have different skills and traits. For example, a cheetah uses its speed to catch prey. On the other hand, tigers have colors that let them blend with the environment so that prey cannot see them easily. ☑

READING CHECK

5. Identify What are two traits different predators may have to help them catch prey?

____

____

____

PREY

Prey generally have some way to protect themselves from being eaten. Different types of organisms protect themselves in different ways:

Say It

Discuss In small groups, talk about other animals that escape predators in the four ways described in the text.

1. **Run Away** When a rabbit is in danger, it runs.

2. **Travel in Groups** Some animals, such as musk oxen, travel in herds, or groups. Many fishes, such as anchovies, travel in schools. All the animals in these groups can help one another by watching for predators.

Critical Thinking

6. Infer Why do you think it would be difficult for predators to attack animals in a herd?

____

____

____

____

____

When musk oxen sense danger, they move close together to protect their young.

3. **Show Warning Colors** Some organisms have bright colors that act as a warning. The colors warn predators that the prey might be poisonous. A brightly colored fire salamander, for example, sprays a poison that burns.

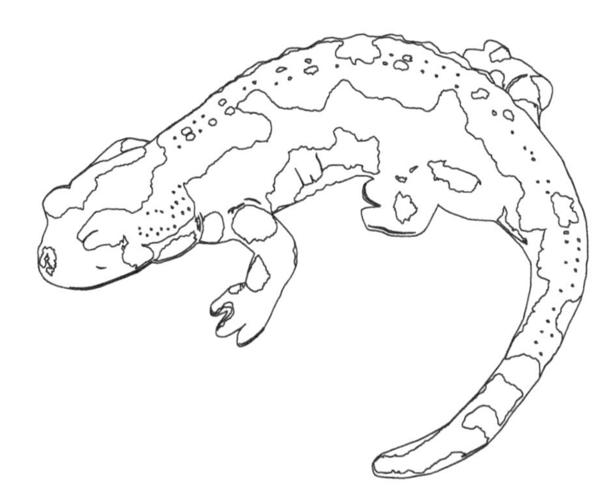

TAKE A LOOK

7. Color A fire salamander has a black body with bright orange or yellow spots. Use colored pencils to give this salamander its warning colors.

4. Use Camouflage Some organisms can hide from predators by blending in with the background. This is called *camouflage*. A rabbit's natural colors, for example, may help it blend in with dead leaves or shrubs so that it cannot be seen. Some animals may look like twigs, stone, or bark.

What Is Symbiosis?

Some species have very close interactions with other species. A close association between two or more species is called **symbiosis**. Each individual in a symbiotic relationship may be helped, hurt, or not affected by another individual. Often, one species lives on or in another species. Most symbiotic relationships can be divided into three types: mutualism, commensalism, and parasitism. ☑

READING CHECK

8. List List the three types of symbiotic relationships.

____

____

____

MUTUALISM

When both individuals in a symbiotic relationship are helped, it is called **mutualism**. You can see mutualism in the relationship between a bee and a flower.

Organism hurt?	Organism helped?	Example
No one	both organisms	A bee transfers pollen for a flower; a flower provides nectar to a bee.

In a mutualistic relationship, both species benefit.

COMMENSALISM

When one individual in a symbiotic relationship is helped but the other is not affected, this is called **commensalism**.

Organism hurt?	Organism helped?	Example
No one	one of the organisms	A fish called a remora attaches to a shark and eats the shark's leftovers.

The remoras get a free meal, but the shark is not harmed.

PARASITISM

A symbiotic relationship in which one individual is hurt and the other is helped is called **parasitism**. The organism that is helped is called the parasite. The organism that is hurt is called the *host*. ☑

Organism hurt?	Organism helped?	Example
Host	parasite	A flea is a parasite on a dog.

Critical Thinking

9. Compare How does mutualism differ from commensalism?

____

____

____

____

____

____

10. Define In parasitism, is the host helped or hurt?

____

This tomato hornworm is being parasitized by young wasps. Their cocoons are on the caterpillar's back.

TAKE A LOOK

11. Infer How do you think the caterpillar helps the wasps?

____

____

____

What Is Coevolution?

Relationships between organisms change over time. Interactions can even be one reason that organisms change. When a long-term change happens in two species because of their close interactions, the change is called **coevolution**.

One example of coevolution can be seen in some flowers and the organisms that pollinate them. A *pollinator* is an organism, such as a bird, insect, or bat, that carries pollen from one flower to another. Flowers need to attract pollinators to help them reproduce. Different flowers have evolved different ways to attract pollinators. Some use colors or odors. Others use nectar as a food reward for the pollinator.

Some plants can use a variety of pollinators. Others have coevolved with certain pollinators. For example, the bat in the picture below has a long sticky tongue. It uses its tongue to get nectar from deep inside the flower. Only an organism with a way to reach the nectar could be a pollinator for this flower.

Say It

Investigate With a partner, look up the meaning of the suffix *co-*. Discuss how the meaning of this suffix can help you remember what *coevolution* means. Think of some other words that have *co-*.

Name ______________________ Class ______________ Date ______________

Section 3 Review

NSES LS 3a, 3c, 4b, 4d

SECTION VOCABULARY

carrying capacity the largest population that an environment can support at any given time

coevolution the evolution of two species that is due to mutual influence, often in a way that makes the relationship more beneficial to both species

commensalism a relationship between two organisms in which one organism benefits and the other is unaffected

mutualism a relationship between two species in which both species benefit

parasitism a relationship between two species in which one species, the parasite, benefits from the other species, the host, which is harmed

predator an organism that kills and eats all or part of another organism

prey an organism that is killed and eaten by another organism

symbiosis a relationship in which two different organisms live in close association with each other

1. **Identify** What are two resources for which organisms are likely to compete?

____

2. **Explain** What happens to a population when it grows larger than its carrying capacity?

____

____

3. **Infer** Do you think the carrying capacity is the same for all species in an ecosystem? Explain your answer.

____

____

4. **Summarize** Complete the chart below to describe the different kinds of symbiotic relationships.

Example organisms	Type of symbiosis	Organism(s) helped	Organism(s) hurt
Flea and dog			host (dog)
Bee and flower	mutualism		
Remora and shark			none

5. **Apply Concepts** The flowers of many plants provide a food reward, such as nectar, to pollinators. Some plants, however, attract pollinators but provide no reward. What type of symbiosis best describes this relationship? Explain your answer.

____

____

Name ____________________ Class ____________ Date ____________

CHAPTER 19 Cycles in Nature

SECTION 1

The Cycles of Matter

BEFORE YOU READ

After you read this section, you should be able to answer these questions:

- Why does matter need to be recycled?
- How are water, carbon, and nitrogen recycled?

National Science Education Standards

LS 1c, 4b, 4c, 5a

Why Is Matter Recycled on Earth?

The matter in your body has been on Earth since the planet formed billions of years ago. Matter on Earth is limited, so it must be used over and over again. Each kind of matter has its own cycles. In these cycles, matter moves between the environment and living things.

Mnemonic As you read, create a mnemonic device, or memory trick, to help you remember the parts of the water cycle.

What Is the Water Cycle?

Without water there would be no life on Earth. All living things are made mostly of water. Water carries other nutrients to cells and carries wastes away from them. It also helps living things regulate their temperatures. Like all matter, water is limited on Earth. The water cycle lets living things use water over and over.

In the environment, water moves between the oceans, atmosphere, land, and living things. Eventually, all the water taken in by organisms returns to the environment. The movement of water is known as the *water cycle*. The parts of the water cycle are explained in the figure below.

Identify Describe to the class all the things you and your family do in a day that use water. Can you think of any ways you might be able to use less water?

Precipitation is rain, snow, sleet, or hail that falls from clouds to Earth's surface. Most precipitation falls into the ocean. It never touches the land.

Condensation happens when water vapor cools and changes into drops of liquid water. The water drops form clouds in the atmosphere.

Evaporation happens when liquid water on Earth's surface changes into water vapor. Energy from the sun makes water evaporate.

Groundwater is water that flows under the ground. Gravity can make water that falls on the land move into rocks underground.

Runoff is water that flows over the land into streams and rivers. Most of the water ends up in the oceans.

Transpiration happens when plants give off water vapor from tiny holes in their leaves.

TAKE A LOOK

1. Describe How do clouds form?

____

____

____

____

____

STANDARDS CHECK

LS 4c For ecosystems, the major source of energy is sunlight. Energy entering ecosystems as sunlight is transferred by producers into chemical energy through photosynthesis. That energy passes from organism to organism in food webs.

2. Analyze Explain the role of photosynthesis in the carbon cycle.

____
____
____
____
____
____
____
____

What Is the Carbon Cycle?

Besides water, the most common molecules in living things are *organic molecules*. These are molecules that contain carbon, such as sugar. Carbon moves between the environment and living things in the *carbon cycle*.

PHOTOSYNTHESIS AND RESPIRATION

Plants are producers. This means they make their own food. They use water, carbon dioxide, and sunlight to make sugar. This process is called *photosynthesis*. Photosynthesis is the basis of the carbon cycle.

Animals are consumers. This means they have to consume other organisms to get energy. Most animals get the carbon and energy they need by eating plants. How does this carbon return to the environment? It returns when cells break down sugar molecules to release energy. This process is called *respiration*.

DECOMPOSITION AND COMBUSTION

Fungi and some bacteria get their energy by breaking down wastes and dead organisms. This process is called **decomposition**. When organisms decompose organic matter, they return carbon dioxide and water to the environment.

When organic molecules, such as those in wood or fossil fuels, are burned, it is called **combustion**. Combustion releases the carbon stored in these organic molecules back into the atmosphere.

TAKE A LOOK

3. Complete Carbon dioxide in the air is used for _______________
_______________.

4. List What three processes release carbon dioxide into the environment?

____
____
____

The Carbon Cycle

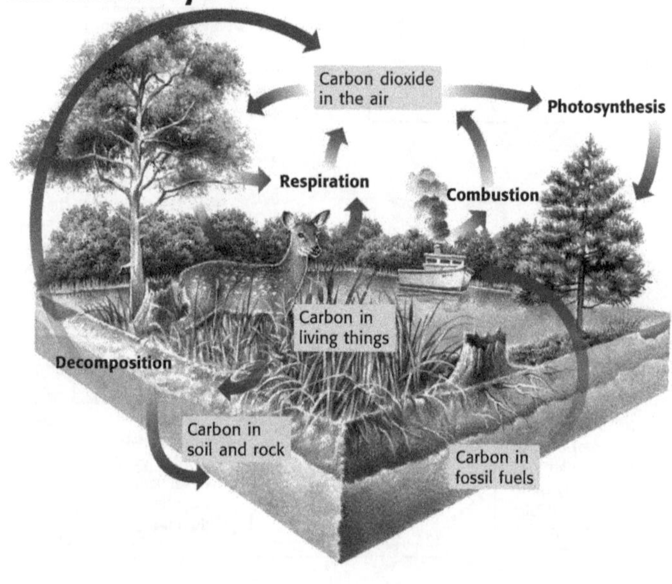

What Is the Nitrogen Cycle?

Nitrogen is also important to living things. Organisms need nitrogen to build proteins and DNA for new cells. Like water and carbon, nitrogen cycles through living things and the environment. This is called the *nitrogen cycle.*

NITROGEN FIXATION

About 78% of Earth's atmosphere is nitrogen gas. Most organisms cannot use nitrogen gas directly. Bacteria in soil can change nitrogen gas into forms that plants can use. This is called *nitrogen fixation.* Other organisms can get the nitrogen they need by eating plants or by eating organisms that eat plants.

Critical Thinking

5. Apply Concepts How is nitrogen fixation important to animals?

____

____

____

____

____

The Nitrogen Cycle

TAKE A LOOK

6. Identify What process releases nitrogen into the soil?

____

How Are the Cycles of Matter Connected?

Other forms of matter on Earth also cycle through the environment. These include many minerals that living cells need, such as calcium and phosphorus. When an organism dies, every substance in its body will be recycled in the environment or reused by other organisms. ☑

All of the cycles of matter are connected. For example, water carries some forms of carbon and nitrogen through the environment. Many nutrients pass from soil to plants to animals and back. Living things play a part in each of the cycles.

READING CHECK

7. Explain What happens to the substances in an organism's body when the organism dies?

____

____

____

____

____

Name ______________________ Class ______________ Date ______________

Section 1 Review

NSES LS 1c, 4b, 4c, 5a

SECTION VOCABULARY

combustion the burning of a substance **condensation** the change of state from a gas to a liquid **decomposition** the breakdown of substances into simpler molecular substances	**evaporation** the change of state from a liquid to a gas **precipitation** any form of water that falls to Earth's surface from the clouds

1. Identify In the water cycle, what makes water evaporate?

____

2. Summarize Draw arrows to show how carbon cycles through the environment and living things.

Air

3. Explain Why does matter need to be recycled?

____

4. Explain Why is water so important to life on Earth?

____

____

____

5. Define What is nitrogen fixation?

____

____

6. Define What are organic molecules?

____

Name ______________________ Class ______________ Date ______________

CHAPTER 19 Cycles in Nature

SECTION 2

Ecological Succession

BEFORE YOU READ

After you read this section, you should be able to answer these questions:

- How do communities of living things form?
- Why do the type of organisms in a community change over time?

National Science Education Standards
LS 1a, 4d

What Is Succession?

In the spring of 1988, much of Yellowstone National Park was a forest. The trees grew close together. Large areas were in shade, and few plants grew under the trees.

That summer, fires burned much of the forest and left a blanket of gray ash on the forest floor. Most of the trees were dead, though some of them were still standing.

The following spring, the forest floor was green. Some of the dead trees had fallen over, and many small, green plants, such as grasses, were growing.

Why were grasses the first things to grow? After the fire, the forest floor was sunny and empty. Nonliving parts of ecosystems, such as water, light, and space, are called *abiotic factors*. When the trees were dead, grasses had the abiotic factors they needed, and their populations grew quickly.

In a few years, larger plants began growing in some areas, and the grasses could not grow without sunlight. Within 10 years, the trees were starting to grow back. The trees began to shade out those plants.

When one type of community replaces another type of community, this is called **succession**. The grasses and other species that are the first to live or grow in an area are called **pioneer species**. ☑

STUDY TIP

Organize As you read, make a table comparing primary succession and secondary succession.

Math Focus

1. Calculate Percentages The fires in Yellowstone National Park in 1988 burned 739,000 acres. The park has 2.2 million acres total. What percentage of the park burned?

____

READING CHECK

2. Define What is a pioneer species?

____

____

____

Critical Thinking

3. Analyze What makes lichens good pioneer species?

____

____

____

____

____

PRIMARY SUCCESSION

Sometimes, a small community starts to grow in an area where living things have never grown before. The area is only bare rock and there is no soil. Over a very long time, a community can develop. The change from bare rock to a community of organisms is called *primary succession.*

Lichens are pioneer species on bare rock. A lichen's structure allows it to function on bare rock. Lichens don't have roots, and they get their water from the air. This means they do not need soil. Most other organisms, however, cannot move into the area without soil.

Lichens produce acid that breaks down the rock they are living on. The rock particles, mixed with the remains of dead lichens, become the first soil.

After many years, there is enough soil for mosses to grow. The mosses eventually replace the lichens. Tiny organisms and insects begin to live there. When they die, their remains add to the soil.

Over time, the soil gets deeper, and ferns replace mosses. The ferns may be replaced later by grasses and wildflowers. If there is enough soil, shrubs and small trees may grow. After hundreds of years, the soil may be deep enough and rich enough to support a forest community.

TAKE A LOOK

4. Identify Which kind of plants are generally the last to appear in an area going through primary succession?

____

Succession of Lichen and Plant Species in a Forest

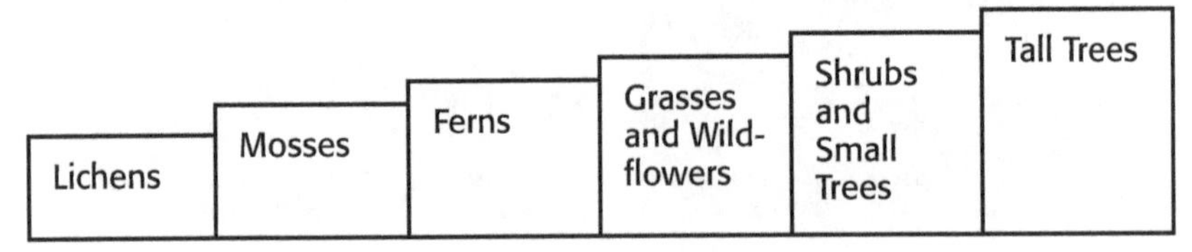

Remember that a community is made up of all the living things in an area. It includes the plants that can live with the abiotic factors there at the time. It also includes the animals that can use the resources there at the time.

When the abiotic factors and resources change, so does the community. For example, a population of cottontail rabbits will get bigger as more small plants grow in the soil over the rock. Later, there will be fewer small plants, when more trees grow and block the sun. Then, there will be fewer rabbits. However, the populations of animals that need trees, such as squirrels, will increase.

SECONDARY SUCCESSION

Sometimes, a community is destroyed by a natural disaster, such as a flood or fire. Sometimes, humans or animals alter an environment. For example, a farmer may stop growing crops in a field. In either case, if there is soil and the area is left alone, the natural community can grow back. The plant species change in a series of stages called *secondary succession*. Secondary succession happens in areas where living things already exist. ☑

READING CHECK

5. Describe Where does secondary succession happen?

____

The figure below shows secondary succession in a farm field that used to be a forest.

First Year Weeds start to grow.

Second Year New weeds appear. Their seeds may have been blown to the field by the wind, or insects may have carried them.

In 5 to 15 Years Small conifer trees, such as pines and firs, grow among the weeds. After about 100 years, the weeds are gone and a forest has formed.

After 100 Years or More As older conifer trees die, they may be replaced by hardwood trees. Oak and maple will grow if the temperature and precipitation are right.

TAKE A LOOK

6. Identify In this example, what are the first kind of plants to grow in secondary succession?

____

7. Identify What are the first kind of trees that may grow in an area?

____

MATURE COMMUNITIES AND BIODIVERSITY

As succession goes on, a community can end up having one well-adapted plant species. This is called a *climax species*. However, in many places, a community is more likely to include many species. The variety of species that live in an area is called its *biodiversity*.

Name ______________________ Class ______________ Date ______________

Section 2 Review

NSES LS 1a, 4d

SECTION VOCABULARY

pioneer species a species that colonizes an uninhabited area and that starts a process of succession	**succession** the replacement of one type of community by another at a single location over a period of time

1. **Define** What are abiotic factors? Give three examples.

____

____

2. **Compare** What is the difference between primary and secondary succession?

____

____

____

3. **Apply Concepts** Secondary succession generally happens faster than primary succession. Why do you think this happens?

____

____

____

4. **Apply Ideas** Consider a species of animal that eats grass and a species of animal that eats nuts. Which species do you think would have a larger population in a mature forest? Explain your answer.

____

____

____

5. **Analyze** Why, in general, can't tall trees be pioneer species?

____

____

____

6. **Define** What is biodiversity?

____

7. **Describe** When soil first forms over bare rock, what is it made of?

____

____

Name ______________________ Class ______________ Date ______________

CHAPTER 20 The Earth's Ecosystems

SECTION 1

Land Biomes

BEFORE YOU READ

After you read this section, you should be able to answer these questions:

- What are eight kinds of land biomes?
- What kinds of organisms live in each land biome?

National Science Education Standards

LS 1a, 3a, 3c, 3d, 4b, 4c, 4d, 5a, 5b

What Are Biomes?

Imagine that you have to explain the difference between a desert and a forest to someone. What would you say? You might say that a forest gets more rain than a desert. You might also say that different organisms live in forests than in deserts. Both of these features, as well as many others, make these two environments different from each other.

Rainfall is an *abiotic factor*, or nonliving part, of the environment. Organisms and their interactions are *biotic factors*, or living parts, of the environment. A large area with similar biotic and abiotic factors is called a **biome**. A biome is made up of many related ecosystems.

There are many different biomes on Earth. The figure below shows a map of the major land biomes. *Land biomes* are biomes that are found on land.

STUDY TIP

Summarize As you read about each type of land biome, write down its important features. When you finish reading this section, write a summary of the features of each type of land biome into your notebook.

Critical Thinking

1. Compare How is a biome different from an ecosystem?

____

____

Polar ice
Tundra
Coniferous forest
Tropical rain forest
Temperate deciduous forest
Temperate grassland
Savanna
Desert
Chaparral
Mountains

This map shows some of the major land biomes on Earth.

TAKE A LOOK

2. Identify On which two continents are most savanna biomes found?

____

____

What Are Three Kinds of Forest Biomes?

Three forest biomes are coniferous forests, temperate deciduous forests, and tropical rain forests. All of these forests receive plenty of rain, and the temperatures are not extreme. The kind of forest that forms in an area depends on the area's climate. ☑

READING CHECK

3. Identify What are two features that forest biomes have in common?

____

____

TEMPERATE DECIDUOUS FORESTS

Have you seen trees with leaves that change color and then fall off the trees? If so, you have seen deciduous trees. The word *deciduous* comes from a Latin word that means "to fall off." Deciduous trees shed their leaves to save water during the winter. Temperate deciduous forests contain many deciduous trees. As the figure below shows, many different organisms live in deciduous forests.

When deciduous trees lose their leaves, more light can reach the ground. Therefore, the ground in deciduous forests gets a lot of sunlight during part of the year. Small trees and shrubs can grow on the forest floor.

STANDARDS CHECK

LS 4b Populations of organisms can be categorized by the functions they serve in an ecosystem. Plants and some microorganisms are producers—they make their own food. All animals, including humans, are consumers, which obtain their food by eating other organisms. Decomposers, primarily bacteria and fungi, are consumers that use waste materials and dead organisms for food. Food webs identify the relationship among producers, consumers, and decomposers in an ecosystem.

Word Help: function
use or purpose

4. Identify What are two producers in a temperate deciduous forest?

____

____

CONIFEROUS FORESTS

The coniferous forest gets its name from conifers, the main type of tree that grows there. *Conifers* are trees that produce seeds in cones. They have special needle-shaped leaves covered in a thick, waxy coating. These features help the tree conserve water. The waxy coating also protects the needles from being damaged by cold weather. Most conifers are evergreens. They stay green all year and do not lose all their leaves at once. ☑

In coniferous forests, decomposition is slow. The ground may be covered by a thick layer of needles. The trees prevent much sunlight from reaching the ground. Because there is little light there, not many plants live under the conifer trees. The figure below shows some of the organisms that live in coniferous forests.

READING CHECK

5. Identify Where do conifers produce their seeds?

____

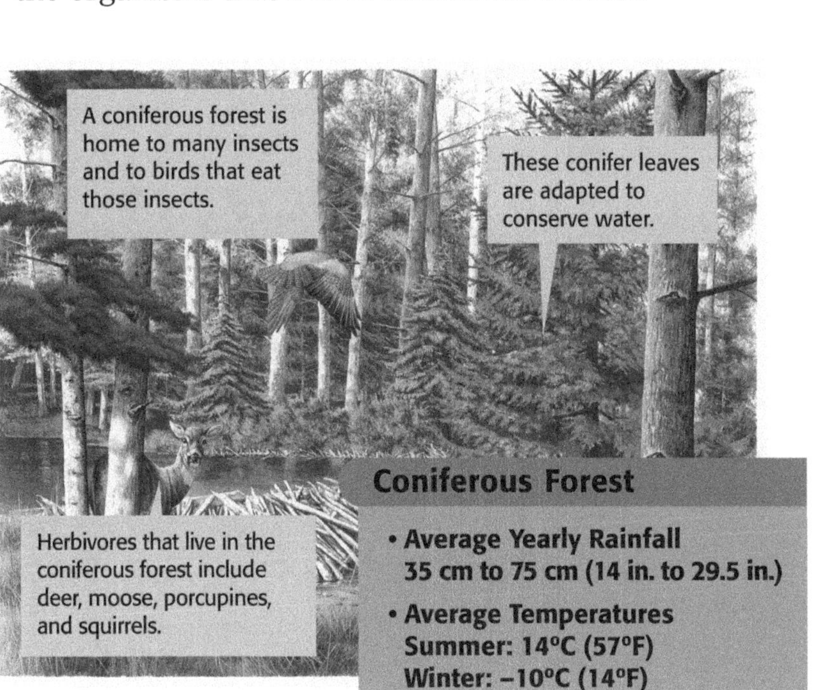

Coniferous Forest

- **Average Yearly Rainfall**
  35 cm to 75 cm (14 in. to 29.5 in.)
- **Average Temperatures**
  Summer: 14°C (57°F)
  Winter: –10°C (14°F)

Critical Thinking

6. Compare Why does the floor of a temperate deciduous forest have more small plants than the floor of a coniferous forest?

____

____

____

____

____

TAKE A LOOK

7. Explain Why is it important for trees in a coniferous forest to conserve water?

____

____

____

TROPICAL RAIN FORESTS

Tropical rain forest biomes contain the greatest variety of plants and animals on Earth. The warm temperatures and high rainfall allow a lot of plants to grow. Trees grow very tall and dense. Their leaves prevent much light from reaching the forest floor, so not many small plants live on the forest floor.

The many plants in a rainforest support many different kinds of animals. Most of the animals in a rain forest live in the trees. Birds such as toucans are omnivores that eat fruits, reptiles, and other birds. Carnivores, such as harpy eagles, eat other animals, such as howler monkeys. Howler monkeys are herbivores that eat fruits, nuts, and leaves. ☑

READING CHECK

8. Describe Where in a rain forest are most animals found?

____

____

You may think that the soil in a rain forest is very rich in nutrients because of all the plants that live there. However, most of the nutrients in the tropical rain forest are found in plants, not in soil. The soil is so thin that many trees grow roots above-ground for support.

TAKE A LOOK

9. Infer Why do many vines grow on tree branches instead of on the forest floor?

____

____

____

____

What Are Two Kinds of Grassland?

A *grassland* is a biome made up mainly of grasses, small flowering plants, and a few trees. The two main kinds of grassland are temperate grasslands and savannas.

TEMPERATE GRASSLANDS

In temperate grasslands, the summers are warm and the winters are cold. The soils of temperate grasslands are very rich in nutrients, so many kinds of grasses grow there. Fires, droughts, and grazing prevent trees and shrubs from growing.

Many seed-eating animals, such as prairie dogs and mice, live in this kind of grassland. They use camouflage and burrows to hide from predators, such as coyotes. ☑

10. Identify What are two ways small animals hide from predators?

____

____

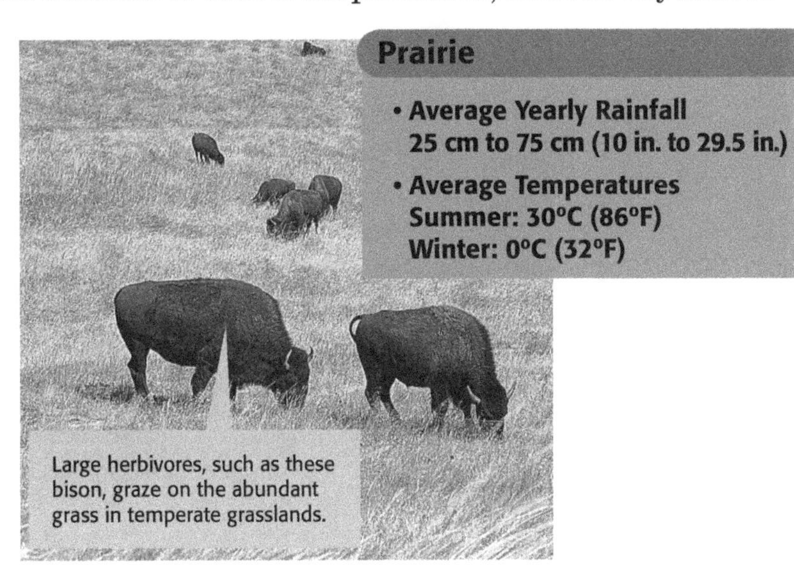

Large herbivores, such as these bison, graze on the abundant grass in temperate grasslands.

SAVANNAS

The **savanna** is a grassland that has a lot of rainfall during some seasons and very little rainfall in others. During the dry season, savanna grasses dry out and turn yellow. However, the roots can live for many months without water.

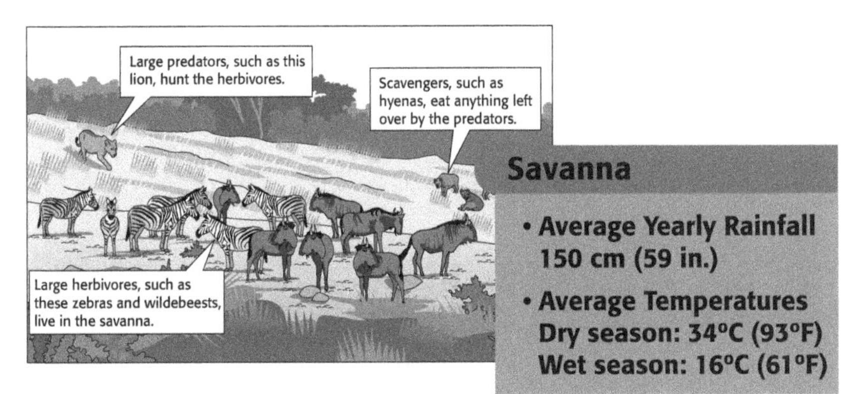

TAKE A LOOK

11. Compare How does the amount of rainfall in a temperate grassland compare with the amount of rainfall in a savanna?

____

____

What Is a Desert?

Deserts are very dry biomes, and most are very hot. The organisms that can live in a desert have special features that let them survive in the dry climate.

Many desert plants have roots that spread near the surface. This allows them to take up water quickly after a rain, before it evaporates.

Desert animals also have ways to survive the hot, dry desert conditions. Some live underground, where it is cooler. They come out only at night, when air temperatures are lower. Others, such as the fringe-toed lizard, bury themselves in the loose sand to escape the heat and avoid predators. ☑

READING CHECK

12. List What are two ways that burrowing under the ground helps the fringe-toed lizard?

____

____

Math Focus

13. Calculate About how many inches of rain does a desert get every year? Show your work.

1 in. = 2.54 cm

____

____

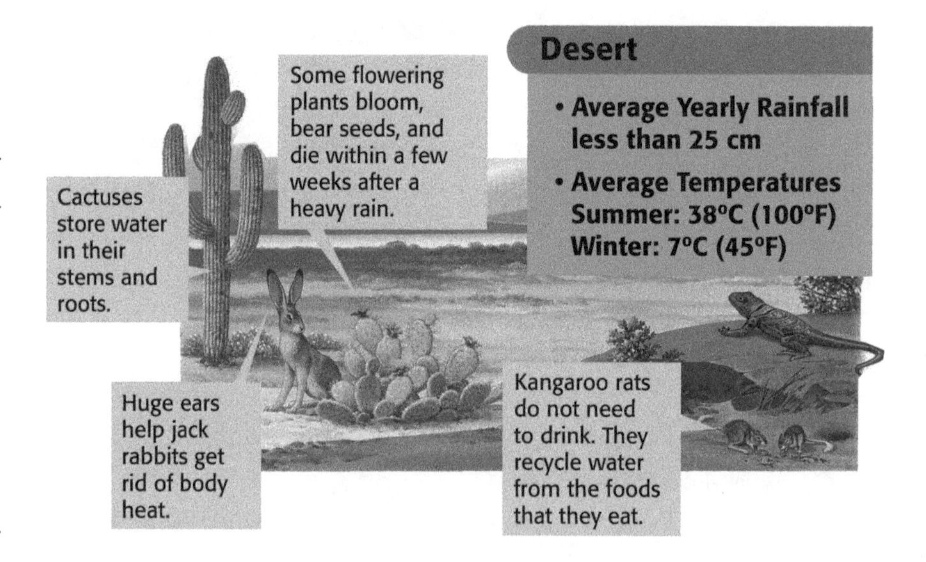

What Is a Tundra?

Imagine a place on Earth that is too cold for trees to grow. A **tundra** is a biome that has very cold temperatures and little rainfall. Two kinds of tundra are polar tundra and alpine tundra.

POLAR TUNDRA

Polar tundra is found near the North and South Poles. In polar tundra, the layer of soil below the surface stays frozen all year long. This layer is called *permafrost.*

During the short, cool summers, only the water in the soil at the surface melts. This surface soil is too shallow for most plants. Only plants with shallow roots, such as grasses and small shrubs, are common. Mosses and lichens grow beneath these plants. Growing close to the ground helps the plants resist the wind and the cold. ☑

Animals of the tundra also have ways to live in this biome. In the winter, food is hard to find and the weather is very harsh. Some animals, such as bears, sleep through much of the winter. Other animals, like the caribou, travel long distances to find food. Many animals have extra layers of fat to keep them warm.

During the summer, the soil above the permafrost becomes muddy from melting ice and snow. Insects, such as mosquitoes, lay eggs in the mud. Birds that prey on these insects are carnivores. Other carnivores, such as wolves, prey on herbivores, such as caribou and musk oxen.

READING CHECK

14. Explain How does growing close to the ground helps tundra plants?

____

____

____

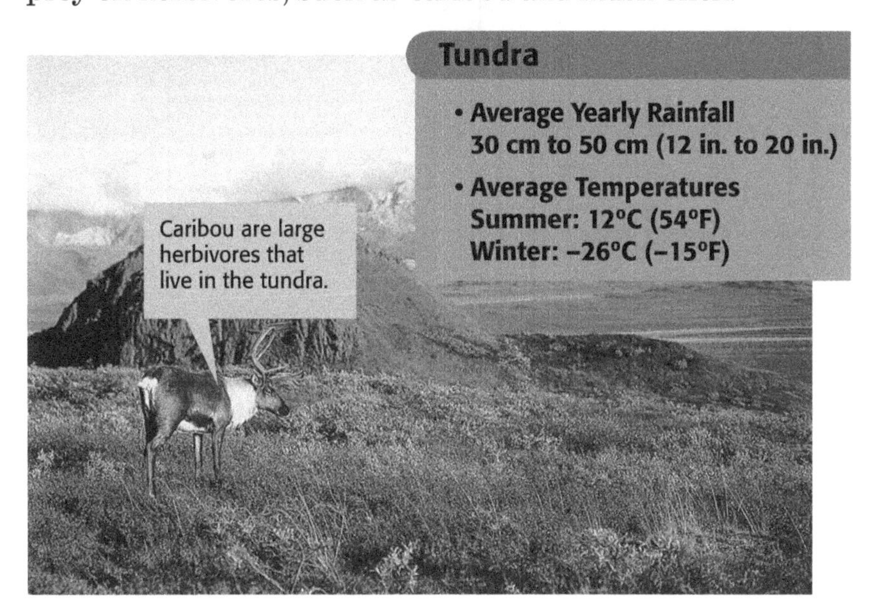

TAKE A LOOK

15. Describe How are the ecological roles of a caribou in a tundra and a zebra in a savanna similar?

____

____

____

ALPINE TUNDRA

Alpine tundra is similar to polar tundra. Alpine tundra has permafrost. However, alpine tundra is found at the tops of tall mountains. Above an elevation called the *tree line*, trees cannot grow on a mountain. Alpine tundra is found above the tree line. It gets a lot of sunlight and a moderate amount of rainfall.

Share Experiences Have you ever been to a very cold place? In a group, discuss what it was like.

Name ______________________ Class ______________ Date ______________

Section 1 Review

NSES **LS 1a, 3a, 3c, 3d, 4b, 4c, 4d, 5a, 5b**

SECTION VOCABULARY

biome a large region characterized by a specific type of climate and certain types of plant and animal communities

desert a region that has little or no plant life, long periods without rain, and extreme temperatures: usually found in hot climates

savanna a grassland that often has scattered trees and that is found in tropical and subtropical areas where seasonal rains, fires, and drought happen

tundra a treeless plain found in the Arctic, in the Antarctic, or on the tops of mountains that is characterized by very low winter temperatures and short, cool summers

1. **Explain** The tundra has been called a "frozen desert." Explain why this is a good name for the tundra.

____

____

____

2. **Compare** Compare the temperate grassland and the savanna by filling in the blank spaces in the table below.

	Temperate grassland	**Savanna**
Abiotic factors		constant warmth with seasonal rains
Types of producers	grass with a few trees	
Types of consumers		

3. **List** What are some of the adaptations that allow desert plants to live in such a hot, dry environment?

____

____

____

4. **Compare** How is alpine tundra different from polar tundra?

____

____

____

Name ______________________ Class ______________ Date ______________

CHAPTER 20 The Earth's Ecosystems

SECTION 2

Marine Ecosystems

BEFORE YOU READ

After you read this section, you should be able to answer these questions:

- What abiotic factors affect marine ecosystems?
- What are the major zones found in the ocean?
- What organisms are found in marine ecosystems?

National Science Education Standards

LS 1a, 3d, 4a, 4b, 4c, 4d

What Are Marine Ecosystems?

Oceans cover almost three-fourths of Earth's surface! Scientists call the ecosystems in the ocean *marine ecosystems*. Marine ecosystems, like all ecosystems, are affected by abiotic factors.

Compare Create a table comparing the abiotic factors and organisms for each marine ecosystem.

TEMPERATURE

One abiotic factor in marine ecosystems is the temperature of the water. The water near the surface is much warmer that the rest of the ocean because it is heated by the sun. Deep ocean water is much colder.

Water temperatures at the surface are also affected by latitude. Water near the equator is generally warmer than water closer to the poles. The water at the surface is also warmer in summer than winter. ☑

1. Identify Where is the warmest surface ocean water?

____

Temperature affects the animals in marine ecosystems. For example, fish that live near the poles have a chemical in their blood that keeps them from freezing. Most animals that live in coral reefs need warm water to live.

Ocean Temperature and Depth

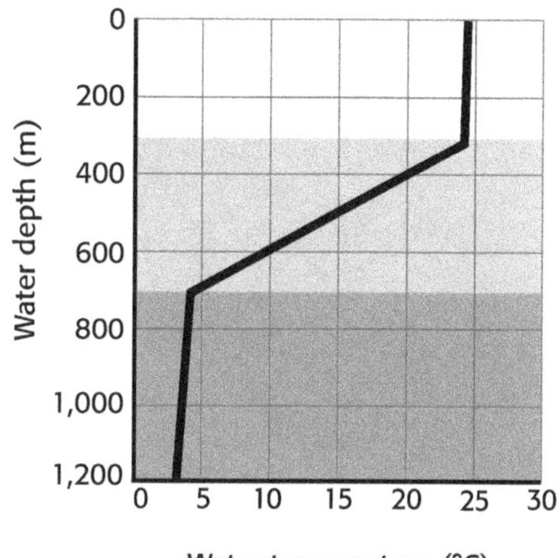

Math Focus

2. Read a Graph About how much colder is ocean water at 600 m depth than at 400 m depth?

____

WATER DEPTH AND SUNLIGHT

Two other abiotic factors that affect marine ecosystems are water depth and sunlight. The average depth of the oceans is 4,000 m, but sunlight does not reach deeper than 200 m. Producers that carry out photosynthesis, such as algae, can live only in water less than about 200 m deep. ☑

Plankton are tiny organisms that float near the surface of the ocean. Many kinds of plankton are producers. These *phytoplankton* use photosynthesis to make their own food. They are the base of most ocean food chains. Large consumers, such as whales, feed on these tiny producers.

READING CHECK

3. List What are the three main abiotic factors that affect marine ecosystems?

____

____

____

Marine ecosystems support many different organisms. Both large humpback whales and tiny phytoplankton live near the surface of the ocean.

TAKE A LOOK

4. Identify In the picture, which organism is the producer and which is the consumer?

____

____

____

What Are the Major Zones in the Ocean?

Scientists divide the ocean into zones. The divisions are based on things such as water depth, the amount of sunlight, and water temperature.

Say It

Share Experiences In a group, discuss the abiotic factors and the living organisms you have seen or might see at the beach.

THE INTERTIDAL ZONE

The intertidal zone is where the ocean meets the shore. The organisms of the intertidal zone are covered with water at high tide and exposed to air at low tide.

The Intertidal Zone Sea grasses, periwinkle snails, and herons are common in an intertidal mudflat. Sea stars and anemones often live on rocky shores, while clams, crabs, snails, and conchs are common on sandy beaches.

TAKE A LOOK

5. Explain Why is it difficult for many sea creatures to live in the intertidal zone?

____

____

THE NERITIC ZONE

The neritic zone is further from shore. In this zone, the water becomes deeper as the ocean floor starts to slope downward. This water is warmer than deep ocean water and receives a lot of sunlight. Corals and producers thrive in this zone. Sea turtles, sea urchins, and fishes are some of the consumers of this zone.

The Neritic Zone Although phytoplankton are the major producers in this zone, seaweeds are common, too. Sea turtles and dolphins live in the neritic zone. Other animals, such as corals, sponges, and colorful fishes, contribute to this vivid landscape.

TAKE A LOOK

6. Identify What are the two main kinds of producers in the neritic zone?

____

____

THE OCEANIC ZONE

In the oceanic zone, the sea floor drops off quickly. The oceanic zone extends from the surface to the deep water of the open ocean. Phytoplankton live near the surface, where there is sunlight. ☑

Consumers such as fishes, whales, and sharks live in the oceanic zone. Some of the animals live in deep waters, where there is no sunlight. These animals feed on each other and on material that sinks from the surface waters.

READING CHECK

7. Explain Why do phytoplankton need to live near the surface?

____

____

The Oceanic Zone Many unusual animals are adapted for the deep ocean. Whales and squids can be found in this zone. Also, fishes that glow can be found in very deep, dark water.

TAKE A LOOK

8. Explain How can the consumers that live in deep waters survive if there are no producers present?

____

____

____

THE BENTHIC ZONE

The benthic zone is the ocean floor. It does not get any sunlight and is very cold. Fishes, worms, and crabs have special features to live in this zone. Many of them feed on material that sinks from above.

Some organisms, such as angler fish, eat smaller fish. Other organisms, such as bacteria, are decomposers and help break down dead organisms.

TAKE A LOOK

9. Describe What abiotic factors do organisms that live in the benthic zone need to adapt to?

____

____

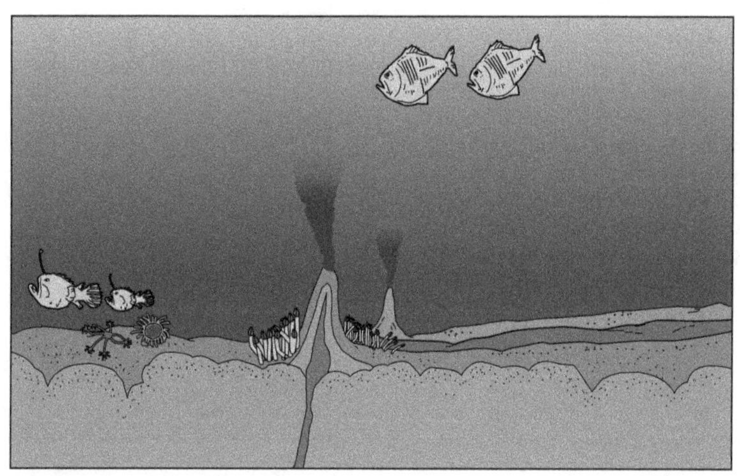

The Benthic Zone Organisms such as bacteria, worms, and sea urchins thrive on the sea floor.

What Are Some Marine Ecosystems?

Life on Earth depends on the ocean. The water that evaporates from the ocean becomes most of the rain and snow that falls on land. The ocean affects world climates and wind patterns. People depend on the ocean for food.

Many different kinds of organisms live in the ocean. They live in the many ecosystems in the different zones of the ocean.

Critical Thinking

10. Predict Consequences How would humans be affected if there were no oceans?

____

____

____

____

THE SARGASSO SEA

Floating mats of algae in the middle of the Atlantic Ocean make up the base of the Sargasso Sea ecosystem. Many animals live in this ecosystem. Most of them are omnivores that can eat many different organisms.

POLAR ICE

The icy waters near the poles are rich in nutrients that support large numbers of phytoplankton. These producers can support many types of consumers. One of these is a small shrimplike organism called krill. Larger consumers, such as fish, eat krill. These consumers, in turn, serve as food for other consumers, such as seals.

INTERTIDAL ECOSYSTEMS

Organisms in intertidal ecosystems must be able to live both underwater and in the air. Those that live in mudflats and beaches may dig into the ground during low tide.

On rocky shores, organisms have adaptations to keep from being swept away by crashing waves. For example, seaweeds use structures called *holdfasts* to attach themselves to rocks. Other organisms, such as barnacles, attach themselves to rocks with a special glue. Sea stars feed on these organisms. ☑

READING CHECK

11. Describe How do organisms in intertidal ecosystems protect themselves from being washed away by waves?

____

____

____

ESTUARIES

An **estuary** is an area where fresh water flows into the ocean. The water in an estuary is a mix of fresh water from rivers and salt water from the ocean. Organisms that live in estuaries must be able to survive the changing amounts of salt in the water.

The fresh water that flows into an estuary is rich in nutrients washed from the land. The nutrients in the water support large numbers of producers, such as algae. The algae support many consumers, such as fish and shellfish.

CORAL REEFS

Coral reefs are named for the small animals called *corals* that form the reefs. Many of these tiny animals live together in a colony, or group. When the corals die, their hard skeletons remain. New corals grow on the remains.

Over time, layers of skeletons build up and form a rock-like structure called a *reef*. The reef is a home for many marine animals. These organisms include fishes, sponges, sea stars, and sea urchins. Because so many kinds of organisms live there, coral reefs are some of the most diverse ecosystems on Earth. ☑

READING CHECK

12. Explain How is a coral reef both a living and a nonliving structure?

____

____

____

____

____

Coral reefs are very diverse marine ecosystems.

Name ______________________ Class ______________ Date ______________

Section 2 Review

NSES LS 1a, 3d, 4a, 4b, 4c, 4d

SECTION VOCABULARY

estuary an area where fresh water mixes with salt water from the ocean	**plankton** the mass of mostly microscopic organisms that float or drift freely in freshwater and marine environments

1. **Describe** What unique abiotic factor do organisms in an estuary have to adapt to? What causes this abiotic factor?

____

____

____

2. **Describe** What are some of the different kinds of producers found in marine ecosystems?

____

____

____

3. **Apply Concepts** Complete this food chain that shows the flow of energy through a polar ice ecosystem.

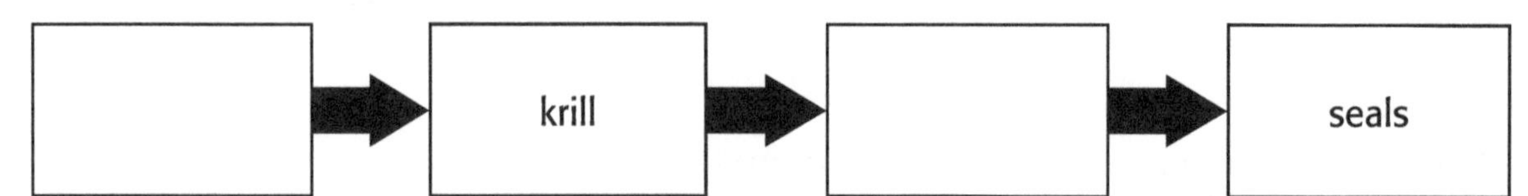

4. **Explain** Why are there few producers below 200 m in the ocean?

____

____

____

____

5. **Identify** What abiotic factors make the neritic zone a good home for many different organisms?

____

____

____

____

Name ______________________ Class ______________ Date ______________

CHAPTER 20 The Earth's Ecosystems

SECTION 3

Freshwater Ecosystems

BEFORE YOU READ

After you read this section, you should be able to answer these questions:

- What organisms live in stream and river ecosystems?
- What are the three zones in a pond or lake?
- What are two kinds of wetlands?

National Science Education Standards
LS 1a, 3d, 4a, 4b, 4c, 4d

What Are Stream and River Ecosystems?

One important abiotic factor that affects freshwater ecosystems is how quickly the water is moving. In rivers and streams, the water is moving faster than in other freshwater ecosystems.

The water in streams may come from melted ice or snow. It may also come from a spring. A *spring* is a place where water from under the ground flows to the surface.

Each stream of water that joins a larger stream is called a *tributary*. As more tributaries join a stream, it becomes stronger and wider. A very strong, wide stream is called a *river*.

Answer Questions Before reading this section, write the three Before You Read questions on a piece of paper. As you read, write down the answers to the questions.

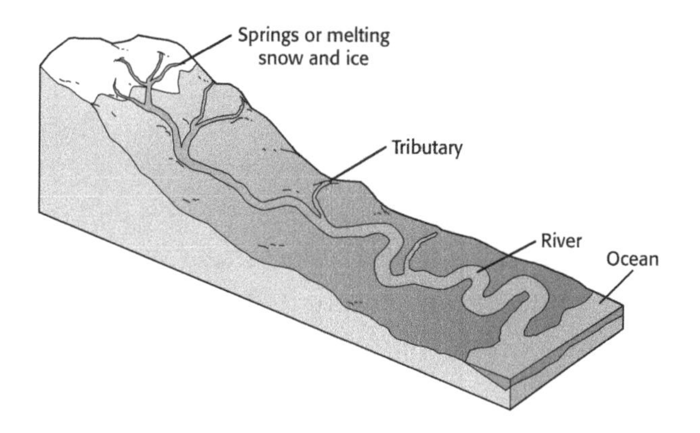

TAKE A LOOK

1. Describe What happens to the size of a stream when a tributary flows into it?

____

Stream and river ecosystems are full of life. Plants live along the edges of streams and rivers. Fish live in the open waters. Clams and snails live in the mud at the bottom.

Organisms that live in fast-moving water have to keep themselves from being washed away. Some producers, such as algae and moss, are attached to rocks. Consumers, such as tadpoles, use suction to hold themselves to rocks. Other consumers, such as crayfish, hide under rocks.

What Are Pond and Lake Ecosystems?

The water in ponds and lakes is not moving very much compared with rivers and streams. As a result, they have different types of ecosystems. Like marine ecosystems, pond and lake ecosystems are affected by water depth, sunlight, and temperature.

LIFE NEAR THE SHORE

The area of water near the edge of a pond or lake is called the **littoral zone**. Sunlight reaches the bottom, which allows producers such as algae to grow in this zone. Plants, such as cattails and rushes, grow here too, farther from shore.

Many consumers, such as tadpoles and some insects, eat the algae and plants. Some consumers, such as snails and insects, make their homes in plants. Consumers that live in the mud include clams and worms. Other consumers, such as fishes, also live in this zone. ☑

2. Identify Which consumers of the littoral zone are herbivores?

____

LIFE AWAY FROM THE SHORE

The area of a lake or pond away from the littoral zone near the surface is called the **open-water zone**. This zone is as deep as sunlight can reach. Producers such as phytoplankton grow well here. This zone is home to bass, lake trout, and other consumers.

Beneath the open-water zone is the **deep-water zone**, where no sunlight reaches. Photosynthetic organisms cannot live in this zone. Scavengers, such as catfish and crabs, live here and feed on dead organisms that sink from above. Decomposers, such as fungi and bacteria, also help to break down dead organisms. ☑

3. Explain Why can't producers live in the deep-water zone?

____

____

TAKE A LOOK

4. Identify Fill in the blank spaces in the figure with the correct words.

What Is a Wetland?

A **wetland** is an area of land that is sometimes under water or whose soil contains a lot of water. Wetlands help control floods. During heavy rains, wetlands soak up large amounts of water. This water sinks into the ground and helps refill underground water supplies. ☑

Wetlands contain many different plants and animals. There are two main types of wetlands: marshes and swamps.

READING CHECK

5. Define What is a wetland?

____

A **marsh** is a treeless wetland. Marshes form along the shores of lakes, ponds, rivers, and streams.

Grasses and other small plants are the main producers in marsh ecosystems.

Consumers such as turtles, frogs, and birds live in marshes.

A **swamp** is a wetland in which trees and vines grow. Swamps form in low-lying areas and near slow-moving rivers.

Trees and vines are important producers in swamp ecosystems.

Birds, fishes, and snakes are consumers that live in swamps.

TAKE A LOOK

6. Compare What is a quick way to tell the difference between a marsh and a swamp?

____

How Can an Ecosystem Change?

Did you know that a pond or lake can disappear? The water flowing into the lake carries sediment. The sediment, along with dead leaves and other materials, sinks to the bottom of the lake.

Bacteria decompose the material at the bottom of the lake. The decay process uses up some of the oxygen in the water. As the amount of oxygen in the water goes down, fewer fish and other organisms can live in it.

Over time, the pond or lake is filled with sediment. New kinds of plants grow in the new soil. Shallow places fill in first, so plants grow closer and closer to the center of the pond or lake. What is left of the pond or lake becomes a wetland. As the soils dry out and the oxygen levels increase, forest plants can grow. In this way, a pond or lake can become a forest.

Critical Thinking

7. Apply Concepts Why is the amount of oxygen in pond water an abiotic factor?

____

Name ______________________ Class ______________ Date ______________

Section 3 Review

NSES **LS 1a, 3d, 4a, 4b, 4c, 4d**

SECTION VOCABULARY

deep-water zone the zone of a lake or pond below the open-water zone, where no light reaches

littoral zone the shallow zone of a lake or pond where light reaches the bottom and nurtures plants

marsh a treeless wetland ecosystem where plants such as grasses grow

open-water zone the zone of a pond or lake that extends from the littoral zone and that is only as deep as light can reach

swamp a wetland ecosystem in which shrubs and trees grow

wetland an area of land that is periodically under water or whose soil contains a great deal of moisture

1. **Compare** Why are the kinds of producers in the littoral zone of a lake different from the producers in the open-water zone?

____

____

____

2. **Compare** How are the producers in a swamp different from those in a marsh?

____

____

3. **Identify** Give two examples of consumers in wetlands.

____

____

4. **Describe** What abiotic factors do organisms living in rivers and streams have to adapt to?

____

____

5. **Describe** Fill in the blank spaces in the flow chart below to show how a pond can become a forest.

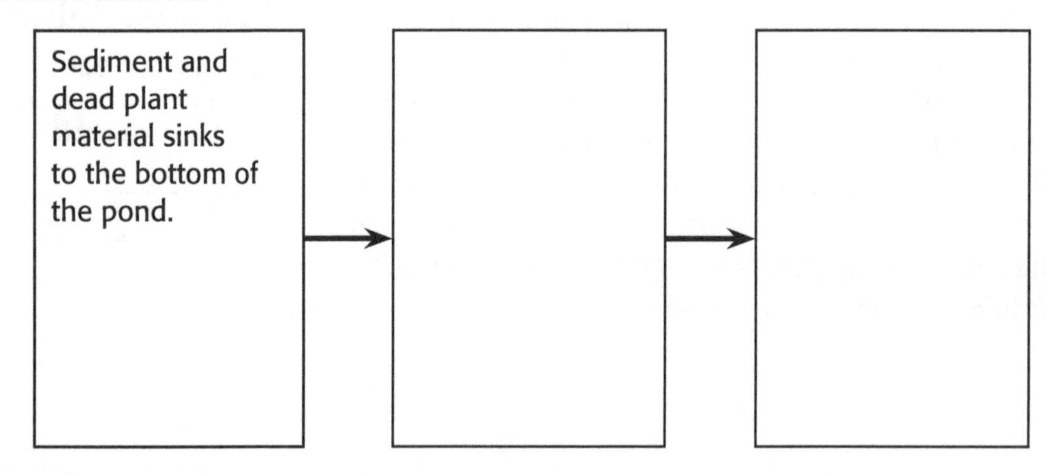

SECTION 1

Environmental Problems

BEFORE YOU READ

After you read this section, you should be able to answer these questions:

- What is pollution?
- What are some other environmental problems?

National Science Education Standards
LS 3a, 4d

What Is Pollution?

In the late 1700s, people started to depend on machines more and more. This is known as the *Industrial Revolution*. As people used more machines, they put larger amounts of harmful substances into the air, water, and soil. Machines today don't make as many harmful substances as machines many years ago. However, there are now more sources of pollution than there once were.

STUDY TIP
Underline As you read, underline any unfamiliar words. Find out what these words mean before you move on to the next section.

Pollution is an unwanted change in the environment caused by substances such as wastes, or energy, such as radiation. Anything that causes pollution is called a *pollutant*. Pollutants can harm living things. Natural events such as volcanic eruptions make pollutants. Humans make many other pollutants. ☑

READING CHECK
1. Explain Is all pollution caused by humans? Explain your answer.

____

____

____

GARBAGE

Americans throw away more trash than people of any other country. Most trash goes into landfills, like the one shown below. Some kinds of trash, such as medical waste and lead paint, are very dangerous. They are called *hazardous wastes*, and include things that can catch fire, eat through metal, explode, or make people sick. ☑

READING CHECK
2. Identify Give two examples of hazardous wastes.

____

____

Many industries, such as hospitals, oil refineries, paint manufacturers, power plants, and paper mills, produce hazardous wastes. People need to dispose of hazardous wastes in places set aside for them.

Every year, Americans throw away 200 million metric tons of garbage. Much of the garbage ends up in landfills.

TAKE A LOOK

3. Summarize As you read, complete this chart to describe different types of pollution.

Type of pollution	Examples or sources	Harmful effects
	nuclear power plants	
Greenhouse gases		

CHEMICALS

People use chemicals for many things. Some chemicals treat diseases. Others are used to make plastics and to preserve food. Sometimes, the same chemicals that help people may also harm the environment. For example, fertilizers and pesticides can make plants grow bigger and faster. However, they may also pollute the soil and water.

CFCs and PCBs are two examples of chemical pollutants. *CFC*s were once used in spray cans, refrigerators, and plastics. Scientists found that CFCs were destroying the ozone layer. The ozone layer protects Earth by absorbing harmful ultraviolet radiation from the sun. Even though CFCs were banned years ago, they are still found in the atmosphere.

*PCB*s were once used in appliances and paints. They are poisonous and may cause cancer. PCBs are now banned. However, they are still found all over the Earth.

Critical Thinking

4. Infer New refrigerators do not use CFCs. However, CFCs are still being released into the atmosphere. Explain why this is happening.

____

____

____

NUCLEAR WASTES

Nuclear power plants provide electricity. They also produce radioactive wastes. Radioactive wastes are hazardous. They give off radiation, which can cause cancer or radiation poisoning. They may take thousands of years to break down into less harmful materials.

NOISE

Some pollutants affect your senses. These include loud noises, such as airplanes taking off, and even loud music. Noise isn't just annoying. It can affect your ability to think clearly. It can also damage your hearing.

GREENHOUSE GASES

Earth's atmosphere is made up of a mixture of gases, including carbon dioxide. The atmosphere acts as a blanket. It keeps Earth warm enough for life to exist.

Since the Industrial Revolution, the amount of carbon dioxide in the atmosphere has increased. Carbon dioxide and many pollutants act like a greenhouse, trapping heat around the Earth. Many scientists think the increase in carbon dioxide has caused global temperatures to go up. If temperatures continue to rise, ice at Earth's poles could melt. This would cause the level of the world's oceans to rise and flood many areas of land along coasts.

Critical Thinking

5. Analyze Ideas Are greenhouse gases always harmful? Explain your answer.

____
____
____
____
____
____

What Is Resource Depletion?

Resources are depleted when they are used up without being replaced. Some of Earth's resources can be replaced as quickly as they are used. Others, however, can never be replaced.

RENEWABLE RESOURCES

A **renewable resource** is one that can be replaced as quickly as it is used. Solar and wind energy, as well as some kinds of trees, are renewable resources. Fresh water is generally a renewable resource because it is replaced every time it rains. However, some areas are using up water faster than it can replaced. This may cause water to become a nonrenewable resource.

NONRENEWABLE RESOURCES

A **nonrenewable resource** is one that cannot be replaced or can only be replaced over many thousands of years. Minerals and fossil fuels, such as oil, coal, and natural gas, are nonrenewable resources. Nonrenewable resources are depleted as they are used because they cannot be replaced. In addition, removing some resources from Earth may lead to oil spills, loss of habitats, and damage from mining. All are harmful to the environment.

Renewable resources	Nonrenewable resources

TAKE A LOOK

6. Identify Use the text to complete the chart with examples of both types of resources.

What Are Exotic Species?

People can carry plant seeds, animal eggs, and adult organisms from one part of the world to another. An animal or plant that is brought into a new environment is an *exotic species*. Sometimes an exotic species cannot survive outside of its natural environment. However, sometimes an exotic species does well in its new home.

Exotic species can cause problems. One reason for this is the organism does not have its natural predators in the new environment. This can allow populations of exotic species to grow out of control. An exotic species may compete with native species for resources. They may also kill the native species. ☑

READING CHECK

7. Identify What are two effects exotic species can have on native species?

____

____

____

____

Northern snakehead fish are an exotic species in the United States. These fish can move across land as they look for water. They can survive out of water for up to four days!

STANDARDS CHECK

LS 4d The number of organisms an ecosystem can support depends on the resources available and abiotic factors, such as the quantity of light and water, range of temperatures, and soil composition. Given adequate biotic and abiotic resources and no disease or predators, populations (including humans) increase at rapid rates. Lack of resources and other factors, such as predation and climate, limit the growth of populations in specific niches in the ecosystem.

8. Predict What will happen to resources as the human population continues to grow?

____

____

____

____

____

Why Is Human Population Growth a Problem?

Advances in medicine, such as immunizations, have helped people live longer. Advances in farming have let farmers grow food to feed more people. Because of this, the number of people on Earth has grown very quickly over the past few hundred years.

Eventually there could be too many people on Earth. **Overpopulation** happens when a population gets so large that individuals cannot get the resources they need. For example, one day there may not be enough food or water on Earth to support the growing human population.

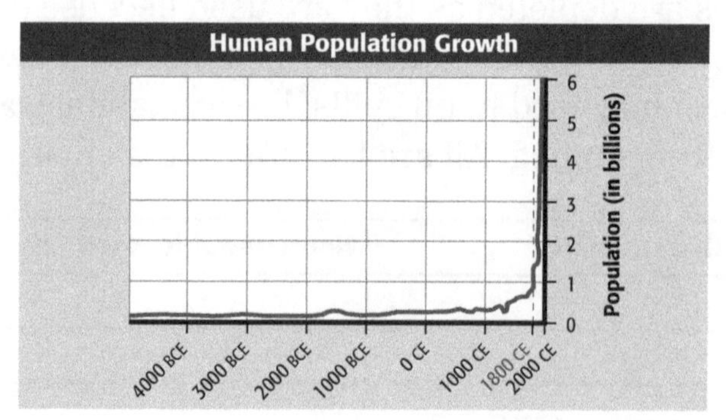

What Is Habitat Destruction?

An organism's *habitat* is where it lives. Every habitat has its own variety of organisms. This is known as **biodiversity**. If a habitat is damaged or destroyed, biodiversity is lost.

Habitats can be destroyed in different ways. Topsoil may erode, or wash away, when people clear land for crops or buildings. Also, chemicals may pollute streams and rivers. Organisms living in these areas may be left without food and shelter, and may die.

FOREST HABITATS

Trees provide humans with oxygen, wood, food, rubber, and paper. For many of these products trees must be cut down. Sometimes all the trees in a forest are cut down. This is called *deforestation*. People can plant new trees to replace ones that they cut. However, trees take many years to grow. ☑

READING CHECK

9. Define What is deforestation?

____

____

____

Tropical rain forests have some of the highest biodiversity on Earth. However, people clear many acres of rainforest for farm land, roads, and lumber. After a forest is cleared, the biodiversity of the area is lost.

Deforestation can lead to soil erosion. This means that soil washes away.

MARINE HABITATS

When people think of pollution in marine habitats, many think of oil spills. An oil spill is an example of *point source pollution*, or pollution that comes from one source.

Unlike oil spills, some pollution comes from many different sources. This is called *nonpoint-source pollution*. For example, chemicals on land wash into rivers, lakes, and oceans. These chemicals can harm or kill organisms that live in marine habitats.

Critical Thinking

10. Apply Concepts Dumping plastics into the oceans is another form of pollution. It can kill many marine animals. Is dumping plastics in oceans an example of point-source or nonpoint-source pollution? Explain your answer.

____

____

____

Name ______________________ Class ______________ Date ______________

Section 1 Review

NSES LS 3a, 4d

SECTION VOCABULARY

biodiversity the number and variety of organisms in a given area during a specific period of time

nonrenewable resource a resource that forms at a rate that is much slower than the rate at which the resource is consumed

overpopulation the presence of too many individuals in an area for the available resources

pollution an unwanted change in the environment caused by substances or forms of energy

renewable resource a natural resource that can be replaced at the same rate at which the resource is consumed

1. Apply Concepts Use the vocabulary terms above to complete the Concept Map.

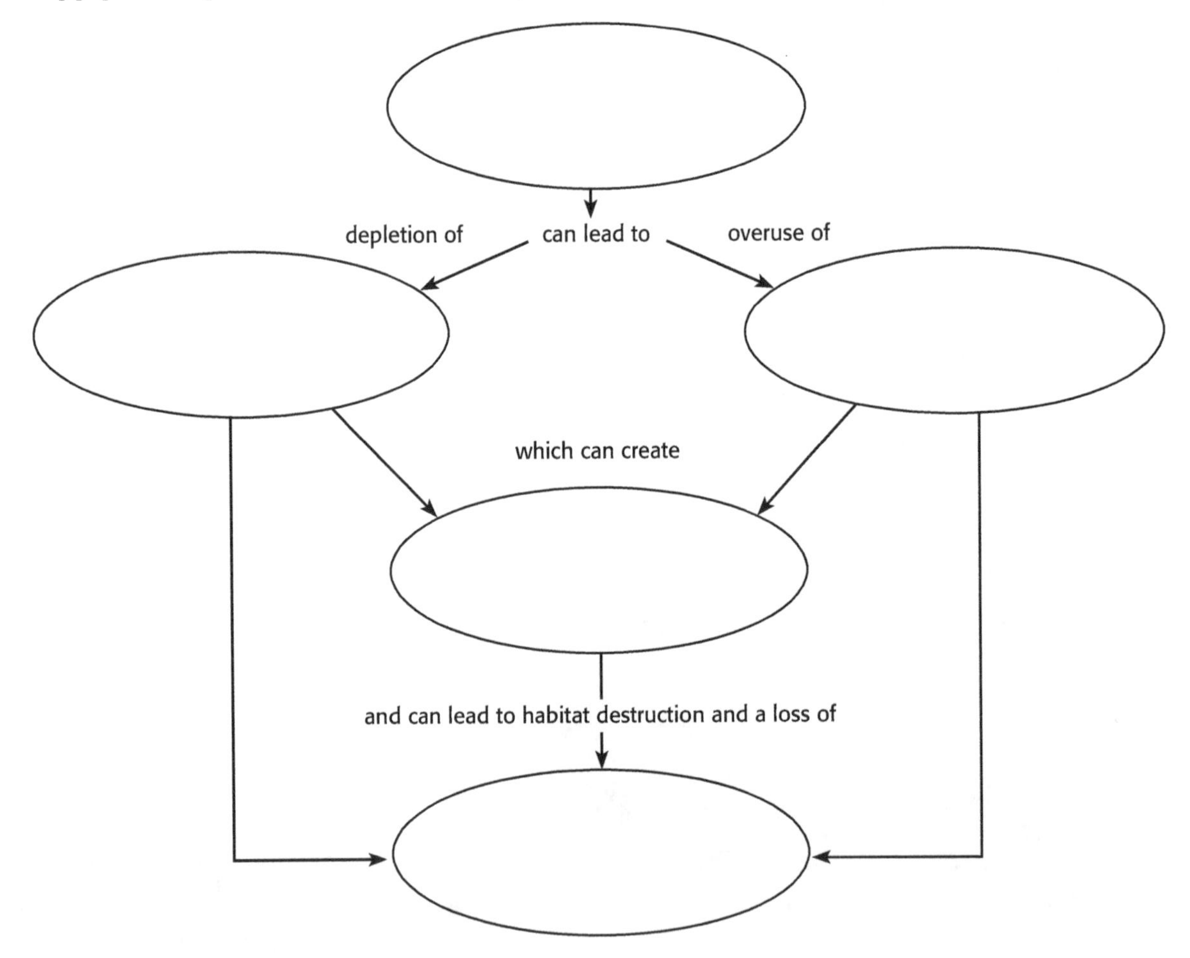

2. Define What is resource depletion?

____

3. Analyze Ideas Is it possible for a renewable resource to become nonrenewable? Explain your answer and give an example.

____

____

____

Name ______________________ Class ______________ Date ____________

CHAPTER 21 Environmental Problems and Solutions

SECTION 2

Environmental Solutions

BEFORE YOU READ

After you read this section, you should be able to answer these questions:

- What is conservation?
- What are some things people can do to protect the environment?

National Science Education Standards

LS 5c

How Can People Conserve Resources?

As the human population grows, it will need more resources. People will need food, health care, transportation, and ways to get rid of wastes. All of these needs will affect the Earth. If people don't use resources wisely, people will continue to pollute the air, soil, and water. More natural habitats could be lost and many species could die out. However, there are many things people can do to protect the environment.

Conservation is the protection and wise use of natural resources. When you conserve resources, you use fewer of them. Conservation not only keeps us from using up resources, but also reduces waste, pollution, and habitat destruction. For example, using a bike instead of a car can save fuel and help prevent air pollution.

There are three ways to conserve resources: **R**educe, **R**euse, and **R**ecycle. These are known as the three Rs. ☑

Organize As you read, make a Spider Map to show different ways you can help protect the environment.

READING CHECK

1. List What are the three Rs of conservation?

____

____

____

Reduce

Reuse

Recycle

TAKE A LOOK

2. Infer How is the girl with the bag reducing her use of resources?

____

____

____

____

READING CHECK

3. Identify What is the best way to conserve resources?

____

REDUCE

The best way to conserve resources is to use fewer of them. Using fewer resources also helps reduce pollution and the amount of waste we produce. ☑

Almost one third of the waste produced by some countries is packaging material, such as boxes, plastic bags, and paper. Wrapping products in less paper and plastic reduces waste. For example, fast-food restaurants used to serve sandwiches in large plastic containers. Today, many wrap food in thin paper instead. Paper is biodegradable. *Biodegradable* means that living organisms, such as bacteria, can break it down.

In addition to reducing waste, people need to reduce the amount of nonrenewable resources they use. For example, scientists are looking for sources of energy that can replace fossil fuels, such as oil and gas. One way to use fewer nonrenewable resources is to use renewable resources, such as solar and wind energy.

TAKE A LOOK

4. Identify Are the people living in this house using a renewable or nonrenewable resource for energy?

____

The people who live in this house use solar panels to get energy from the sun.

Scientists are also studying energy sources such as wind, tides, and waterfalls. Car companies have developed automobiles that use electricity and hydrogen for fuel. Driving these cars uses less gas and oil, and produces less pollution.

REUSE

Every time you reuse a plastic bag, one bag fewer needs to be made. This also means that one bag fewer will end up in a landfill. Reusing products is an important way to conserve resources.

You can reuse many things. For example, clothes can be passed down and worn by someone else. Sports equipment can be repaired instead of being thrown away. Builders can reuse materials such as wood, bricks, and tiles to make new structures. Even old tires can be used again. They can be reused for playground surfaces or even in new homes.

Critical Thinking

5. Identify Relationships How are reducing and reusing related?

____

____

____

____

____

This home is made of reused tires and aluminum cans

American homes use about 100 billion liters of water each day. When you wash your hands or take a shower, most of the water goes down the drain. This is called *wastewater*. No new water is ever made on Earth. Wastewater eventually returns to Earth as rain in the water cycle. However, this takes a very long time. Because of this, many communities are learning how to reuse, or *reclaim*, wastewater. ☑

Reclaiming wastewater allows people to use the same water for more than one thing. One way to do this is to clean the water so that it can be used again. Some organisms can be used to filter water and take out many of the wastes. These organisms include plants and filter-feeding animals, such as clams. The reclaimed water may not be pure enough to drink. However, it can be used to water crops, lawns, and golf courses.

READING CHECK

6. Explain What does reclaiming wastewater mean?

____

____

RECYCLE

Recycling is the recovery of materials from waste. This means that new products can be made from old products. A material that can be recycled is *recyclable*. Some recyclable items, such as paper, are used to make the same kind of product. Other recyclable items are made into different products. For example, yard wastes, such as dead leaves and cut grass, can be recycled into a natural fertilizer.

Say It

Investigate Find out where in your neighborhood you can take materials to be recycled. Does your city or town pick up recyclables? What sort of things can you recycle in your community? Share with the class what you learn.

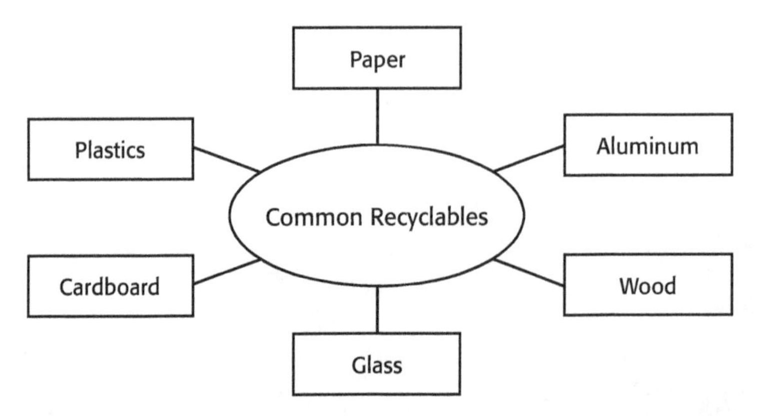

Recycling newspapers could save millions of trees. Recycling aluminum uses 95% less energy than turning raw ore into aluminum. Glass can be recycled many times to make new bottles and jars. This uses less energy than making new glass.

Burning some wastes can produce electricity. Using garbage to make electricity is an example of *resource recovery*. When companies make electricity from their waste products, they save money and conserve resources. However, some people are concerned that burning wastes for electricity pollutes the air. ☑

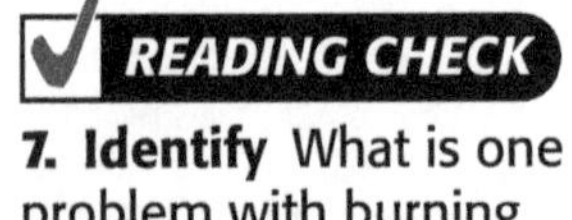

7. Identify What is one problem with burning garbage to make electricity?

____

____

A waste-to-energy plant can provide electricity for many homes and businesses.

How Can People Help Maintain Biodiversity?

Biodiversity is important because each species has a unique role in an ecosystem. Losing one species could affect the entire ecosystem. Biodiversity also protects against the spread of disease. For example, if a disease spreads in a forest that has only one species, the entire forest may be destroyed. If there are many different species, a disease is not likely to affect them all. ☑

READING CHECK

8. Identify Give two reasons that maintaining biodiversity is important.

____

____

____

____

____

____

PROTECTING SPECIES

One way to maintain biodiversity is to protect individual species. In the United States, a law called the *Endangered Species Act* was designed to do just that. An endangered species is one that generally has few individuals left. Without protection, an endangered species may become extinct. Scientists identify endangered species and put that species on a special list. The law forbids activities that would harm any species on this list.

The law also requires recovery programs for endangered species. This means that scientists work both to protect existing individuals of an endangered species, and to increase the population. To help some species recover, scientists remove individuals from the wild to breed them. This is known as *captive breeding*. It can help some species produce more offspring than they would in the wild. ☑

READING CHECK

9. Explain What is the purpose of a species recovery program?

____

____

____

Individuals that are bred in captivity are often released back into the wild. However, this is not a solution for all endangered species. Many species are endangered because their habitats have been destroyed. They may have no habitat left to return to. These species can only survive if their habitat is protected as well.

Captive-breeding programs have helped the population of California condors to grow.

PROTECTING HABITATS

Scientists want to keep species from becoming endangered. They don't want to wait until a species is almost extinct before they try to protect it.

Plants, animals, and microorganisms depend on each other. Each is part of a huge, interconnected web of organisms. To protect the web, whole habitats need to be protected, not just individual species.

Describe Have you ever visited a nature reserve or national park? Tell the class about your visit. If you have never been to a reserve or national park, tell the class about the kind of park you would like to visit.

Nature preserves are large areas of public land set aside for wildlife. They can protect important habitats.

ENVIRONMENTAL STRATEGIES

Laws have been passed to help protect resources and preserve habitats. By following these laws, people can help the environment. People can also use the following environmental strategies.

- **Reduce pollution.** Recycle as much as possible and buy recycled products. Do not dump wastes in forests, lakes, rivers, or oceans. Participate in a local cleanup project.
- **Reduce pesticide use.** Use pesticides that work only on harmful insects. Use natural pesticides whenever possible.
- **Protect habitats.** Preserve entire habitats. Use fewer resources, such as wood and paper, that can destroy habitats.
- **Learn about local issues.** Research how laws or projects will affect your area, and let people know about your concerns.
- **Develop alternative energy sources.** Use fewer nonrenewable resources, such as coal and oil. Use more renewable resources, such as solar and wind power.

Critical Thinking

10. Infer How could recycling help protect habitats?

____

____

____

____

____

____

____

____

EPA

The United States Environmental Protection Agency (EPA) is a government organization. Its job is to help protect the environment. The EPA keeps people informed about environmental problems and helps enforce environmental laws.

WHAT YOU CAN DO

Reduce, reuse, and recycle. Protect the Earth. These are jobs for everyone. Children and adults can help clean up the environment. As they improve their environment, they improve their quality of life.

How You Can Help the Environment

1. Volunteer at a local preserve or nature center, and help other people learn about conservation.
2. Give away your old toys.
3. Use recycled paper.
4. Fill up both sides of a sheet of paper.
5. Start an environmental awareness club at your school or in your neighborhood.
6. Recycle glass, plastics, paper, aluminum, and batteries.
7. Don't buy any products made from an endangered plant or animal.
8. Turn off electrical devices when you are not using them.
9. Wear hand-me-downs.
10. Share books with friends, or use the library.
11. Walk, ride a bicycle, or use public transportation.
12. Carry a reusable cloth shopping bag to the store.
13. Use a lunch box, or reuse your paper lunch bags.
14. Turn off the water while you brush your teeth.
15. Buy products made from biodegradable and recycled materials.
16. Use cloth napkins and kitchen towels.
17. Buy things in packages that can be recycled.
18. Use rechargeable batteries.
19. Make a compost heap.

TAKE A LOOK

11. Apply Concepts Next to each item in the figure, write whether the activity is a way to reduce, reuse, or recycle. Some items might have more than one answer.

12. Identify Circle five things in this figure that you could start to do today to help the environment.

Name ______________________ Class ______________ Date ______________

Section 2 Review

NSES LS 5c

SECTION VOCABULARY

conservation the preservation and wise use of natural resources	**recycling** the process of recovering valuable or useful materials from waste or scrap

1. **Define** What does biodegradable mean?

____

____

2. **Describe** Decribe one way that water can be reclaimed.

____

____

____

3. **Identify** What are two roles of the EPA?

____

____

4. **Explain** How would driving a car that uses hydrogen for fuel be helpful to the environment?

____

____

____

5. **List** Name six materials that can be recycled in many communities.

____

____

6. **Explain** What is the purpose of the Endangered Species Act?

____

____

7. **Apply Concepts** How is using cloth napkins an example of both reusing and reducing?

____

____

____

Name ______________________ Class ______________ Date ______________

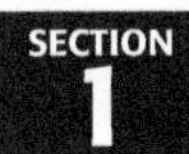

CHAPTER 22 Body Organization and Structure

SECTION 1

Body Organization

BEFORE YOU READ

After you read this section, you should be able to answer these questions:

- What is homeostasis?
- How is the human body organized?
- What are the 11 different human organ systems?

National Science Education Standards
LS 1a, 1d, 3a

How Is the Body Organized?

The different parts of your body all work together to maintain, or keep, the conditions in your body stable. Your body works to keep itself stable even when things outside your body change. This is called **homeostasis**. For example, your body temperature needs to stay the same even when temperatures outside are very cold or very hot. If your body could not keep its inside conditions the same, many processes in your body would not work.

Conditions inside and outside your body are always changing. Your body can maintain homeostasis because each cell does not have to do everything your body needs. Instead, your body is organized into different levels. The parts at each level work together to help your body maintain homeostasis.

There are four levels of organization in the body: cells, tissues, organs, and organ systems. Cells are the smallest level of organization. A group of similar cells working together forms a **tissue**. Your body has four main kinds of tissue: epithelial, nervous, muscle, and connective.

STUDY TIP

Discuss Read this section silently. When you finish reading, work with a partner to answer any questions you may have about the section.

STANDARDS CHECK

LS 3a All organisms must be able to obtain and use resources, grow, reproduce, and maintain stable internal conditions while living in a constantly changing external environment.

Word Help: resource
anything that can be used to take care of a need

Word Help: environment
the surrounding natural conditions that affect an organism

1. Define What is homeostasis?

____

____

Four Kinds of Tissue

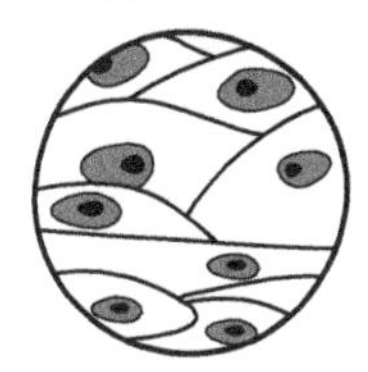

Epithelial tissue covers and protects other tissues.

Nervous tissue sends electrical signals through the body.

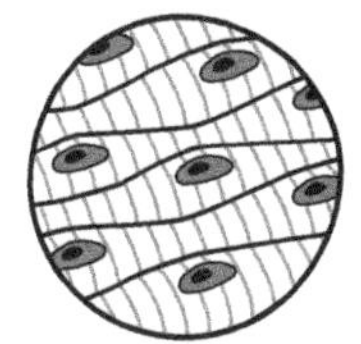

Muscle tissue is made of cells that contract and relax to produce movement.

Connective tissue joins, supports, protects, insulates, nourishes, and cushions organs. It also keeps organs from falling apart.

ORGANS

When different kinds of tissues work together, they can do more than any one tissue can do alone. A group of two or more tissues working together to do a job is an **organ**. For example, your stomach is an organ that helps you digest your food. None of the stomach's tissues could digest food alone. ☑

READING CHECK

2. Define What is an organ?

____

____

____

Four Kinds of Tissue in the Stomach

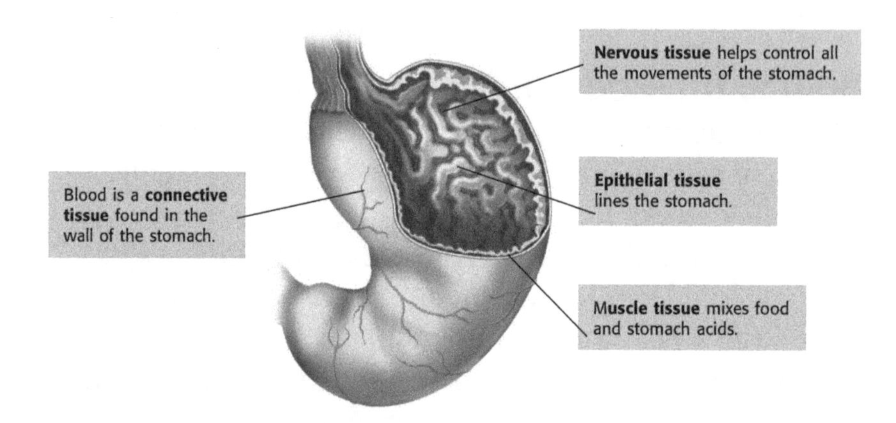

ORGAN SYSTEMS

Organs that work together to do a job make up an *organ system*. For example, your stomach works with other organs in the digestive system, such as the intestines, to digest food. Organ systems can do jobs that one organ alone cannot do. Each organ system has a special function.

There are 11 different organ systems that make up the human body. No organ system works alone. For example, the respiratory system and cardiovascular system work together to move oxygen through your body.

Critical Thinking

3. Apply Concepts How does the stomach work as part of an organ system?

____

____

____

____

____

Integumentary System Your skin, hair, and nails protect the tissue that lies beneath them.

Muscular System Your muscular system works with the skeletal system to help you move.

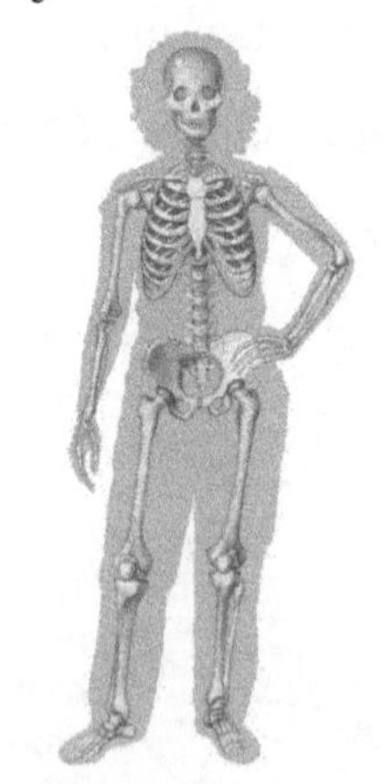

Skeletal System Your bones provide a frame to support and protect your body parts.

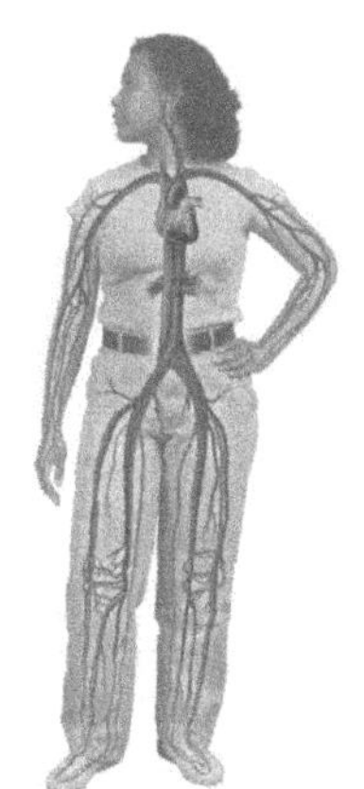

Cardiovascular System Your heart pumps blood through all of your blood vessels.

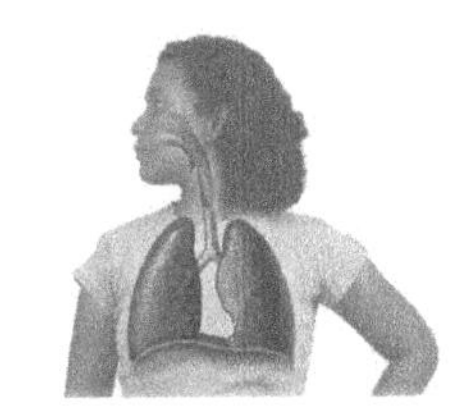

Respiratory System Your lungs absorb oxygen and release carbon dioxide.

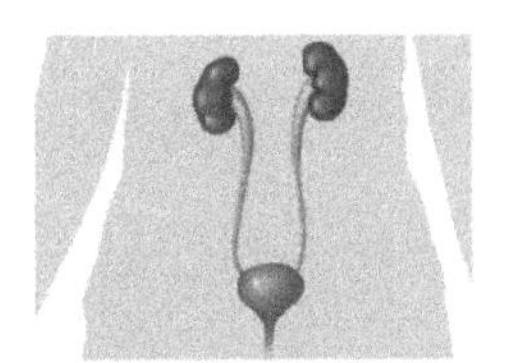

Urinary System Your urinary system removes wastes from the blood and regulates your body's fluids.

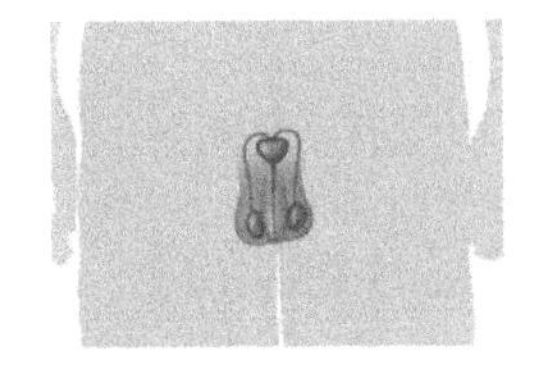

Male Reproductive System The male reproductive system produces and delivers sperm.

Female Reproductive System The female reproductive system produces eggs and nourishes and protects the fetus.

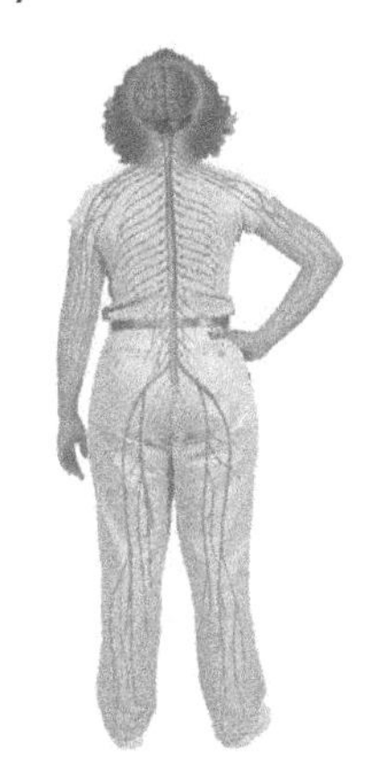

Nervous System Your nervous system receives and sends electrical messages throughout your body.

TAKE A LOOK

4. Identify Which organ system includes your lungs?

____

5. Identify Which organ system is different in males and females?

____

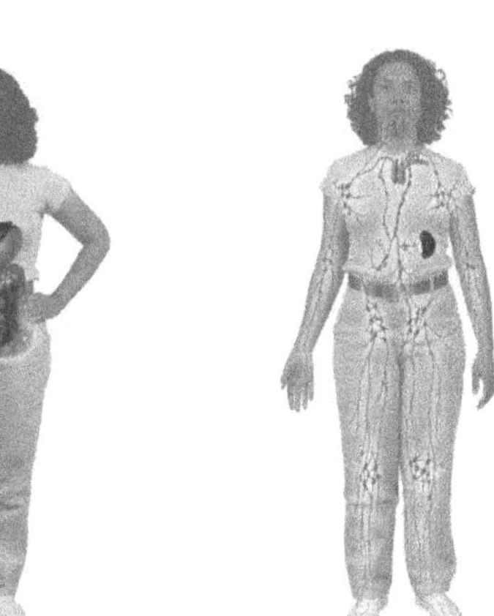

Digestive System Your digestive system breaks down the food you eat into nutrients that your body can absorb.

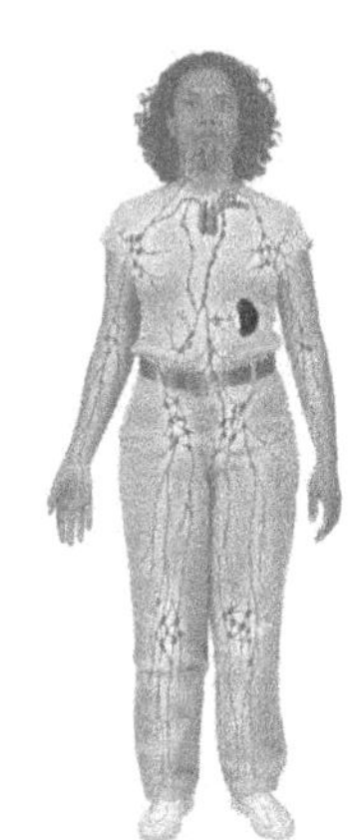

Lymphatic System The lymphatic system returns leaked fluids to blood vessels and helps get rid of bacteria and viruses.

Endocrine System Your glands send out chemical messages. Ovaries and testes are part of this system.

Say It

Discuss With a partner, see how many organs you can name from each organ system.

Name ______________________ Class ______________ Date ______________

Section 1 Review

NSES LS 1a, 1d, 3a

SECTION VOCABULARY

homeostasis the maintenance of a constant internal state in a changing environment **organ** a collection of tissues that carry out a specialized function of the body	**tissue** a group of similar cells that perform a common function

1. **Compare** How is an organ different from a tissue?

____

____

____

2. **List** Name five organ systems in the human body.

____

____

3. **Explain** Why is it important for your body to maintain homeostasis?

____

____

____

4. **Infer** What organ systems must work together to help a person eat and digest a piece of pizza? Give at least three systems.

____

____

5. **Infer** What organ systems must work together to help a person play a soccer game? Give at least four systems.

____

____

6. **Apply Concepts** Can an organ do the same job as an organ system? Explain your answer.

____

____

____

7. **Identify Relationships** How is the lymphatic system related to the cardiovascular system?

____

____

SECTION 2 The Skeletal System

BEFORE YOU READ

After you read this section, you should be able to answer these questions:

- What are the major organs of the skeletal system?
- What are the functions of the skeletal system?
- What are the three kinds of joints in the body?

National Science Education Standards

LS 1d, 1e

What Are Bones?

Many people think that bones are dry and brittle, but your bones are actually living organs. Bones are the major organs of the skeletal system. The **skeletal system** is made up of bones, cartilage, and connective tissue.

STUDY TIP

Organize As you read this section, make a chart listing the functions of bones and the tissue or bone structure that does each job.

What Are the Functions of the Skeletal System?

An average adult human skeleton has 206 bones. Bones have many jobs. For example, they help support and protect your body. They work with your muscles so you can move. Bones also help your body maintain homeostasis by storing minerals and making blood cells. The skeletal system does the following jobs for your body:

- It protects other organs. For example, your rib cage protects your heart and lungs.
- It stores minerals that help your nerves and muscles work properly. Long bones store fat that can be used as energy.
- Skeletal muscles pull on bones to cause movement. Without bones, you would not be able to sit, stand, or run.
- Some bones make blood cells. *Marrow* is a special material that makes blood cells.

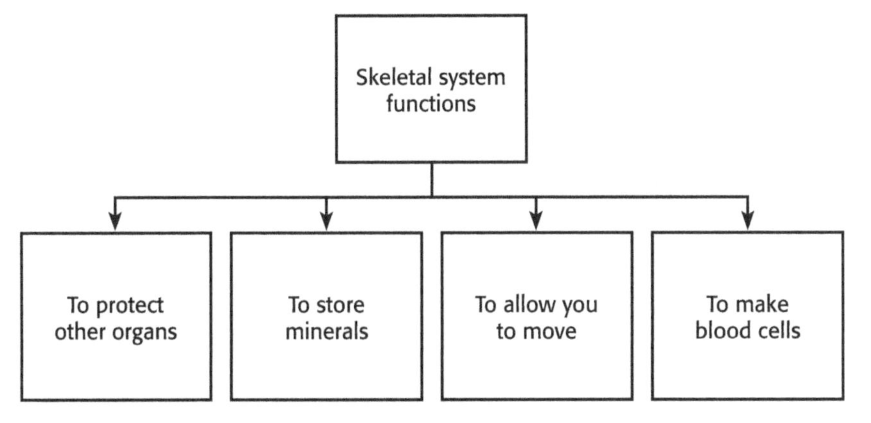

Critical Thinking

1. Predict Name one organ system, other than the skeletal system, that would be affected if you had no bones. Explain your answer.

____

____

____

____

What Is the Structure of a Bone?

A bone may seem lifeless. Like other organs, however, bone is a living organ made of several different tissues. Bone is made of connective tissue and minerals. Living cells in the bone deposit the minerals.

BONE TISSUE

If you look inside a bone, you will see two kinds of bone tissue: spongy bone and compact bone. Spongy bone has many large open spaces that help the bone absorb shocks. Compact bone has no large open spaces, but it does have tiny spaces filled with blood vessels. ☑

READING CHECK

2. Identify What are the two kinds of bone tissue?

____

____

MARROW

Some bones contain a tissue called marrow. There are two types of marrow. Red marrow makes red and white blood cells. Yellow marrow stores fat.

CARTILAGE

Did you know that most of your skeleton used to be soft and rubbery? Most bones start out as a flexible tissue called *cartilage*. When you were born, you didn't have much true bone. As you grow, your cartilage is replaced by bone. However, bone will never replace cartilage in a few small areas of your body. For example, the end of your nose and the tops of your ears will always be made of cartilage. ☑

READING CHECK

3. Define What is cartilage?

____

____

____

Bone Tissues

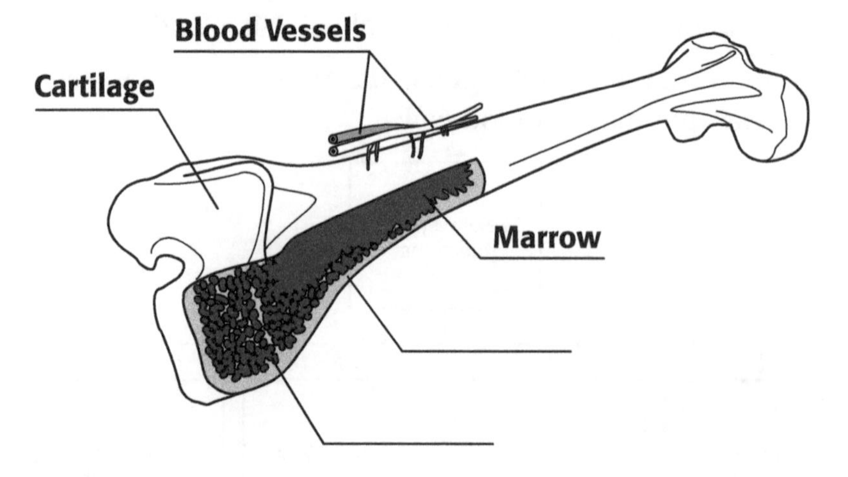

TAKE A LOOK

4. Label Fill in the missing labels for tissues that are found in this bone.

What Is a Joint?

A place where two or more bones meet is called a **joint**. Some joints, called *fixed joints*, do not let bones move very much. Many of the joints in the skull are fixed joints. However, most joints let your bones move when your muscles *contract*, or shorten. Joints can be grouped based on how the bones in the joint move. ☑

READING CHECK

5. Define What is a joint?

____

____

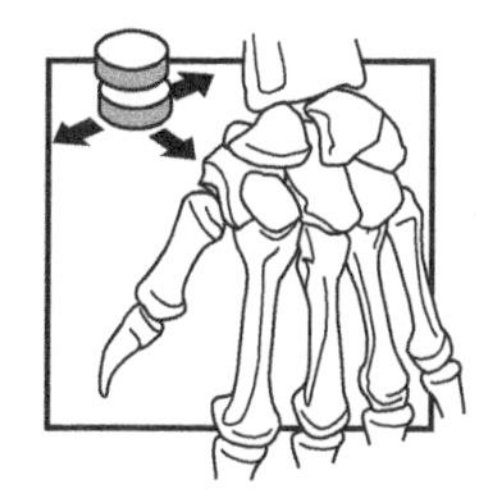

Gliding Joint Gliding joints let bones in the wrist slide over each other. This type of joint makes a body part flexible.

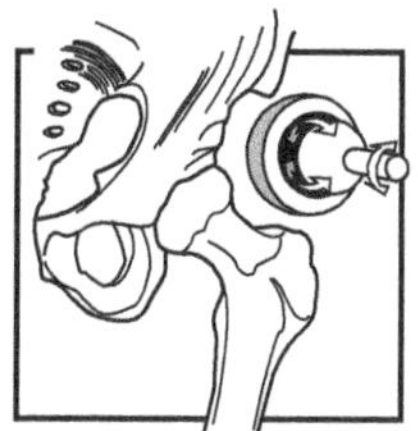

Ball-and-Socket Joint In the same way that a video-game joystick lets you move your character around, the shoulder lets your arm move freely in all directions.

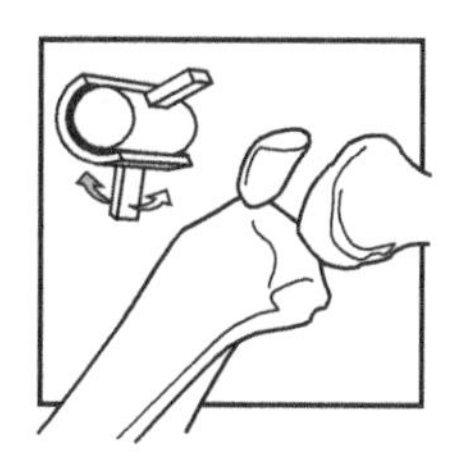

Hinge Joint A hinge lets a door open and close. Your knee joint lets your leg bend in only one direction.

TAKE A LOOK

6. List What are the three types of joints in the human body?

____

____

____

Joints can handle a lot of wear and tear because of how they are made. Joints are held together by ligaments. *Ligaments* are strong bands of connective tissue. Cartilage covers the ends of many bones and helps cushion the areas where bones meet. ☑

READING CHECK

7. Identify How does cartilage help protect bones and joints?

____

____

____

SKELETAL SYSTEM INJURIES AND DISEASES

Sometimes, parts of the skeletal system are injured. For example, bones may be broken. Joints and ligaments can also be injured. Many of these injuries happen when too much stress is placed on the skeletal system.

There are also some diseases that affect the skeletal system. For example, the disease *osteoporosis* makes bones brittle and easy to break. Some diseases make bones soft or affect bone marrow. *Arthritis* is a disease that makes joints stiff, so they are painful and hard to move.

Name ______________________ Class ______________ Date ______________

Section 2 Review

NSES LS 1d, 1e

SECTION VOCABULARY

joint a place where two or more bones meet	**skeletal system** the organ system whose primary function is to support and protect the body and to allow the body to move

1. List What are four functions of the skeletal system?

____

____

2. Identify What three things make up the skeletal system?

____

____

3. Describe Fill in the chart below to describe the three types of joints. Give an example of each.

Type of joint	Example
	wrist
hinge	

4. Compare What is the difference between red marrow and yellow marrow?

____

____

5. Explain What happens to the cartilage in your body as you grow up?

____

____

6. Identify What are two diseases that can affect the skeletal system?

____

7. Describe Describe a joint and its structure.

____

____

____

8. Explain What causes most injuries to the skeletal system?

____

Name ______________________ Class ______________ Date ____________

CHAPTER 22 Body Organization and Structure

SECTION 3

The Muscular System

BEFORE YOU READ

After you read this section, you should be able to answer these questions:

- What are the three kinds of muscle tissue?
- How do skeletal muscles work?
- How can exercise help keep you healthy?

National Science Education Standards
LS 1d, 1e

What Is the Muscular System?

The **muscular system** is made up of the muscles that let you move. There are three kinds of muscle in your body: smooth muscle, cardiac muscle, and skeletal muscle.

STUDY TIP

Circle As you read this section, circle any new science terms. Make sure you know what these words mean before moving to the next chapter.

Skeletal muscle makes bones move.

Smooth muscle moves food through the digestive system.

Cardiac muscle pumps blood around the body.

Muscle action can be voluntary or involuntary. Muscle action that you can control is *voluntary*. Muscle action that you cannot control is *involuntary*. For example, cardiac muscle movements in your heart are involuntary. They happen without you having to think about it. Skeletal muscles, such as those in your eyelids, can be both voluntary and involuntary. You can blink your eyes anytime you want, but your eyes also blink automatically.

Critical Thinking

1. Apply Concepts Your diaphragm is a muscle that helps you breathe. Do you think this muscle is voluntary or involuntary? Explain.

____
____
____
____
____

Kind of muscle	Where in your body is it found?	Are its actions voluntary or involuntary?
Cardiac	heart	involuntary
Smooth	digestive tract, blood vessels	involuntary
Skeletal	attached to bones and other organs	both

How Do Skeletal Muscles Work?

Skeletal muscles let you move. When you want to move, signals travel from your brain to your skeletal muscle cells. The muscle cells then contract, or get shorter. ☑

READING CHECK

2. Explain What causes skeletal muscle cells to contract?

____

____

HOW MUSCLES AND BONES WORK TOGETHER

Strands of tough connective tissue connect your skeletal muscles to your bones. These strands are called tendons. When a muscle that connects two bones contracts, the bones are pulled closer to each other. For example, tendons attach the biceps muscle to bones in your shoulder and forearm. When the biceps muscle contracts, your forearm bends toward your shoulder.

PAIRS OF MUSCLES

Your skeletal muscles often work in pairs to make smooth, controlled motions. Generally, one muscle in the pair bends part of the body. The other muscle straightens that part of the body. A muscle that bends part of your body is called a *flexor*. A muscle that straightens part of your body is an *extensor*. ☑

READING CHECK

3. Complete A muscle that bends part of your body is a

______________________.

In the figure below, the biceps muscle is the flexor. When the biceps muscle contracts, the arm bends. The triceps muscle is the extensor. When it contracts, the arm straightens out.

TAKE A LOOK

4. Identify On the figure, label the flexor muscle and the extensor muscle.

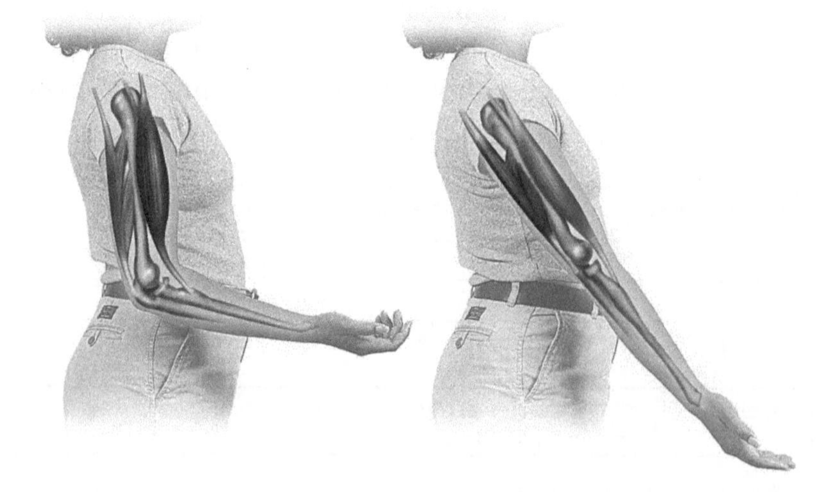

How Can You Keep Your Muscles Healthy?

Muscles get stronger when you exercise them. Strong muscles can help other organs to work better. For example, when your heart is strong, it can pump more blood to the rest of your organs. More blood brings more oxygen and nutrients to your organs.

Certain kinds of exercises can give muscles more strength and endurance. More endurance means that your muscles can work longer before they get tired.

Resistance exercise is a good way to make skeletal muscles stronger. During *resistance exercise*, the muscles work against the resistance, or weight, of an object. Some resistance exercises use weights. Others, such as sit-ups, use your own body weight as resistance.

Aerobic exercise can increase skeletal muscle strength and endurance. Aerobic exercise can also make your heart muscles stronger. During *aerobic exercise*, the muscles work steadily for a fairly long period of time. Jogging, skating, swimming, and walking are all aerobic exercises.

Type of exercise	Description	Example
Resistance		weight-lifting, sit-ups
	Muscles work steadily for a long time.	

MUSCLE INJURY

Most muscle injuries happen when people try to do too much exercise too quickly. For example, a *strain* is an injury in which a muscle or tendon is overstretched or torn. To avoid muscle injuries, you should start exercising slowly. Don't try to do too much too fast.

Exercising too much can also harm your muscles and tendons. For example, if you exercise a tendon that has a strain, the tendon cannot heal. It can become swollen and painful. This condition is called *tendonitis*.

Some people try to make their muscles stronger by taking drugs called *anabolic steroids*. These drugs can cause serious health problems. They can cause high blood pressure and can damage the heart, liver, and kidneys. They can also cause bones to stop growing.

Math Focus

5. Calculate A student is doing resistance exercise. After one week, she can lift a weight of 2 kg. After four weeks, she can lift a weight of 3 kg. By what percentage has the weight that she can lift increased?

____

TAKE A LOOK

6. Describe Complete the table to describe types of exercise.

Discuss In a small group, talk about some of the ways that exercise can help keep you healthy.

Name ______________________ Class ______________ Date ______________

Section 3 Review

NSES LS 1d, 1e

SECTION VOCABULARY

muscular system the organ system whose primary function is movement and flexibility	

1. **List** What three kinds of muscle make up the muscular system?

____

____

2. **Identify** Which kind of muscle movement happens without you having to think about it? Give two kinds of muscle that show this kind of movement.

____

____

3. **Describe** How are muscles attached to bones?

____

4. **Explain** How do muscles cause bones to move?

____

____

5. **Describe** What happens to muscle when you exercise it?

____

6. **Compare** How is aerobic exercise different from resistance exercise?

____

____

____

7. **Identify** What are two kinds of injuries to the muscular system?

____

8. **Compare** How is a flexor different from an extensor?

____

____

9. **Explain** What kinds of problems can anabolic steroids cause?

____

____

____

Name ______________________ Class ______________ Date ______________

CHAPTER 22 Body Organization and Structure

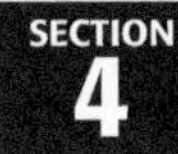

The Integumentary System

BEFORE YOU READ

After you read this section, you should be able to answer these questions:

- What is the integumentary system?
- What are the functions of the skin, hair, and nails?

National Science Education Standards

LS 1c, 1d, 1e, 1f, 3a, 3b

What Is the Integumentary System?

Your **integumentary system** is made up of your skin, hair, fingernails, and toenails. The Latin word *integere* means "to cover." Your integumentary system covers your body and helps to protect it. Your integumentary system also helps your body to maintain homeostasis.

Compare As you read, make a table comparing skin, hair, and nails. In the table, describe each structure and list its functions.

THE SKIN

Your skin is the largest organ in your body. It is an important part of the integumentary system. The skin has four main functions.

- Skin protects your body. It keeps water inside your body, and it keeps many harmful particles outside your body.
- Skin keeps you in touch with the world. Nerve endings in your skin let you feel things around you.
- Skin helps to keep your body temperature from getting too high. Small organs in the skin called *sweat glands* make sweat, which flows onto the skin. When sweat evaporates, your body cools down.
- Skin helps your body get rid of some wastes. Sweat can carry these wastes out of your body.

Critical Thinking

1. Apply Concepts Why are you more likely to get sick if you touch a dirty surface with damaged skin than if you touch it with healthy skin?

____

____

____

____

____

____

____

As you know, skin can be many different colors. The color of your skin is determined by a chemical called *melanin*. If your skin contains a lot of melanin, it is dark. If your skin contains very little melanin, it is light.

Melanin helps to protect your skin from being damaged by the ultraviolet radiation in sunlight. People's skin may darken if they are exposed to a lot of sunlight. This happens because the cells in your skin make extra melanin to help protect themselves from ultraviolet radiation.

LAYERS OF SKIN

Your skin has two main layers: the epidermis and the dermis. The **epidermis** is the outermost layer of skin. It is the layer that you see when you look at your skin. The prefix *epi-* means "above." Therefore, the epidermis lies above the dermis. The **dermis** is the thick layer of skin that lies underneath the epidermis. ☑

READING CHECK

2. List What are the two main layers of skin?

____

____

The epidermis is made of *epithelial tissue*. These tissues are made of many layers of cells. However, on most parts of your body, the epidermis is only a few millimeters thick.

Most of the cells in the epidermis are dead. The dead cells are filled with a protein called *keratin*. Keratin helps to make your skin tough.

The dermis is much thicker than the epidermis. It contains many fibers made of a protein called *collagen*. Collagen fibers make the dermis strong and let the skin bend without tearing. The dermis also contains many small structures, as shown in the figure below.

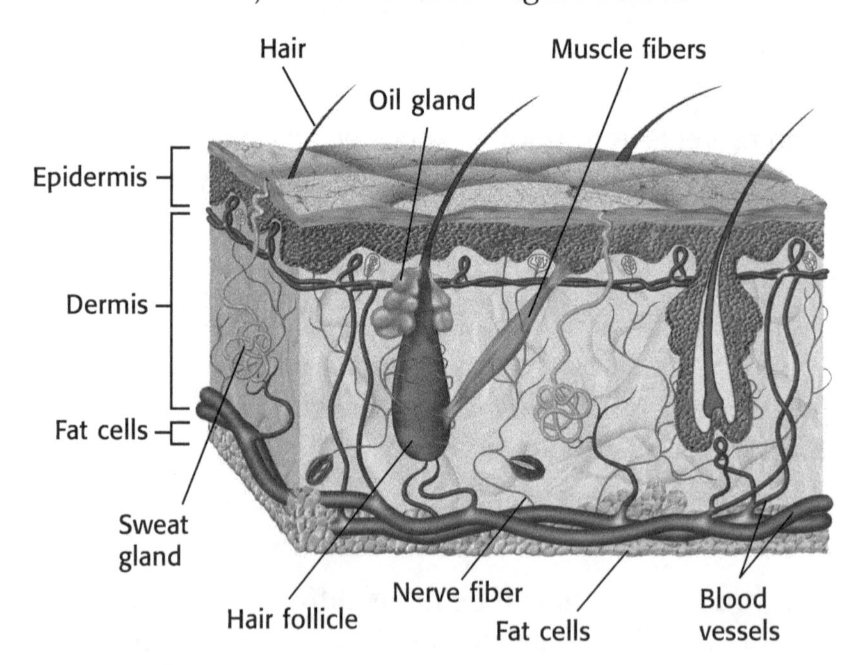

TAKE A LOOK

3. Identify Give three structures that attach to hair follicles.

____

____

____

SKIN INJURIES

Your skin is always coming into contact with the outside world. Therefore, it is often damaged. Fortunately, your skin can repair itself. The figure on the top of the next page shows how a cut in the skin heals.

Some skin problems are caused by conditions inside your body. For example, hormones can cause your skin to make too much oil. The oil can combine with bacteria and dead skin cells to form acne.

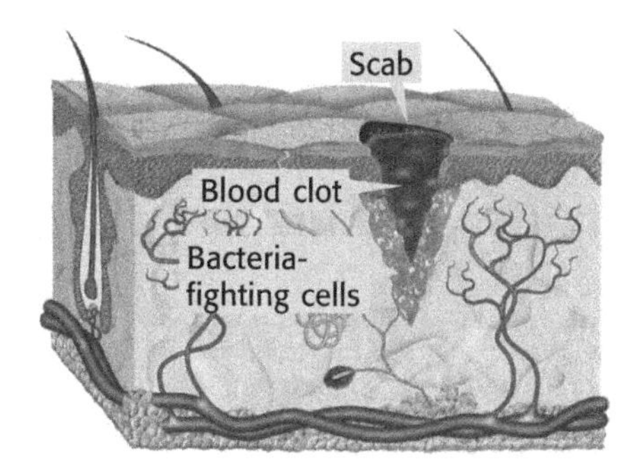

❶ A blood clot forms over a cut to stop bleeding and to keep bacteria from entering the wound. Bacteria-fighting cells then come to the area to kill bacteria.

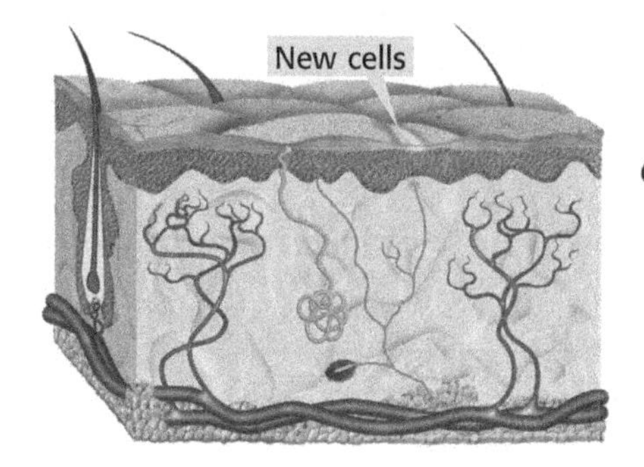

❷ Damaged cells are replaced through cell division. Eventually, all that is left on the surface is a scar.

TAKE A LOOK

4. Explain How do blood clots help protect your body?

____

____

HAIR AND NAILS

Your hair and nails are also important parts of your integumentary system. Like skin, hair and nails are made of both living and dead cells. Hair and nails grow from your skin.

A hair forms at the bottom of a tiny sac called a *hair follicle*. The hair grows as new cells are added at the hair follicle. Older cells get pushed upward. The only living cells in hair are found in the hair follicle. ☑

READING CHECK

5. Describe How does hair grow?

____

____

____

____

Like skin, hair gets its color from melanin. Hair helps protect skin from being damaged. The hair in and around your nose, eyes, and ears helps keep dust and other particles out of your body. Your hair also helps to keep you warm. When you feel cold, tiny muscles cause your hair to stand up. The raised hairs act like a sweater. They trap warm air near your body.

A nail forms at the *nail root*. A nail root is found at the base of the nail. As new cells form in the nail root, the nail grows longer. The hard part of the nail is made of dead cells that are filled with keratin. Nails protect the ends of your fingers and toes. They allow your fingers and toes to be soft and sensitive to touch.

Section 4 Review

NSES LS 1c, 1d, 1e, 1f, 3a, 3b

SECTION VOCABULARY

dermis the layer of skin below the epidermis

epidermis the surface layer of cells on a plant or animal

integumentary system the organ system that forms a protective covering on the outside of the body

1. **Identify** Name three functions of the integumentary system.

____

____

____

2. **Compare** Give three differences between the dermis and the epidermis.

____

____

____

3. **Infer** The epidermis on the palms of your hands and the soles of your feet is thicker than it is on other parts of your body. What do you think is the reason for this?

____

____

____

____

4. **Explain** Why can skin get darker if it is exposed to a lot of sunlight?

____

____

5. **Identify** Give two ways that hair helps to protect your body.

____

____

6. **Infer** Blood clots help to prevent bacteria from entering your body through a cut. Why do bacteria-fighting cells travel to a cut, even though there is a blood clot there?

____

____

Name ______________________ Class ______________ Date ______________

CHAPTER 23 Circulation and Respiration

The Cardiovascular System

BEFORE YOU READ

After you read this section, you should be able to answer these questions:

- What is the cardiovascular system?
- What are some cardiovascular problems?

National Science Education Standards
LS 3b

What Is the Cardiovascular System?

Your heart, blood, and blood vessels make up your **cardiovascular system**. The word *cardio* means heart. The word *vascular* means blood vessels. *Blood vessels* are hollow tubes that your blood flows through. The cardiovascular system is also sometimes called the *circulatory system*. This is because it *circulates*, or moves, blood through your body.

STUDY TIP

Summarize As you read, underline the main ideas in each paragraph. When you finish reading, write a short summary of the section using the ideas you underlined.

The cardiovascular system helps your body maintain homeostasis. *Homeostasis* is the state your body is in when its internal conditions are stable. The cardiovascular system helps maintain homeostasis in many ways:

- it carries oxygen and nutrients to your cells
- it carries wastes away from your cells
- it carries heat throughout your body
- it carries chemical signals called *hormones* throughout your body ☑

READING CHECK

1. Identify What are two functions of the cardiovascular system?

____
____
____
____

THE HEART

Your heart is an organ about the same size as your fist. It is near the center of your chest. There is a thick wall in the middle of your heart that divides it into two halves. The right half pumps oxygen-poor blood to your lungs. The left half pumps oxygen-rich blood to your body.

Each side of your heart has two chambers. Each upper chamber is called an *atrium* (plural, *atria*). Each lower chamber is called a *ventricle*. These chambers are separated by flap-like structures called *valves*. Valves keep blood from flowing in the wrong direction. The closing of valves is what makes the "lub-dub" sound when your heart beats. The figure at the top of the next page shows how blood moves through your heart.

Math Focus

2. Calculate A person's heart beats about 70 times per minute. How many times does a person's heart beat in one day? How many times does it beat in one year?

____

____

____

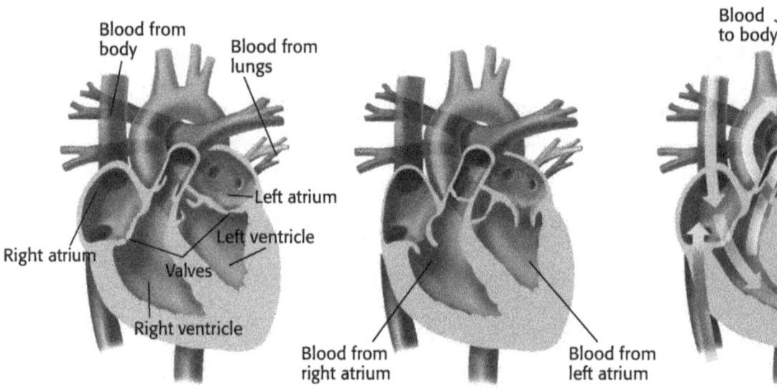

❶ Blood enters the atria first. The left atrium receives blood that has a lot of oxygen in it from the lungs. The right atrium receives blood that has little oxygen in it from the body.

❷ When the atria contract, blood moves into the ventricles.

❸ While the atria relax, the ventricles contract and push blood out of the heart. Blood from the right ventricle goes to the lungs. Blood from the left ventricle goes to the rest of the body.

TAKE A LOOK

3. Identify Where does the left ventricle receive blood from? Where does the right atrium receive blood from?

____

____

____

____

BLOOD VESSELS

Blood travels throughout your body in your blood vessels. There are three types of blood vessels: arteries, capillaries, and veins.

An **artery** is a blood vessel that carries blood away from the heart. Arteries have thick walls that contain a layer of muscle. Each heartbeat pumps blood into your arteries. The blood is under high pressure. Artery walls are strong and can stretch to handle this pressure. Your *pulse* is caused by the pumping of blood into your arteries. ☑

4. Describe What causes your pulse?

____

____

____

A **capillary** is a tiny blood vessel. Capillary walls are very thin. Therefore, substances can move across them easily. Capillaries are also very narrow. They are so narrow that blood cells have to pass through them in single file. Nutrients and oxygen move from the blood in your capillaries into your body's cells. Carbon dioxide and other wastes move from your body's cells into the blood.

A **vein** is a blood vessel that carries blood toward the heart. Veins have valves to keep the blood from flowing backward. When skeletal muscles contract, they squeeze nearby veins and help push blood toward the heart.

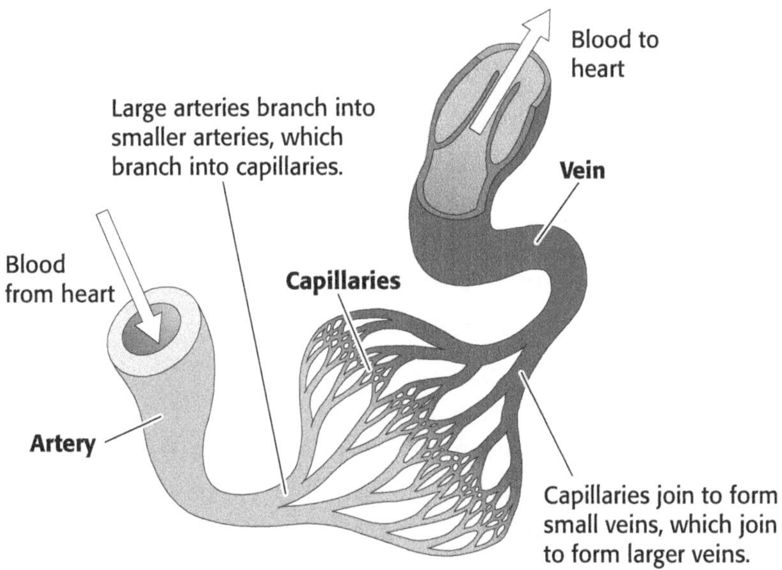

TAKE A LOOK

5. Compare What is one main difference between arteries and veins?

____

____

____

How Does Blood Flow Through Your Body?

Where does blood get the oxygen to deliver to your body? From your lungs! Your heart contracts and pumps blood to the lungs. In the lungs, carbon dioxide leaves the blood and oxygen enters the blood. The oxygen-rich blood then flows back to your heart. This circulation of blood between your heart and lungs is called **pulmonary circulation**. ☑

READING CHECK

6. Define What is pulmonary circulation?

____

____

The oxygen-rich blood returning to your heart from your lungs is then pumped to the rest of your body. The circulation of blood between your heart and the rest of your body is called **systemic circulation**. The figure below shows how blood moves through your body.

Pulmonary circulation

The right ventricle pumps oxygen-poor blood into arteries that lead to the lungs. These are the only arteries in the body that carry oxygen-poor blood.

In the lungs, blood gives off carbon dioxide and takes up oxygen. Oxygen-rich blood flows through veins to the left atrium. These are the only veins in the body that carry oxygen-rich blood.

Oxygen-poor blood travels back to the heart through veins. These veins deliver the blood to the right atrium.

The heart pumps oxygen-rich blood from the left ventricle into arteries. The arteries branch into capillaries.

Systemic circulation

Oxygen, nutrients, and water move into the cells of the body as blood moves through capillaries. At the same time, carbon dioxide and other waste materials move out of the cells and into the blood.

TAKE A LOOK

7. Color Use a blue pen or colored pencil to color the vessels carrying oxygen-poor blood. Use a red pen or colored pencil to color the vessels carrying oxygen-rich blood.

What Are Some Problems of the Cardiovascular System?

Problems in the cardiovascular system can affect other parts of your body. Cardiovascular problems can be caused by smoking, too much cholesterol, stress, physical inactivity, or heredity. Eating a healthy diet and getting plenty of exercise can help to keep your cardiovascular system, and the rest of your body, healthy.

Critical Thinking

8. Infer How can a problem in your cardiovascular system affect the rest of your body?

____

____

____

____

____

____

ATHEROSCLEROSIS

Heart disease is the most common cause of death in the United States. One major cause of heart disease is atherosclerosis. *Atherosclerosis* happens when cholesterol and other fats build up inside blood vessels. This buildup causes the blood vessels to become narrower and less stretchy. When the pathway through a blood vessel is blocked, blood cannot flow through. ☑

READING CHECK

9. Identify What is the most common cause of death in the United States?

____

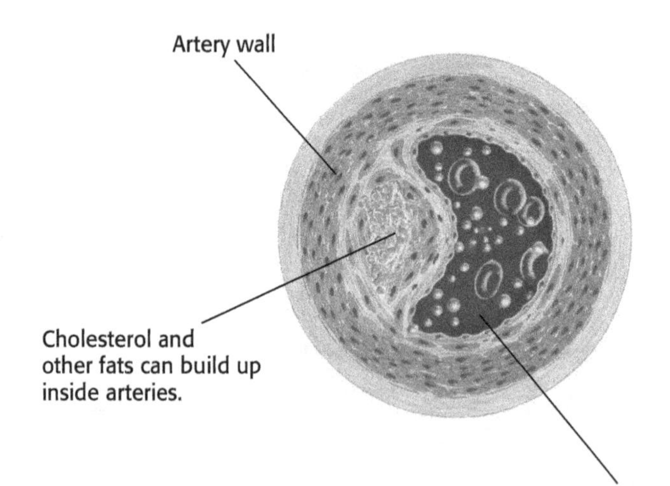

TAKE A LOOK

10. Explain How can too much cholesterol cause problems in your cardiovascular system?

____

____

____

HIGH BLOOD PRESSURE

Hypertension is high blood pressure. Hypertension can make it more likely that a person will have cardiovascular problems. For example, atherosclerosis may be caused by hypertension.

High blood pressure can also cause a stroke. A *stroke* happens when a blood vessel in the brain is blocked or breaks open. Blood cannot flow through the vessel to the brain cells. Without blood, the brain cells cannot get oxygen, so the cells die. ☑

Say It

Discuss Learn about two ways to maintain healthy blood pressure. In a small group, talk about how you can apply these ideas in your life.

READING CHECK

11. Identify What is a stroke?

____

____

HEART ATTACKS AND HEART FAILURE

Hypertension can also cause heart attacks and heart failure. A *heart attack* happens when heart muscle cells do not get enough blood. Arteries that deliver oxygen to the heart may be damaged. Without oxygen from the arteries, heart muscle cells can be damaged. If enough heart muscle cells are damaged, the heart may stop.

Arteries carry blood and oxygen to the heart muscle.

If an artery is blocked, blood and oxygen cannot flow to part of the heart muscle.

Without oxygen from blood, the heart muscle can be damaged. It can become weak or die.

TAKE A LOOK

12. Explain How can blocking an artery in the heart cause heart damage?

____

____

____

Heart failure happens when the heart is too weak to pump enough blood to meet the body's needs. Organs may not receive enough oxygen or nutrients to function correctly. Waste products can build up in the organs and damage them.

Name ______________________ Class ______________ Date ______________

Section 1 Review

NSES LS 3b

SECTION VOCABULARY

artery a blood vessel that carries blood away from the heart to the body's organs

capillary a tiny blood vessel that allows an exchange between blood and cells in tissue

cardiovascular system a collection of organs that transport blood throughout the body; the organs in this system include the heart, the arteries, and the veins

pulmonary circulation the flow of blood from the heart to the lungs and back to the heart through the pulmonary arteries, capillaries, and veins

systemic circulation the flow of blood from the heart to all parts of the body and back to the heart

vein in biology, a vessel that carries blood to the heart

1. Identify What are the three main parts of the cardiovascular system?

____

____

2. Describe Beginning and ending in the left atrium, describe the path that blood takes through your body and lungs.

____

____

____

____

____

3. Compare How is a heart attack different from heart failure?

____

____

____

4. Explain What is the function of valves in the heart and the veins?

____

____

5. Compare How are the arteries that lead from your heart to your lungs different from the other arteries in your body?

____

____

____

Name ______________________ Class ______________ Date ______________

CHAPTER 23 Circulation and Respiration

SECTION 2

Blood

BEFORE YOU READ

After you read this section, you should be able to answer these questions:

- What is blood?
- What is blood pressure?
- What are blood types?

National Science Education Standards

LS 3a, 3b

What Is Blood?

Your cardiovascular system is made up of your heart, your blood vessels, and blood. **Blood** is a connective tissue made up of plasma, red blood cells, platelets, and white blood cells. Blood travels in blood vessels and carries oxygen and nutrients to all parts of your body. An adult human has only about 5 L of blood. All the blood in your body would not even fill up three 2-L soda bottles! ☑

STUDY TIP

Ask Questions As you read this section, write down the questions that you have. Then, discuss your questions with a small group.

READING CHECK

1. Define What is blood?

____

____

____

____

PLASMA

The fluid part of the blood is called plasma. *Plasma* is made up of water, minerals, nutrients, sugars, proteins, and other substances.

PLATELETS

Platelets are pieces of larger cells found in bone marrow. When you get a cut, you bleed because blood vessels have been opened. Platelets clump together in the damaged area to form a plug. They also give off chemicals that cause fibers to form. The fibers and clumped platelets form a blood clot and stop the bleeding.

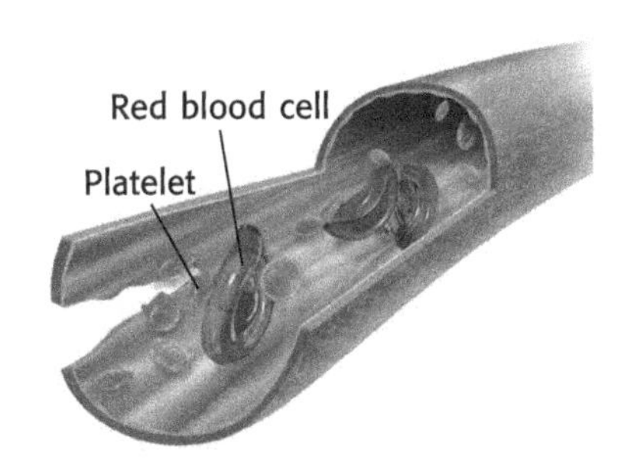

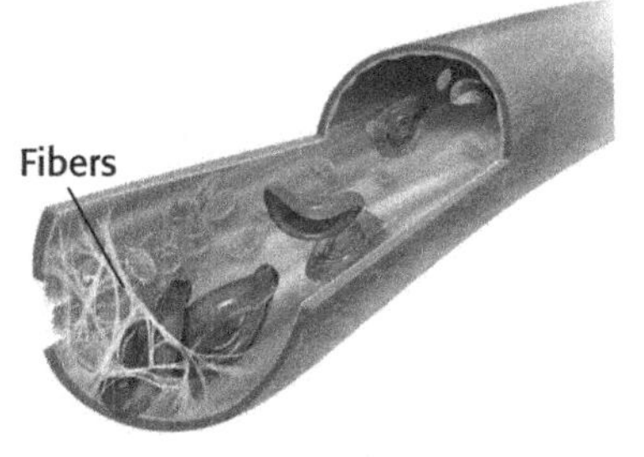

Critical Thinking

2. Infer If a person does not have enough platelets in her blood, what will happen if she gets a cut?

____

____

____

____

____

Math Focus

3. Calculate One cubic millimeter of blood contains 5 million RBCs and 10,000 WBCs. How many times more RBCs are there than WBCs?

____

RED BLOOD CELLS

Most blood cells are *red blood cells*, or RBCs. RBCs carry oxygen to all the cells in your body. Cells need oxygen to do their jobs. *Hemoglobin* is the protein in red blood cells that carries the oxygen. It is what makes RBCs look red.

WHITE BLOOD CELLS

A *pathogen* is a virus, bacteria, or other tiny particle that can make you sick. When pathogens get into your body, *white blood cells*, or WBCs, help kill them. WBCs can fight pathogens by:

- leaving blood vessels to destroy pathogens in tissues
- making chemicals called *antibodies* to help destroy pathogens
- destroying body cells that have died or been damaged

Most WBCs are made in bone marrow. Some mature in the lymphatic system. ☑

4. Identify Where are most white blood cells made?

____

How Does Blood Control Body Temperature?

Your blood also helps keep your body temperature constant. Your blood vessels can open wider or get narrower to control how much heat is lost through your skin.

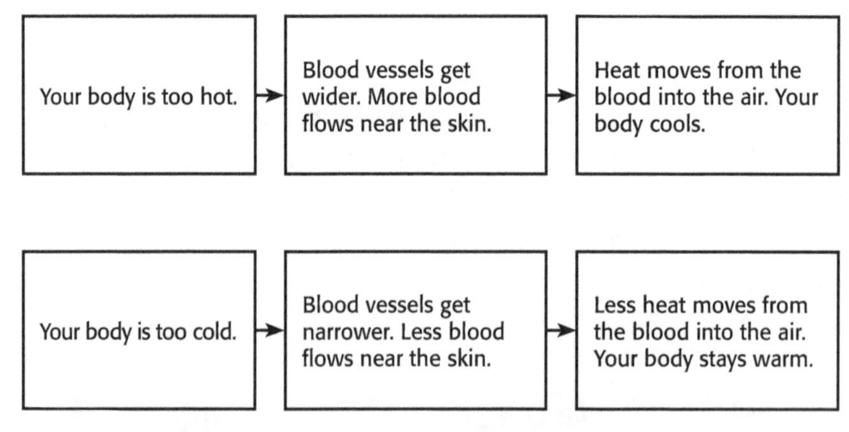

STANDARDS CHECK

LS 3b Regulation of an organism's internal environment involves sensing the internal environment and changing physiological activities to keep conditions within the range required to survive.

5. Explain How do wider or narrower blood vessels help your body stay at a constant temperature?

____

____

____

What Is Blood Pressure?

When your heart beats, it pushes blood out of your heart and into your arteries. The force of the blood on the inside walls of the arteries is called **blood pressure**. Blood pressure is measured in millimeters of mercury (mm Hg).

SYSTOLIC AND DIASTOLIC PRESSURE

Blood pressure is given by two numbers, such as 120/80. The first, or top, number is systolic pressure. *Systolic pressure* is the pressure in arteries when the ventricles contract. The rush of blood causes arteries to bulge and produce a pulse. The second, or bottom, number is diastolic pressure. *Diastolic pressure* is the pressure in arteries when the ventricles relax.

For adults, a blood pressure of 120/80 mm Hg or less is healthy. High blood pressure can cause heart or kidney damage.

What Are Blood Types?

Every person has one of four blood types: A, B, AB, or O. Chemicals called *antigens* on the outside of your RBCs determine whch blood type you have. The plasma of different blood types may have different antibodies. *Antibodies* are chemicals that react with antigens of other blood types as if the antigens were pathogens. ☑

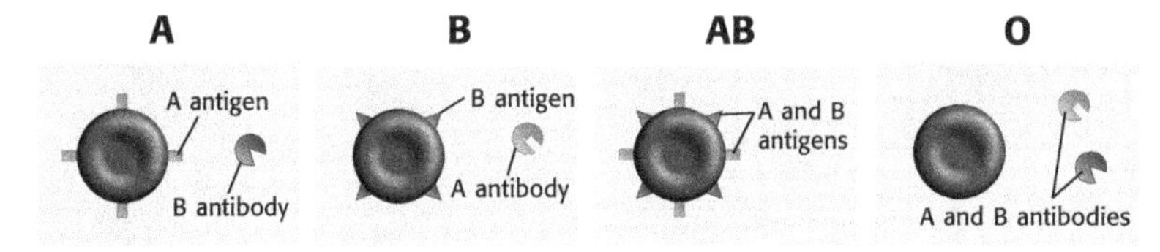

This figure shows which antigens and antibodies may be present in each blood type.

READING CHECK

6. Identify What determines your blood type?

____

____

TAKE A LOOK

7. Identify What kinds of antigens are found on the RBCs of a person with type AB blood?

____

IMPORTANCE OF BLOOD TYPES

A person can lose blood from an injury, illness or surgery. To replace lost blood, a person can receive a blood transfusion. A *transfusion* is when a person is given blood from another person.

However, a person cannot receive blood from just anyone. If someone who is type A gets type B blood, the type B antibodies can make the RBCs clump together. The clumps can block blood vessels. A reaction to the wrong blood type can kill you.

Blood type	Can receive blood from:	Can donate blood to:
A	types A and O	types A and AB
B	types B and O	types B and AB
AB	types A, B, AB, and O	type AB only
O	type O only	types A, B, AB, and O

TAKE A LOOK

8. Identify Which blood type can receive blood from the most other blood types? Which type can donate blood to the most other types?

____

____

Name ______________________ Class ______________ Date ______________

Section 2 Review

NSES LS 3a, 3b

SECTION VOCABULARY

blood the fluid that carries gases, nutrients, and wastes through the body and that is made up of platelets, white blood cells, red blood cells, and plasma	**blood pressure** the force that blood exerts on the walls of arteries

1. Identify What are two functions of white blood cells?

____

____

2. Describe Complete the table to describe the two parts of blood pressure.

Type of pressure	Description	Where it is found in a blood-pressure measurement
systolic		top number
	pressure in the arteries when ventricles relax	

3. List What are three functions of blood?

____

____

4. Infer Why does your face get redder when you are hot?

____

____

____

5. Explain Why is it important that a person with type O blood only receive a blood transfusion from another person with type O blood?

____

____

____

6. Predict If a person has a disease that causes hemoglobin to break down, what can happen to his RBCs?

____

____

Name ______________________ Class ______________ Date ______________

CHAPTER 23 Circulation and Respiration

SECTION 3

The Lymphatic System

BEFORE YOU READ

After you read this section, you should be able to answer these questions:

- What is the function of the lymphatic system?
- What are the parts of the lymphatic system?

National Science Education Standards

LS 1a, 1c, 1d, 1e, 1f

What Does the Lymphatic System Do?

Every time your heart pumps, small amounts of plasma are forced out of the thin walls of the capillaries. What happens to this fluid? Most of it is reabsorbed into your blood through the capillaries. Some of the fluid moves into your lymphatic system.

The **lymphatic system** is the group of vessels, organs, and tissues that collects excess fluid and returns it to the blood. The lymphatic system also helps your body fight pathogens.

STUDY TIP

Describe Work with a partner to quiz each other on the names and functions of each structure in the lymphatic system.

What Are the Parts of the Lymphatic System?

Fluid collected by the lymphatic system is carried in vessels. The smallest of these vessels are called lymph capillaries. Larger lymph vessels are called lymphatic vessels. These vessels, along with bone marrow, lymph nodes, the thymus, and the spleen, make up the lymphatic system.

STANDARDS CHECK

LS 1e The human organism has systems for digestion, reproduction, circulation, excretion, movement, control and coordination, and protection from disease. These systems interact with one another.

1. Identify Relationships How do the lymphatic and circulatory systems work together?

____

____

____

____

____

____

LYMPH CAPILLARIES

Lymph capillaries absorb some of the fluid and particles from between cells in the body. Some of the particles are dead cells or pathogens. These particles are too large to enter blood capillaries. The fluid and particles absorbed into lymph capillaries are called **lymph**.

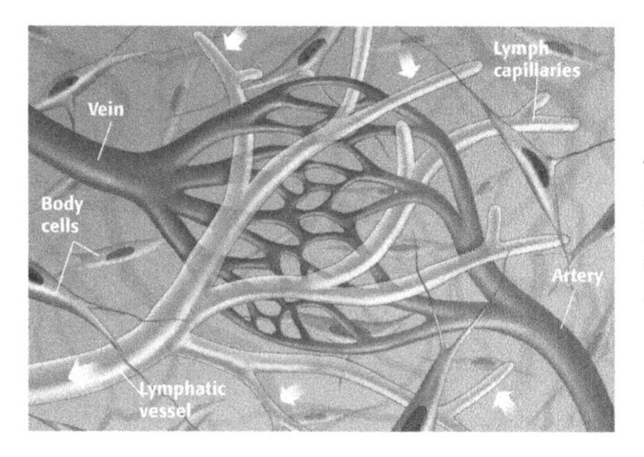

The white arrows show how lymph moves into lymph capillaries and through lymphatic vessels.

LYMPHATIC VESSELS

Lymph capillaries carry lymph into larger vessels called *lymphatic vessels*. Skeletal muscles and valves help push the lymph through the lymphatic system. Lymphatic vessels drain the lymph into large veins in the neck. This returns the fluid to the cardiovascular system. ☑

READING CHECK

2. Identify Where is fluid returned to the cardiovascular system?

____

____

The Lymphatic System

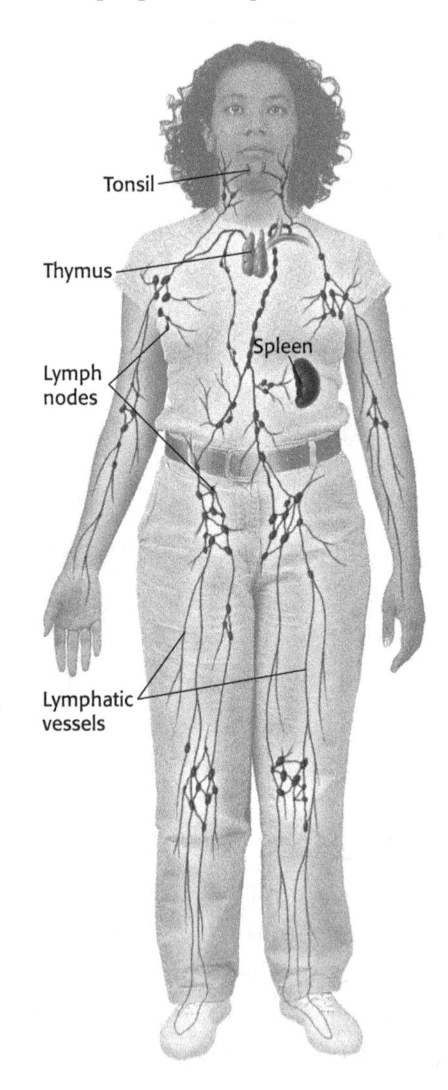

TAKE A LOOK

3. Describe As you read, write on the diagram the function of each labelled structure.

BONE MARROW

Bone marrow is the soft tissue inside bones that makes red and white blood cells. Recall that platelets, which help blood clot, are made in marrow. White blood cells called *lymphocytes* are part of the lymphatic system. They help fight infection. Killer T cell lymphocytes surround and destroy pathogens. B cell lymphocytes make antibodies that cause pathogens to stick together. This marks them for destruction. ☑

READING CHECK

4. Identify What is the function of bone marrow?

____

____

LYMPH NODES

As lymph travels through lymphatic vessels, it passes through lymph nodes. **Lymph nodes** are small masses of tissue that remove pathogens and dead cells from the lymph. When bacteria or other pathogens cause an infection, white blood cells multiply and fill the lymph nodes. This may cause lymph nodes to become swollen and painful.

Critical Thinking

5. Infer Sometimes you can easily feel your lymph nodes when they are swollen. If you had swollen lymph nodes, what could you infer?

____

____

THYMUS

T cells are made in the bone marrow. Before these cells are ready to fight infections, however, they develop further in the **thymus** gland. The thymus is located just above the heart. Mature T cells leave the thymus and travel through the lymphatic system.

SPLEEN

The **spleen** is the largest lymphatic organ. It stores lymphocytes and fights infection. It is a purplish organ located in the upper left side of the abdomen. As blood flows through the spleen, lymphocytes attack or mark pathogens in the blood. The spleen may release lymphocytes into the bloodstream when there is an infection. The spleen also monitors, stores, and destroys old blood cells. ☑

READING CHECK

6. List Name three functions of the spleen.

____

____

____

____

TONSILS

The **tonsils** are lymphatic tissue at the back of the mouth. Tonsils help defend the body against infection by trapping pathogens. Sometimes, however, tonsils can become infected. Infected tonsils may be red, swollen, and sore. They may be covered with patches of white, infected tissue and make swallowing difficult. Tonsils may be removed if there are frequent, severe tonsil infections that make breathing difficult.

Inflamed tonsils

Tonsils help protect your throat and lungs from infection by trapping pathogens.

Name ______________________ Class ______________ Date ______________

Section 3 Review

NSES **LS 1a, 1c, 1d, 1e, 1f**

SECTION VOCABULARY

lymph the fluid that is collected by the lymphatic vessels and nodes

lymph node an organ that filters lymph and that is found along the lymphatic vessels

lymphatic system a collection of organs whose primary function is to collect extracellular fluid and return it to the blood; the organs in this system include lymph nodes and the lymphatic vessels

spleen the largest lymphatic organ in the body; serves as a blood reservoir, disintegrates old red blood cells, and produces lymphocytes and plasmids

thymus the main gland of the lymphatic system; it releases mature T lymphocytes

tonsils organs that are small, rounded masses of lymphatic tissue located in the pharynx and in the passage from the mouth to the pharynx

1. Describe How does the lymphatic system fight infection?

____

____

____

2. Summarize Complete the Process Chart below to show how fluid travels between the cardiovascular system and the lymphatic system.

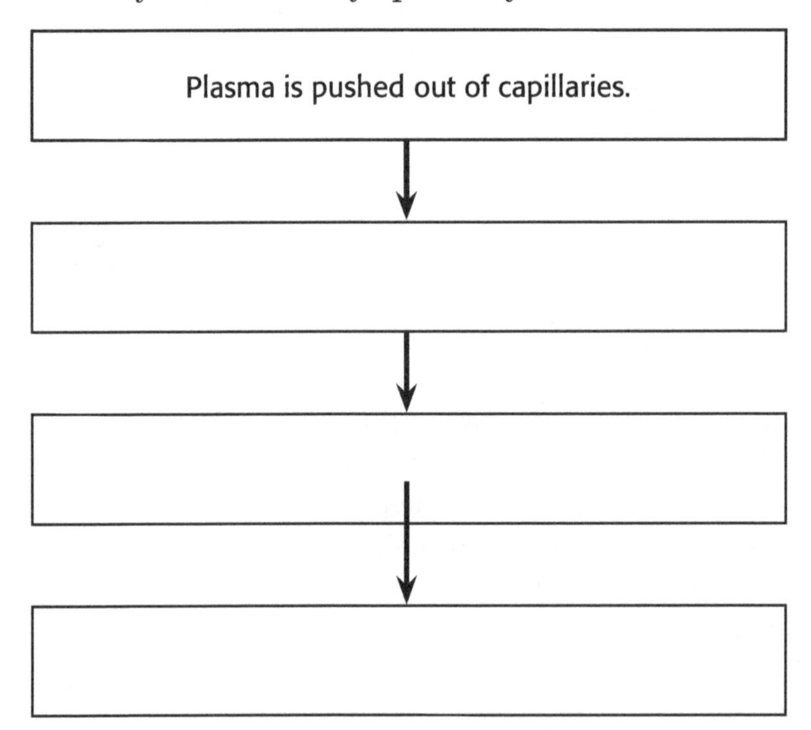

3. List What are three things that can be found in lymph?

____

4. Analyze Why is it important that lymphatic tissue is spread throughout the body?

____

____

Name ______________________ Class ______________ Date ____________

CHAPTER 23 Circulation and Respiration

SECTION 4

The Respiratory System

BEFORE YOU READ

After you read this section, you should be able to answer these questions:

- What is the respiratory system?
- What are some respiratory disorders?

What Is the Respiratory System?

Breathing: you do it all the time. You're doing it right now. You probably don't think about it unless you can't breathe. Then, it becomes very clear that you have to breathe in order to live. Why is breathing important? Breathing helps your body get oxygen. Your body needs oxygen in order to get energy from the foods you eat.

The words *breathing* and *respiration* are often used to mean the same thing. However, breathing is only one part of respiration. **Respiration** is the way the body gains and uses oxygen and gets rid of carbon dioxide. ☑

Respiration is divided into two parts. The first part involves inhaling and exhaling, or breathing. The second part is cellular respiration. *Cellular respiration* involves the chemical reactions that let you get energy from food.

The **respiratory system** is the group of organs and structures that take in oxygen and get rid of carbon dioxide. The nose, throat, lungs, and passageways that lead to the lungs make up the respiratory system.

Compare Make a chart showing the features of the different parts of the respiratory system.

READING CHECK

1. Define What is respiration?

____

____

____

____

Parts of the Respiratory System

TAKE A LOOK

2. List What are the parts of the respiratory system?

____

____

____

____

THE NOSE, PHARYNX, LARYNX, AND TRACHEA

Your *nose* is the main passageway into and out of the respiratory system. You breathe air in through your nose. You also breathe air out of your nose. Air can also enter and leave through your mouth. ☑

READING CHECK

3. Describe What is the main function of the nose?

____

____

____

From the nose or mouth, air flows through the **pharynx**, or throat. Food and drink also move through the pharynx on the way to the stomach. The pharynx branches into two tubes. One tube, the *esophagus*, leads to the stomach. The other tube leads to the lungs. The larynx sits at the start of this tube.

The **larynx** is the part of the throat that contains the vocal cords. The *vocal cords* are bands of tissue that stretch across the larynx. Muscles connected to the larynx control how much the vocal cords are stretched. When air flows between the vocal cords, the cords vibrate. These vibrations make sound.

The larynx guards the entrance to a large tube called the **trachea**, or windpipe. The trachea is the passageway for air traveling from the larynx to the lungs. ☑

READING CHECK

4. Identify What is the trachea?

____

____

THE BRONCHI AND ALVEOLI

Inside your chest, the trachea splits into two branches called **bronchi** (singular, *bronchus*). One bronchus connects to each lung. Each bronchus branches into smaller and smaller tubes. These branches form a smaller series of airways called *bronchioles*. In the lungs, each bronchiole branches to form tiny sacs that are called **alveoli** (singular, *alveolus*).

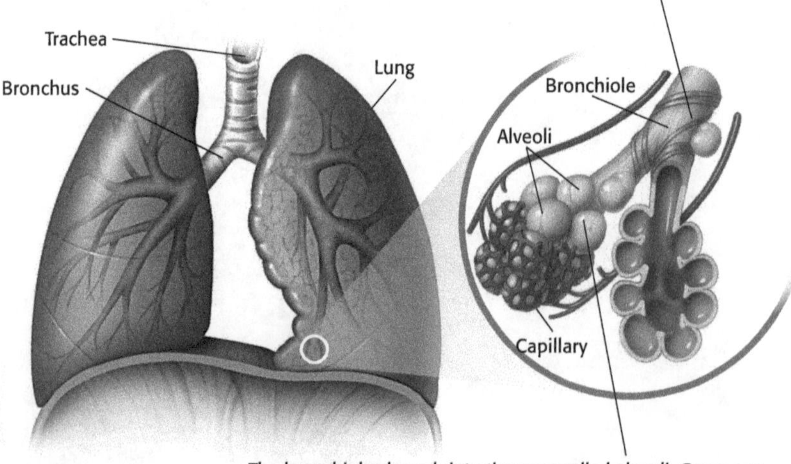

The bronchioles branch into tiny sacs called alveoli. Gases can move between the alveoli and the blood that is in the capillaries.

TAKE A LOOK

5. Infer Which do you have the most of in your lungs: bronchi, bronchioles, or alveoli?

____

How Does Breathing Work?

Your lungs have no muscles of their own. Instead, your diaphragm and rib muscles do the work that helps you breathe. The *diaphragm* is a dome-shaped muscle underneath the lungs. When the diaphragm contracts and moves down, you inhale. At the same time, some of your rib muscles contract and lift your rib cage. The volume of your chest gets larger. As a result, air is sucked in.

Exhaling is this process in reverse. Your diaphragm relaxes, your rib muscles relax, and air moves out.

Critical Thinking

6. Predict Consequences What would happen to a person whose diaphragm could not contract?

____

____

BREATHING AND CELLULAR RESPIRATION

In cellular respiration, cells use oxygen to release the energy that is stored in molecules of a sugar. This sugar is called *glucose*. When cells break down glucose, they give off carbon dioxide.

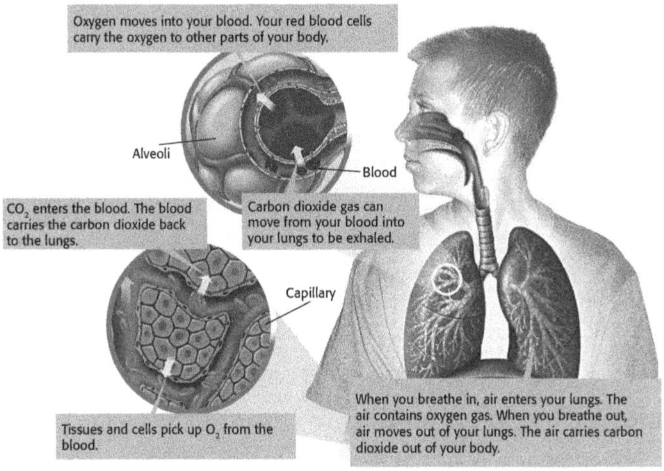

TAKE A LOOK

7. Explain How does oxygen gas get from the air into the cells in your body?

____

____

____

____

____

What Are Some Respiratory Disorders?

People who have *respiratory disorders* have trouble getting the oxygen they need. Their cells cannot release all the energy they need from the food they eat. Therefore, these people may feel tired all the time. They may also have problems getting rid of carbon dioxide. The carbon dioxide can build up in their cells and make them sick.

Respiratory Disorder	What it is
Asthma	A disorder that causes bronchioles to narrow, making it hard to breathe.
Emphysema	A disorder caused when alveoli are damaged.
Severe Acute Respiratory Syndrome (SARS)	A disorder caused by a virus that makes it hard to breathe.

Name ______________________ Class ______________ Date ______________

Section 4 Review

SECTION VOCABULARY

alveolus any of the tiny air sacs of the lungs where oxygen and carbon dioxide are exchanged

bronchus one of the two tubes that connect the lungs with the trachea

larynx the area of the throat that contains the vocal cords and produces vocal sounds

pharynx in flatworms, the muscular tube that leads from the mouth to the gastrovascualr cavity; in animals with a digestive tract, the passage from the mouth to the larynx and esophagus

respiration the exchange of oxygen and carbon dioxide between living cells and their environment; includes breathing and cellular respiration

respiratory system a collection of organs whose primary function is to take in oxygen and expel carbon dioxide; the organs of this system include the lungs, the throat, and the passageways that lead to the lungs

trachea in insects, myriapods, and spiders, one of the network of air tubes; in vertebrates, the tube that connects the larynx to the lungs

1. List What are three respiratory disorders?

____

2. Define What is cellular respiration?

____

____

3. Compare How is respiration different from breathing?

____

____

____

4. Explain The nose is the main way for air to get into and out of your body. How can a person still breathe if his or her nose is blocked?

____

5. Describe How do vocal cords produce sound?

____

____

6. Explain What are two ways that a respiratory disorder can make a person sick?

____

____

____

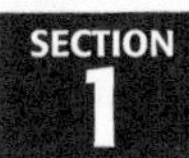

SECTION 1

The Digestive System

BEFORE YOU READ

After you read this section, you should be able to answer these questions:

- What are the parts of the digestive system?
- How does each part of the digestive system work?

National Science Education Standards
LS 1a, 1d, 1e, 1f, 3b

What Are the Parts of the Digestive System?

The **digestive system** is a group of organs that break down, or digest, food so your body can get nutrients. The main organs of the digestive system make one long tube through the body. This tube is called the *digestive tract*. The digestive tract includes the mouth, pharynx, esophagus, stomach, small intestine, and large intestine.

The digestive system has several organs that are not part of the digestive tract. The liver, gallbladder, pancreas, and salivary glands add materials to the digestive tract to help break down food. However, food does not go into these organs. ☑

STUDY TIP

Organize As you read, make combination notes about each digestive organ. Write the function of the organ in the left column of the notes. Draw or describe the structure in the right column.

READING CHECK

1. Identify What is the name of the tube that food passes through?

____

The Digestive System

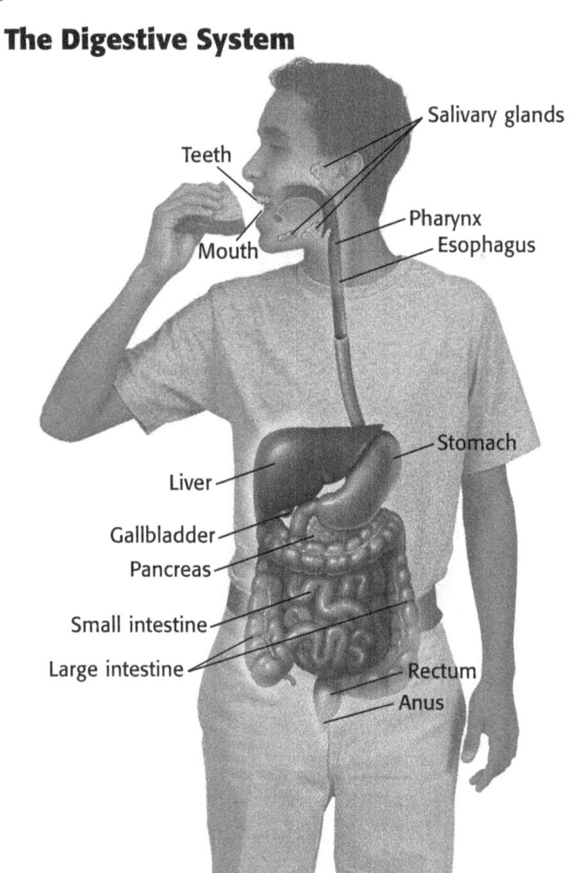

TAKE A LOOK

2. Color Use a colored pencil to shade all of the digestive organs that food passes through.

STANDARDS CHECK

LS 1e The human organism has systems for digestion, respiration, reproduction, circulation, excretion, movement, control and coordination, and protection from disease. These systems interact with one another.

3. Infer How does the circulatory system work with the digestive system?

____

____

____

____

____

____

How Is Food Broken Down?

The sandwich you eat for lunch has to be broken into tiny pieces to be absorbed into your blood. First, food is crushed and mashed into smaller pieces. This is called *mechanical digestion.* However, the food is still too large to enter your blood. Next, the small pieces of food are broken into their chemical parts, or molecules. This is called *chemical digestion.* The molecules can now be taken in and used by the body's cells.

Most food is made up of three types of nutrients: carbohydrates, proteins, and fats. The digestive system uses proteins called *enzymes* to break your food into molecules. Enzymes act as chemical scissors to cut food into smaller particles that the body can use.

What Happens to Food in the Mouth and Stomach?

MOUTH

Digestion begins in the mouth where food is chewed. You use your teeth to mash and grind food. Chewing creates small, slippery pieces of food that are easy to swallow. As you chew, the food mixes with a liquid called *saliva.* Saliva is made in salivary glands in the mouth. Saliva has enzymes that start breaking down starches into simple sugars.

Critical Thinking

4. Infer Why do you think you should chew your food well before you swallow it?

____

____

____

____

____

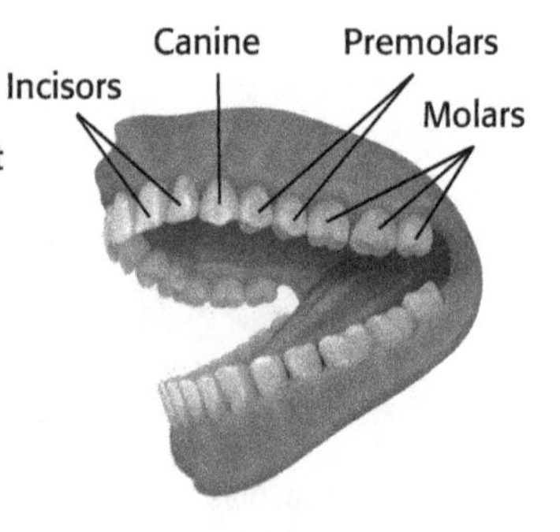

Most adults have 32 permanent teeth. Each type of permanent tooth has a different role in breaking up food.

Math Focus

5. Compute Ratios Young children get a first set of 20 teeth called *baby teeth.* These teeth usually fall out and are replaced by 32 permanent teeth. What is the ratio of baby teeth to permanent teeth?

____

____

ESOPHAGUS

Once the food has been chewed to a soft mush, it can be swallowed. The tongue pushes the food into the *pharynx.* The *pharynx* is the part of the throat that makes food go to the esophagus and air go to the lungs. The **esophagus** is a long, straight tube that leads from the pharynx to the stomach. Muscle contractions, called *peristalsis*, squeeze food in the esophagus down to the stomach.

STOMACH

The stomach is a muscular, saclike organ. The stomach uses its muscles to continue mechanical digestion. It squeezes and mashes food into smaller and smaller pieces.

The stomach also has glands that make enzymes and acid. These chemicals help break down food into nutrients. Stomach acid also kills most bacteria in the food. After a few hours of chemical and mechanical digestion, the food is a soupy mixture called *chyme.* ☑

READING CHECK

6. Identify What kinds of digestion occur in the stomach?

____

____

The stomach squeezes and mixes food for hours before it releases the mixture into the small intestine.

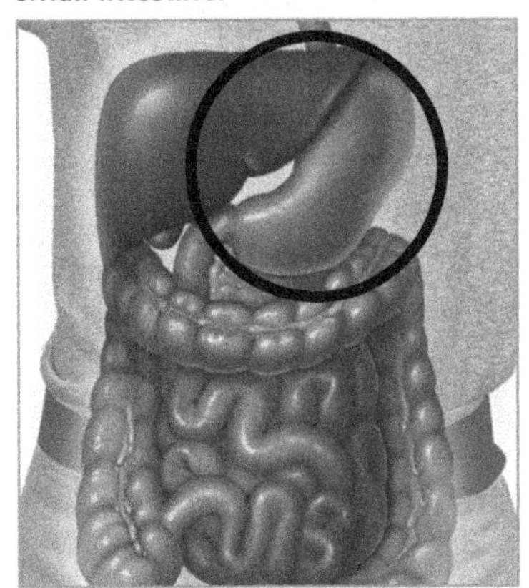

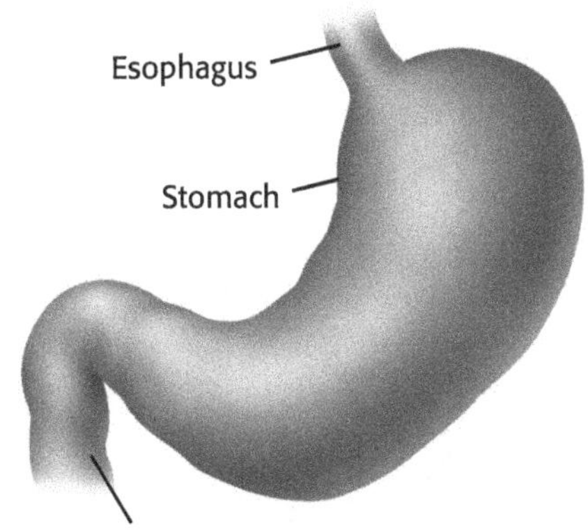

What Happens to Food in the Small and Large Intestines?

Most chemical digestion takes place after food leaves the stomach. Chyme slowly leaves the stomach and enters the small intestine. The pancreas, liver, and gallbladder add enzymes and other fluids to the small intestine to help finish digestion. The large intestine absorbs water and gets rid of waste.

PANCREAS

The **pancreas** is an organ located between the stomach and small intestine. Food does not enter the pancreas. Instead, the pancreas makes a fluid that flows into the small intestine. The chart below shows the chemicals that make up this pancreatic fluid, and the function of each.

Chemical in pancreatic fluid	Function
sodium bicarbonate	to protect the small intestine from acid in the chyme
enzymes	to chemically digest chyme
hormones	to control blood sugar levels

TAKE A LOOK

7. Identify What is the role of enzymes in pancreatic fluid?

____

____

____

LIVER

The **liver** is a large, reddish brown organ found on the right side of the body under the ribs. The liver helps with digestion in the following ways.

- It makes bile.
- It stores extra nutrients.
- It breaks down toxins, such as alcohol.

GALLBLADDER

Bile is made in the liver, and stored in a small, saclike organ called the **gallbladder**. The gallbladder squeezes bile into the small intestine when there is food to digest. Bile breaks fat into very small droplets so that enzymes can digest it. ☑

READING CHECK

8. Identify What is the function of bile?

____

____

____

The liver, gall bladder, and pancreas are linked to the small intestine. However, food does not pass through these organs.

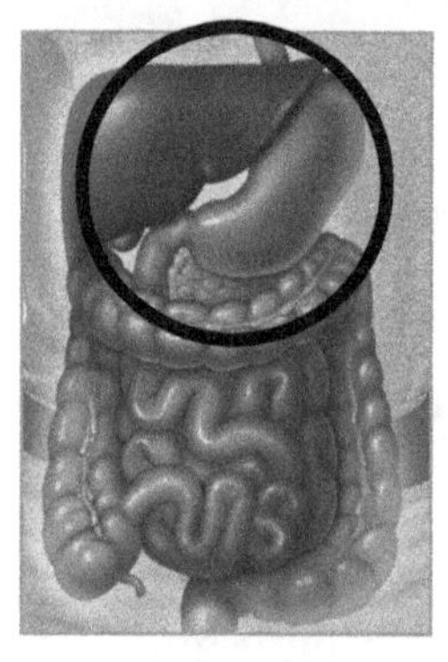

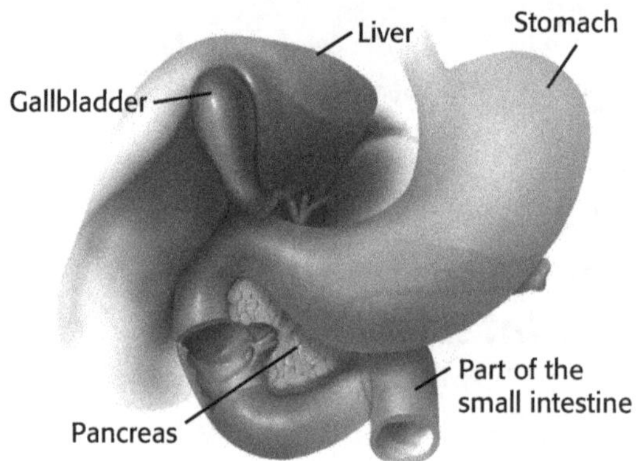

TAKE A LOOK

9. Identify Which organs in this diagram does food pass through?

____

____

SMALL INTESTINE

The **small intestine** is a long, thin, muscular tube where nutrients are absorbed. If you stretched out your small intestine, it would be much longer than you are tall—about 6 m! If you flattened out the surfaces of the small intestine, it would be larger than a tennis court.

The inside wall of the small intestine is covered with many small folds. The folds are covered with cells called *villi.* Villi absorb nutrients. The large number of folds and villi increase the surface area of the small intestine. A large surface area helps the body get as many nutrients from food as possible. ☑

READING CHECK

10. Explain Why is a large surface area in the small intestine important?

____

____

____

____

____

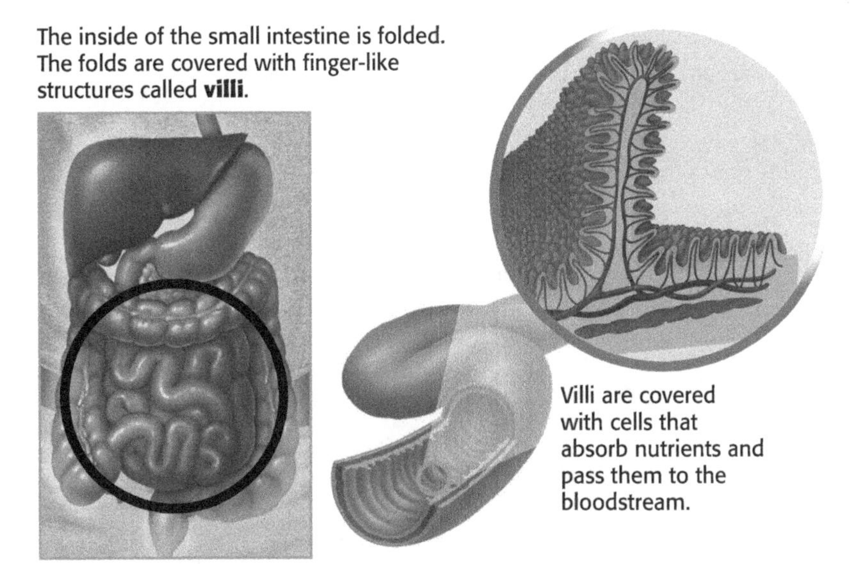

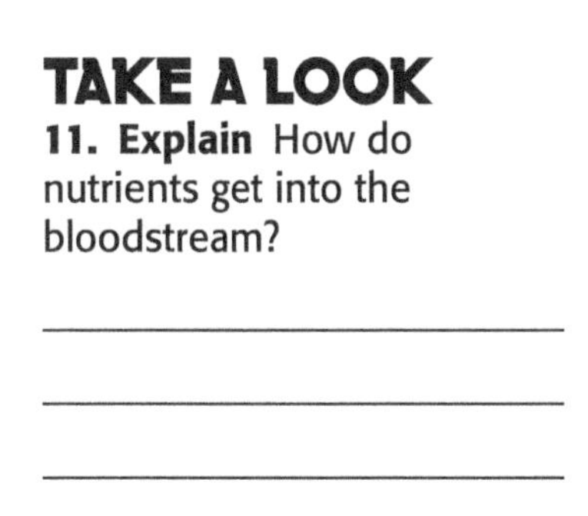

TAKE A LOOK
11. Explain How do nutrients get into the bloodstream?

____

____

____

____

Most nutrient molecules are taken into the blood from the small intestine. However, the body can't use everything you eat. The soupy mixture of water and food that cannot be absorbed moves into the large intestine.

LARGE INTESTINE

The **large intestine** is the last part of the digestive tract. It stores, compacts, and rids the body of waste. The large intestine is wider than the small intestine, but shorter. It takes most of the water out of the mixture from the small intestine. By removing water, the large intestine changes the liquid into mostly solid waste called *feces*, or *stool*.

The *rectum* is the last part of the large intestine. The rectum stores feces until your body can get rid of them. Feces leave the body through an opening called the *anus*. It takes about 24 hours for food to make the trip from your mouth to the end of the large intestine.

You can help keep your digestive system healthy by eating whole grains, fruits, and vegetables. These foods contain a carbohydrate called *cellulose*, or *fiber*. Humans cannot digest fiber. However, fiber keeps the stool soft and keeps materials moving well through the large intestine.

Name With a partner, name as many foods as you can that are sources of fiber.

Section 1 Review

NSES LS 1a, 1d, 1e, 1f, 3b

SECTION VOCABULARY

digestive system the organs that break down food so that it can be used by the body

esophagus a long, straight tube that connects the pharynx to the stomach

gallbladder a sac-shaped organ that stores bile produced by the liver

large intestine the wider and shorter portion of the intestine that removes water from mostly digested food and that turns the waste into semisolid feces, or stool

liver the largest organ in the body; it makes bile, stores and filters blood, and stores excess sugars as glycogen

pancreas the organ that lies behind the stomach and that makes digestive enzymes and hormones that regulate sugar levels

small intestine the organ between the stomach and the large intestine where most of the breakdown of food happens and most of the nutrients from food are absorbed

stomach the saclike, digestive organ between the esophagus and the small intestine and that breaks down food by the action of muscles, enzymes, and acids

1. List What organs in the digestive system are not part of the digestive tract?

____

____

2. Describe What is the function of saliva?

____

____

3. Compare How is chemical digestion different from mechanical digestion?

____

____

____

4. List Name three places in the digestive tract where chemical digestion takes place.

____

5. Explain How does the structure of the small intestine help it absorb nutrients?

____

____

____

6. Apply Concepts How would digestion change if the liver didn't make bile?

____

____

____

Name ______________________ Class ______________ Date ______________

CHAPTER 24 The Digestive and Urinary Systems

SECTION 2

The Urinary System

BEFORE YOU READ

After you read this section, you should be able to answer these questions:

- What is the function of the urinary system?
- How do the kidneys filter the blood?
- What are common problems with the urinary system?

National Science Education Standards
LS 1a, 1c, 1d, 1e, 1f, 3a, 3b

What Does the Urinary System Do?

As your cells break down food for energy, they produce waste. Your body must get rid of this waste or it could poison you! As blood moves through the body and drops off oxygen, it picks up the waste from cells. The **urinary system** is made of organs that take wastes from the blood and send them out of the body.

The chart below describes the main organs of the urinary system.

STUDY TIP
Underline As you read, underline any unfamiliar words. Use the glossary or a dictionary to find out what these words mean. Write the definitions in the margins of the text.

Organ	What it looks like	Function
Kidneys	a pair of organs	to clean the blood and produce urine
Ureters	pair of thin tubes leading from the kidneys to the bladder	to carry urine from the kidneys to the bladder
Urinary bladder	a sac	to store urine
Urethra	tube leading from the bladder to outside your body	to carry urine out of the body

TAKE A LOOK
1. Identify Which organs clean the blood?

____

The Urinary System

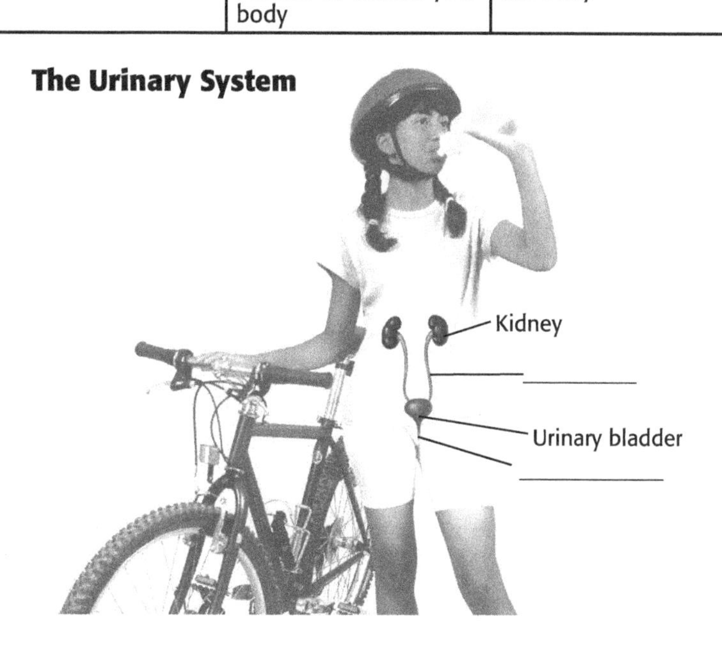

TAKE A LOOK
2. Identify Use the chart above to help you label the organs of the urinary system.

How Do the Kidneys Clean the Blood?

The **kidneys** are a pair of organs that clean the blood. Each kidney is made of many small filters called **nephrons**. Nephrons remove waste from the blood. One of the most important substances that nephrons remove is urea. *Urea* is formed when your cells use protein for energy. The figure below shows how kidneys clean the blood.

Math Focus

3. Calculate Your kidneys filter about 2,000 L of blood each day. Your body has about 5.6 L of blood. About how many times does your blood cycle through your kidneys each day?

____

How the Kidneys Filter Blood

1. Blood enters the kidney through an artery and moves into nephrons.
2. Water, sugar, salts, urea, and other waste products move out of the blood and into the nephron.
3. The nephron returns most of the water and nutrients to the blood. The wastes are left in the nephron.
4. The cleaned blood leaves the kidney and goes back to the body.
5. The waste left in the nephron is a yellow fluid called *urine*. Urine leaves the kidneys through tubes called *ureters*.
6. The urine is stored in the bladder. Urine leaves the body through the urethra.

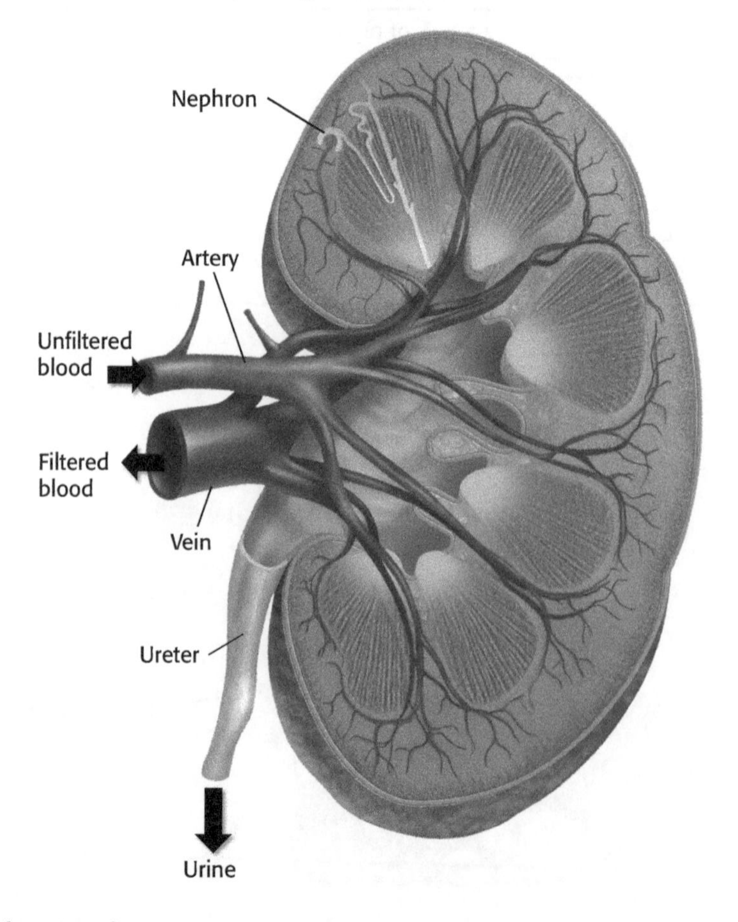

TAKE A LOOK

4. Explain What happens to blood after it is cleaned in the kidneys?

____

____

____

5. Identify What structures carry urine from the kidneys?

____

How Does the Urinary System Control Water?

You lose water every day in sweat and urine, and you replace water when you drink. You need to get rid of as much water as you drink. If you don't, your body will swell up. Chemical messengers called *hormones* help control this balance.

One of these hormones is called *antidiuretic hormone* (ADH). ADH keeps your body from losing too much water. When there is not much water in your blood, ADH tells the nephrons to put water back in the blood. The body then makes less urine. If there is too much water in your blood, your body releases less ADH.

Critical Thinking

6. Infer What do you think happens when your body releases less ADH?

____

____

____

____

____

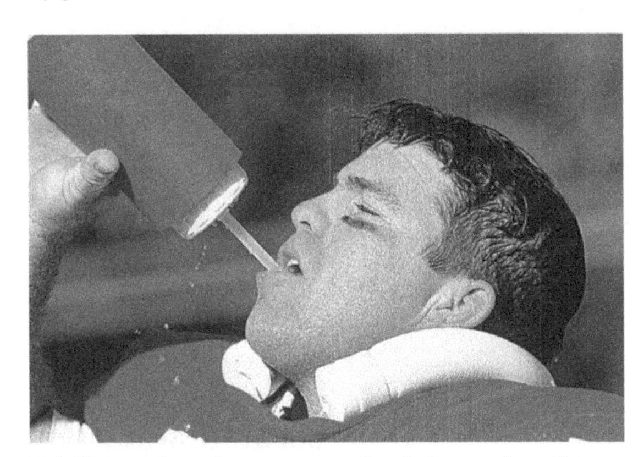

Drinking water when you exercise helps replace the water you lose when you sweat.

Some drinks contain caffeine, which is a diuretic. *Diuretics* cause kidneys to make more urine. This means that a drink with caffeine can actually cause you to lose water.

What Problems Can Happen in the Urinary System?

The chart below shows some common problems of the urinary system.

Problem	Description	Treatment
Bacterial infections	Bacteria can infect the urinary system and cause pain or permanent damage.	Antibiotics
Kidney stones	Wastes can be trapped in the kidney and form small stones. They can stop urine flow and cause pain.	Most can pass out of the body on their own. Some may need to be removed by a doctor.
Kidney disease	Damage to the nephrons can stop the kidneys from working.	A kidney machine can be used to filter the blood.

TAKE A LOOK

7. Explain What problems do kidney stones cause?

____

____

____

Name ______________________ Class ______________ Date ______________

Section 2 Review

NSES LS 1a, 1c, 1d, 1e, 1f, 3a, 3b

SECTION VOCABULARY

kidney one of the pair of organs that filter water and wastes from the blood and that excrete products as urine **nephron** the unit in the kidney that filters blood	**urinary system** the organs that make, store, and eliminate urine.

1. List What are the main organs that make up the urinary system?

____

2. Summarize Complete the process chart to show how blood is filtered.

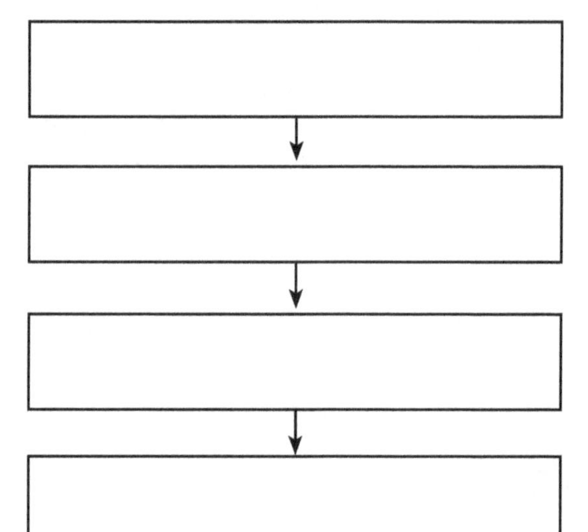

a. Water is put back in the blood.

b. Blood goes into the kidney.

c. Water and waste go into the nephron.

d. Nephrons separate water and waste.

3. Apply Concepts Which of the following has more water: the blood going into the kidney, or the blood leaving it? Explain your answer.

____

____

____

4. Explain How does the urinary system control the amount of water in the body?

____

____

____

5. Infer Is it a good idea to drink beverages with caffeine when you are exercising? Explain your answer.

____

____

____

Name ______________________ Class ______________ Date ______________

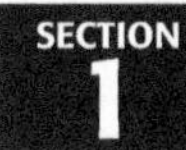

CHAPTER 25 Communication and Control

SECTION 1

The Nervous System

BEFORE YOU READ

After you read this section, you should be able to answer these questions:

- What does the nervous system do?
- What is the structure of the nervous system?

National Science Education Standards

LS 1a, 1d, 1e, 3b, 3c

What Are the Two Main Parts of the Nervous System?

What is one thing that you have done today that did not involve your nervous system? This is a trick question! Your nervous system controls almost everything you do.

The nervous system has two basic functions. First, it collects information and decides what the information means. This information comes from inside your body and from the world outside your body. Second, the nervous system responds to the information it has collected.

The nervous system has two parts: the central nervous system and the peripheral nervous system. The **central nervous system** (CNS) includes the brain and the spinal cord. The CNS takes in and responds to information from the peripheral nervous system. ☑

The **peripheral nervous system** (PNS) includes all the parts of the nervous system except the brain and the spinal cord. The PNS connects all parts of the body to the CNS. Special structures called nerves in the PNS carry information between the body and the CNS.

Organize As you read, make a chart that describes different structures in the nervous system.

READING CHECK

1. Identify What are the two main parts of the nervous system?

____

____

Part of the nervous system	What it includes	What it does
Central nervous system (CNS)		takes in and responds to messages from the PNS
Peripheral nervous system (PNS)		

TAKE A LOOK

2. Summarize Complete the chart to describe the main parts of the nervous system.

Math Focus

3. Calculate To calculate how long an impulse takes to travel a certain distance, you can use the following equation:

$$time = \frac{distance}{speed}$$

If an impulse travels 100 m/s, about how long will it take the impulse to travel 10 meters?

____

How Does Information Move Through the Nervous System?

Special cells called **neurons** carry the information that travels through your nervous system. Neurons carry information in the form of electrical energy. These electrical messages are called *impulses*. Impulses may travel up to 150 m/s!

Like any other cell in your body, a neuron has a nucleus and organelles. The nucleus and organelles are found in the *cell body* of the neuron. However, neurons also have structures called dendrites and axons that are not found in other kinds of cells.

Dendrites are parts of the neuron that branch from the cell body. Most dendrites are very short compared to the rest of the neuron. A single neuron may have many dendrites. Dendrites bring messages from other cells to the cell body.

Axons are longer than dendrites. Some axons can be as long as 1 m! Axons carry information away from the cell body to other cells. The end of an axon is called an *axon terminal*.

TAKE A LOOK

4. Identify Add an arrow to the diagram that shows the direction of impulses at the dendrites.

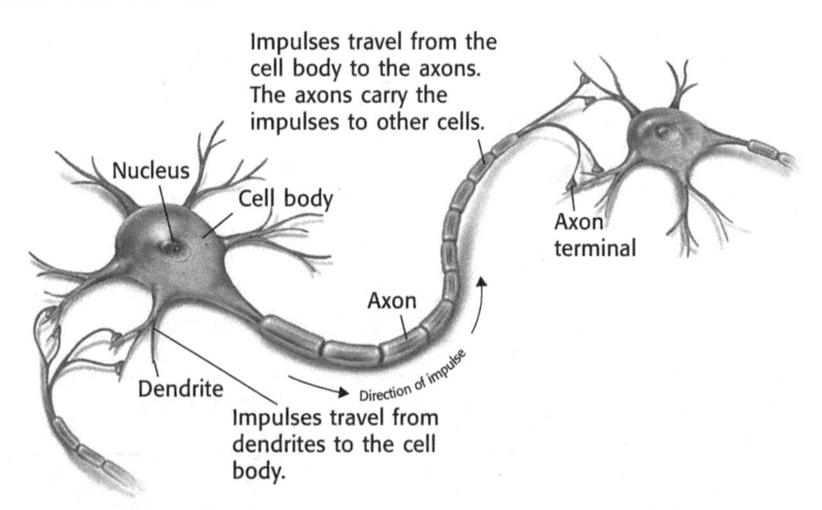

SENSORY NEURONS

Some neurons are sensory neurons. *Sensory neurons* carry information about what is happening in and around your body. Some sensory neurons are called receptors. *Receptors* can detect changes inside and outside your body. For example, receptors in your eyes can sense light. Sensory neurons carry information from the receptors to the CNS.

Identify In a small group, discuss what you think sensory receptors in your eyes, nose, ears, and finger tips respond to.

MOTOR NEURONS

Motor neurons carry impulses from the CNS to other parts of your body. Most motor neurons carry impulses to muscle cells. When muscles cells receive impulses from motor neurons, the muscle cells contract. Some motor neurons carry impulses to glands, such as sweat glands. These messages tell sweat glands when to make sweat.

STANDARDS CHECK

LS 1d Specialized cells perform specialized functions in multicellular organisms. Groups of specialized cells cooperate to form a tissue, such as a muscle. Different tissues are in turn grouped together and form larger functional units called organs. Each type of cell, tissue, and organ has a distinct structure and set of functions that serve the organism as a whole.

Word Help: structure
a whole that is built or put together from parts

Word Help: function
use or purpose

5. Predict What would happen if your nerves stopped working?

____

____

____

____

____

NERVES

In many parts of your body, groups of axons are wrapped together with blood vessels and connective tissue to form bundles. These bundles are called **nerves**. Your central nervous system is connected to the rest of your body by nerves.

Nerves are found everywhere in your PNS. Most nerves contain axons from both sensory neurons and motor neurons. Many nerves carry impulses from your CNS to your PNS. Other nerves carry impulses from your PNS to your CNS.

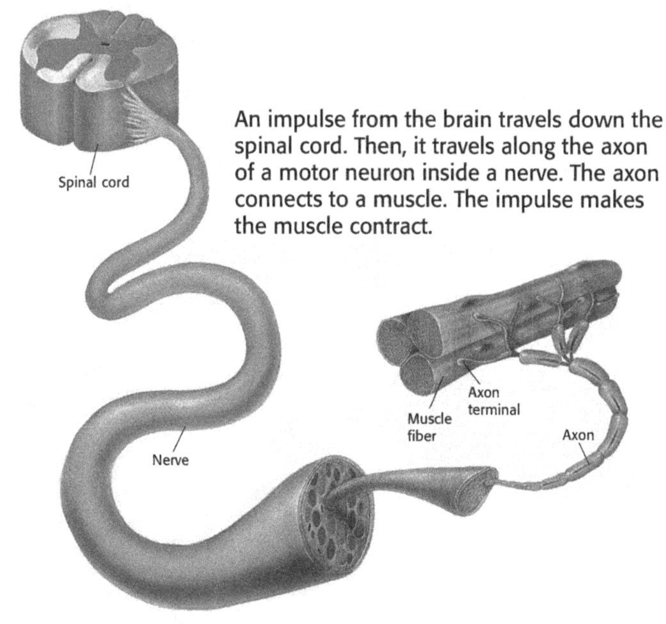

An impulse from the brain travels down the spinal cord. Then, it travels along the axon of a motor neuron inside a nerve. The axon connects to a muscle. The impulse makes the muscle contract.

What Are the Parts of the Peripheral Nervous System?

Recall that the PNS connects the CNS to the rest of the body. Sensory neurons and motor neurons are both found in the PNS. Sensory neurons carry information to the CNS. Motor neurons carry information from the CNS to the PNS. The motor neurons in the PNS make up two groups: the somatic nervous system and the autonomic nervous system. ☑

READING CHECK

6. Identify What are the two main groups of motor neurons in the PNS?

____

____

____

____

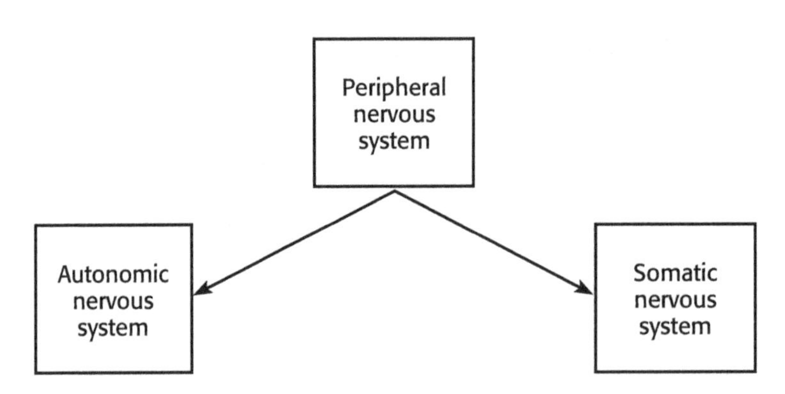

THE SOMATIC NERVOUS SYSTEM

The *somatic nervous* system is made up of motor neurons that you can control. These neurons are connected to skeletal muscles. They control voluntary movements, or movements that you have to think about. These movements include walking, writing, and talking.

Critical Thinking

7. Explain How does eating a piece of pizza involve both the somatic and autonomic nervous systems?

____

____

____

____

____

____

THE AUTONOMIC NERVOUS SYSTEM

The *autonomic nervous system* controls body functions that you do not have to think about. These include digestion and the beating of your heart. The main job of the autonomic nervous system is to keep all of the body's functions in balance.

The autonomic nervous system has two divisions: the *sympathetic nervous system* and the *parasympathetic nervous system*. These two divisions work together to maintain a stable state inside your body. This stable state is called *homeostasis*. The table below shows how the sympathetic and parasympathetic nervous systems work together. ☑

8. Identify What are the two divisions of the autonomic nervous system?

____

____

Organ	Effect of sympathetic nervous system	Effect of parasympathetic nervous system
Eye	makes pupils larger to let in more light	returns pupils to normal size
Heart	raises heart rate to increase blood flow	lowers heart rate to decrease blood flow
Lungs	makes bronchioles larger to get more oxygen into the blood	returns bronchioles to normal size
Blood vessels	makes blood vessels smaller to increase blood pressure	has little or no effect
Intestines	reduces blood flow to stomach and intestines to slow digestion	returns digestion to normal pace

What Are the Parts of the Central Nervous System?

The central nervous system receives information from sensory neurons. It responds by sending messages to the body through motor neurons. The CNS is made of two important organs: the brain and the spinal cord.

The **brain** is the control center of the nervous system. It is also the largest organ in the nervous system. Many processes that the brain controls are involuntary. However, the brain also controls many voluntary processes. The brain has three main parts: the cerebrum, the cerebellum, and the medulla. Each part of the brain has its own job.

THE CEREBRUM

The *cerebrum* is the largest part of your brain. This dome-shaped area is where you think and where most memories are kept. The cerebrum controls voluntary movements. It also lets you sense touch, light, sound, odors, tastes, pain, heat, and cold. ☑

9. Identify What kind of movements does the cerebrum control?

____

____

The cerebrum is made up of two halves called *hemispheres*. The left hemisphere controls most movements on the right side of the body. The right hemisphere controls most movements on the left side of the body. The two hemispheres also control different types of activities, as shown in the figure below. However, most brain activities use both hemispheres.

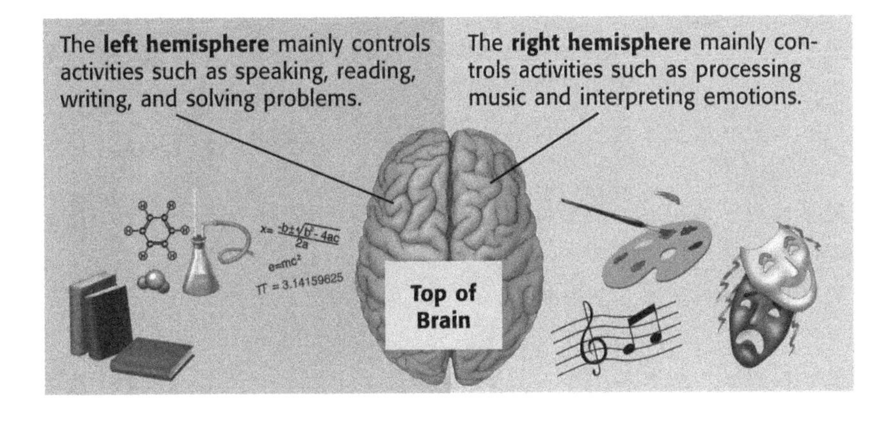

TAKE A LOOK

10. Explain Which hemisphere of your brain are you mainly using as you read this book? Explain your answer.

____

____

____

____

____

____

THE CEREBELLUM

The *cerebellum* is found beneath the cerebrum. The cerebellum receives and processes information from your body, such as from your skeletal muscles and joints. This information lets the brain keep track of your body's position. For example, your cerebellum lets you know when you are upside-down. Your cerebellum also sends messages to your muscles to help you keep your balance. ☑

READING CHECK

11. Identify What are two functions of the cerebellum?

____

____

____

____

THE MEDULLA

The *medulla* is the part of your brain that connects to your spinal cord. It is only about 3 cm long, but you cannot live without it. It controls involuntary processes, such as breathing and regulating heart rate.

Your medulla is always receiving sensory impulses from receptors in your blood vessels. It uses this information to control your blood pressure. If your blood pressure gets too low, your medulla sends impulses that cause blood vessels to tighten. This makes your blood pressure rise. The medulla also sends impulses to the heart to make it beat faster or slower.

TAKE A LOOK

12. Identify Which part of the brain controls breathing?

____

13. Identify Which part of the brain senses smells?

____

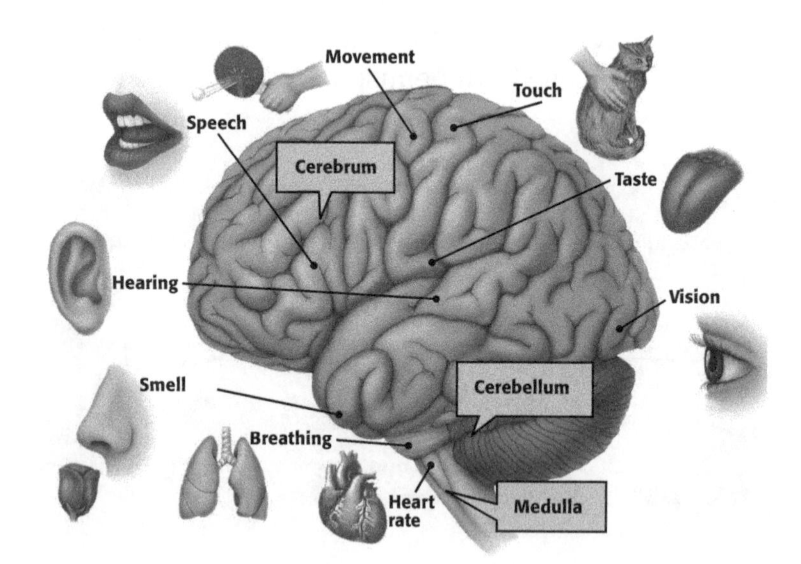

THE SPINAL CORD

Your spinal cord is part of your central nervous system. It is made up of neurons and bundles of axons that send impulses to and from the brain. The spinal cord is surrounded by bones called *vertebrae* (singular, *vertebra*) to protect it.

The axons in your spinal cord let your brain communicate with your PNS. The axons of sensory neurons carry impulses from your skin and muscles to your spinal cord. The impulses travel through the spinal cord to your brain. The brain then processes the impulses and sends signals back through the spinal cord. The impulses travel from the spinal cord to motor neurons. The axons of motor neurons carry the signals to your body.

Critical Thinking

14. Explain What is one way your nervous system and skeletal system work together?

____

____

____

____

____

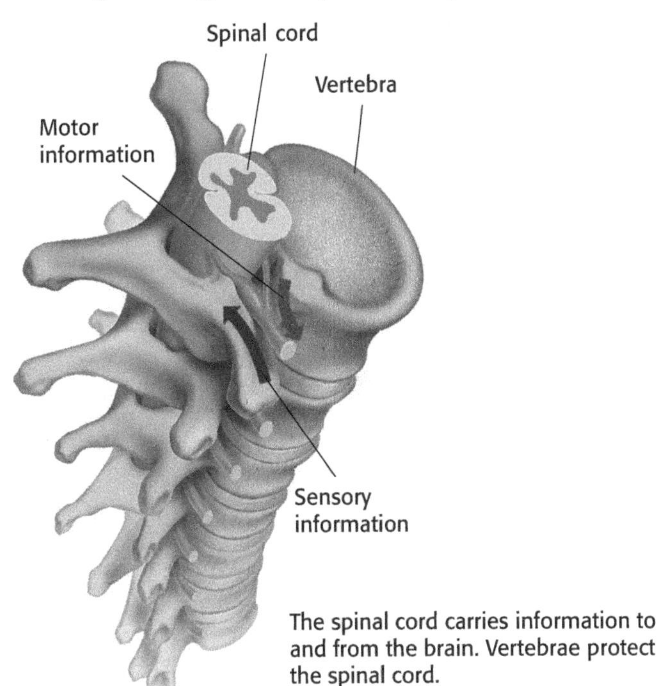

The spinal cord carries information to and from the brain. Vertebrae protect the spinal cord.

SPINAL CORD INJURIES

A spinal cord injury may block the messages to and from the brain. For example, a spinal cord injury might block signals to and from the feet and legs. People with such an injury cannot sense pain, touch, or temperature in their feet or legs. They may not be able to move their legs.

Each year, thousands of people are paralyzed by spinal cord injuries. Most spinal cord injuries in young people happen during sports. You can help prevent such injuries by using proper safety equipment. ☑

READING CHECK

15. Identify How can you help prevent spinal cord injuries while playing sports?

____

____

____

Name ______________________ Class ______________ Date ______________

Section 1 Review

NSES LS 1a, 1d, 1e, 3b, 3c

SECTION VOCABULARY

brain the mass of nerve tissue that is the main control center of the nervous system

central nervous system the brain and the spinal cord

nerve a collection of nerve fibers through which impulses travel between the central nervous system and other parts of the body

neuron a nerve cell that is specialized to receive and conduct electrical impulses

peripheral nervous system all of the parts of the nervous system except for the brain and the spinal cord

1. Compare How do the functions of the CNS and the PNS differ?

____

____

____

2. Summarize Complete the diagram below to show the structure of the nervous system.

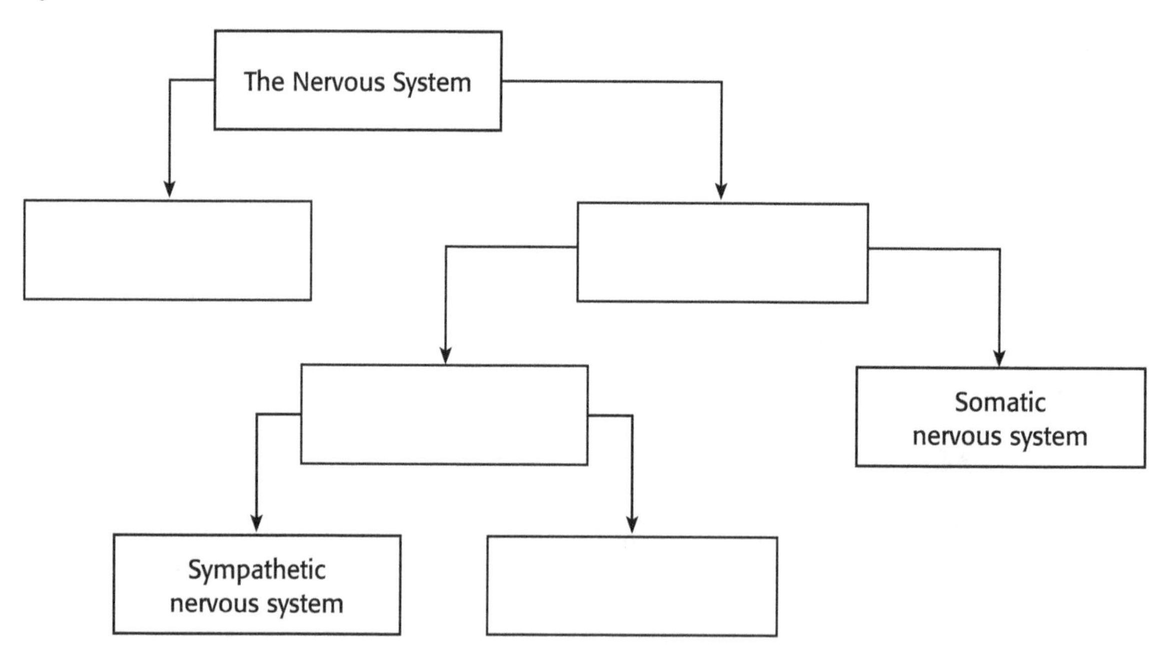

3. Compare How do the functions of dendrites and axons differ?

____

____

4. Explain What can happen to someone with a spinal cord injury?

____

____

____

SECTION 2 Responding to the Environment

BEFORE YOU READ

After you read this section, you should be able to answer these questions:

- How do the integumentary system and nervous system work together?
- What is a feedback mechanism?
- How do your five senses work?

National Science Education Standards
LS 1a, 1d, 1e, 3a

How Does Your Sense of Touch Work?

When a friend taps you on the shoulder or when you feel a breeze, how does your brain know what has happened? Receptors throughout your body gather information about the environment and send this information to your brain.

You skin is part of the integumentary system. The **integumentary system** is an organ system that protects the body. This system also includes hair and nails. Your skin is the main organ that helps you sense touch. It has many different *sensory receptors* that are part of the nervous system. Each kind of receptor responds mainly to one kind of stimulation. For example, *thermoreceptors* respond to temperature changes.

Sensory receptors detect a stimulus and create impulses. These impulses travel to your brain. In your brain, the impulses produce a sensation. A *sensation* is the awareness of a stimulus.

STUDY TIP

List As you read, make a list of the five senses. In your list, include the type of receptors used by those senses.

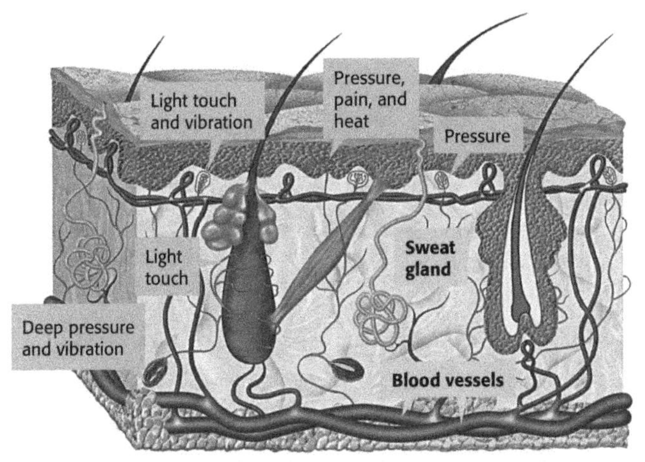

Your skin is made of several layers of tissues and contains many structures. Different kinds of receptors in your skin can sense different things.

TAKE A LOOK

1. List What are three types of sensations that your skin can detect?

____

____

____

REFLEXES

When you step on something sharp, pain receptors in your foot send messages to your spinal cord. The spinal cord sends a message back to move your foot. This immediate reaction that you can't control is a **reflex**. Messages that cause reflexes do not travel all the way to your brain. If you had to wait for your brain to act, you could be badly hurt.

Discuss With a partner, name some other examples of reflexes. What part of the body is involved? When does the reflex happen? How does the reflex protect your body?

FEEDBACK MECHANISMS

Most of the time, the brain decides what to do with the messages from the skin receptors. Your brain helps to control many of your body's functions by using feedback mechanisms. A **feedback mechanism** is a cycle of events in which one step controls or affects another step. In the example below, your brain senses a change in temperature. It tells your sweat glands and blood vessels to react. ☑

2. Complete Feedback mechanisms in your nervous system are controlled by the

______________________.

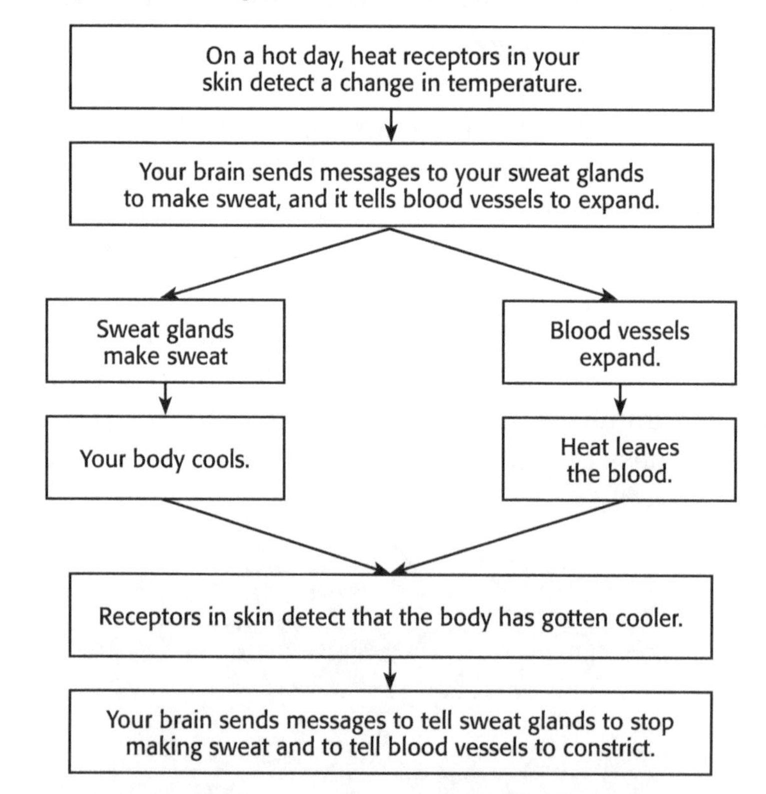

How Does Your Sense of Sight Work?

Sight is the sense that lets you know the size, shape, motion, and color of objects. Light bounces off an object and enters your eyes. Your eyes send impulses to the brain that produce the sensation of sight.

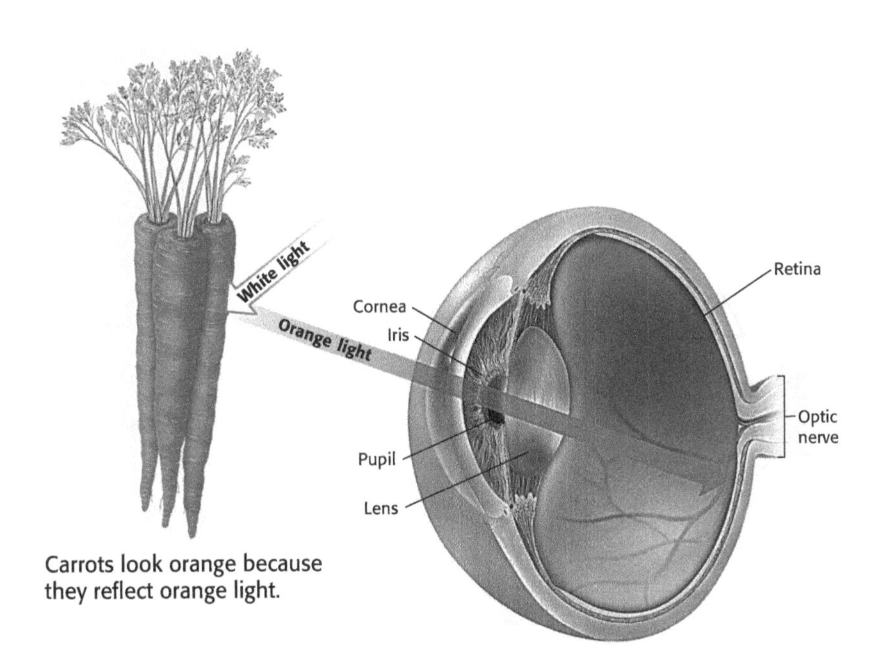

Carrots look orange because they reflect orange light.

TAKE A LOOK

3. Apply Concepts Why do strawberries look red?

____

____

Eyes are complex sensory organs. A clear membrane called the *cornea* protects the eye but lets light through. Light enters an opening called the *pupil.* The light then travels through the lens and hits the retina at the back of the eye.

The **retina** contains neurons called *photoreceptors* that sense light. When light hits a photoreceptor, it sends electrical impulses to the brain. The brain interprets these impulses as light.

The retina has two kinds of photoreceptors: rods and cones. *Rods* are very sensitive to dim light. They are important for night vision and for seeing in black and white. *Cones* are very sensitive to bright light. They let you see colors and fine details. ☑

The impulses from rods and cones travel along axons. The axons form the optic nerve that carries the impulses to your brain.

READING CHECK

4. Identify What are the two kinds of photoreceptors in the retina?

____

____

REACTING TO LIGHT

Your pupil looks like a black dot in the center of your eye. Actually, it is an opening that lets light enter the eye. Around the pupil is a ring of muscle called the *iris.* The iris controls how much light enters your eye. It also gives your eye its color. ☑

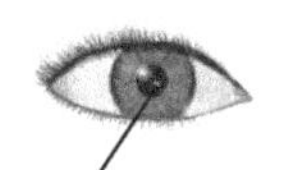

In bright light the iris contracts. This makes the pupil smaller.

In dim light the iris relaxes. This makes the pupil dilate, or get larger

READING CHECK

5. Identify What is the function of the iris?

____

____

____

FOCUSING LIGHT

The lens focuses light onto the retina. The *lens* is an oval-shaped piece of clear, curved material behind the iris. Muscles in the eye change the shape of the lens to focus light on the retina. When you look at something that is close to your eye, the lens becomes more curved. When you look at objects that are far away, the lens gets flatter. ☑

READING CHECK

6. Identify What is the function of the lens?

____

____

How Does Your Sense of Hearing Work?

Sound is produced when something *vibrates*. A drum, for example, vibrates when you hit it. Vibrations produce waves of sound energy. Hearing is the sense that lets you experience sound energy.

Ears are the organs used for hearing. The ear has three main parts: the inner ear, middle ear, and outer ear. The chart below shows the structures and functions of each.

Part of ear	Main structures	Function
Outer ear	ear canal	funnels sound waves into the middle ear
Middle ear	eardrum; three ear bones	sends vibrations to the inner ear
Inner ear	cochlea and auditory nerve	sends impulses to the brain

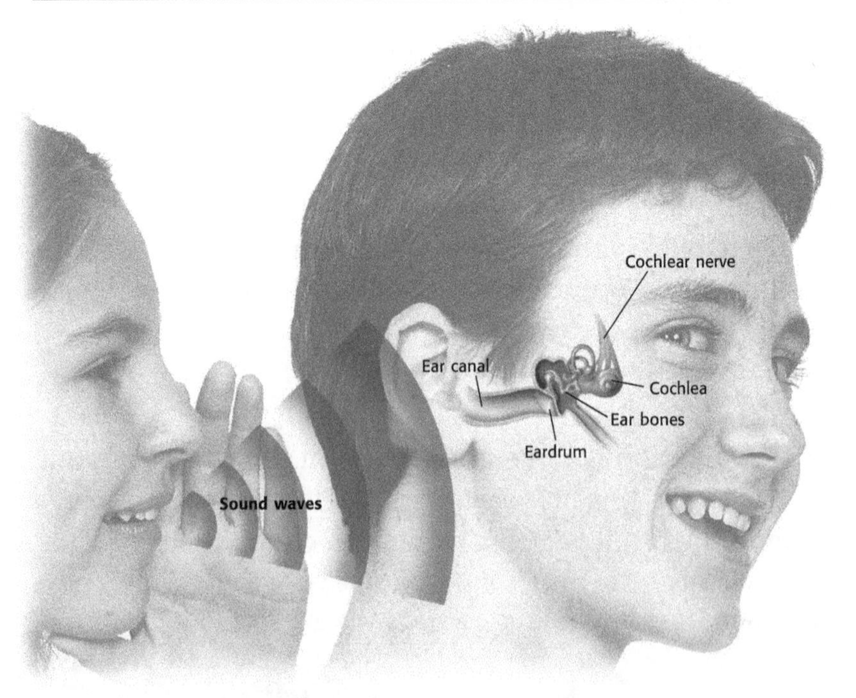

A sound wave travels through the air into the outer ear. The wave produces vibrations in the middle ear and inner ear. These vibrations produce impulses in the cochlear nerve that travel to the brain

TAKE A LOOK

7. Identify Use colored pencils to color the outer ear blue, the middle ear green, and the inner ear red.

SENDING SOUND ENERGY TO YOUR BRAIN

Sound waves create vibrations throughout your ear that your brain can interpret as sound. The outer ear funnels sound waves to the middle ear. Sound waves hit the eardrum and make it vibrate. These vibrations make tiny ear bones vibrate. One of these bones vibrates against the **cochlea**, an organ filled with fluid. The vibrations make waves in the fluid. This causes neurons in the cochlea to send impulses along the auditory nerve to the brain. ☑

READING CHECK

8. Explain What does your brain interpret as sound?

____

How Does Your Sense of Taste Work?

Taste is the sense that lets you detect chemicals and tell one flavor from another flavor. Your tongue is covered with tiny bumps called *papillae* (singular, *papilla*) that contain taste buds. Taste buds have receptors called taste cells that respond to dissolved food molecules in your mouth. Taste cells react to four basic tastes: sweet, sour, salty, and bitter.

How Does Your Sense of Smell Work?

Neurons called *olfactory cells* are located in the nose and have receptors for smell. Your brain combines the information from taste buds and olfactory cells to let you sense flavor.

Critical Thinking

9. Apply Concepts Why do you have a hard time tasting things when you have a cold?

____

____

____

____

____

____

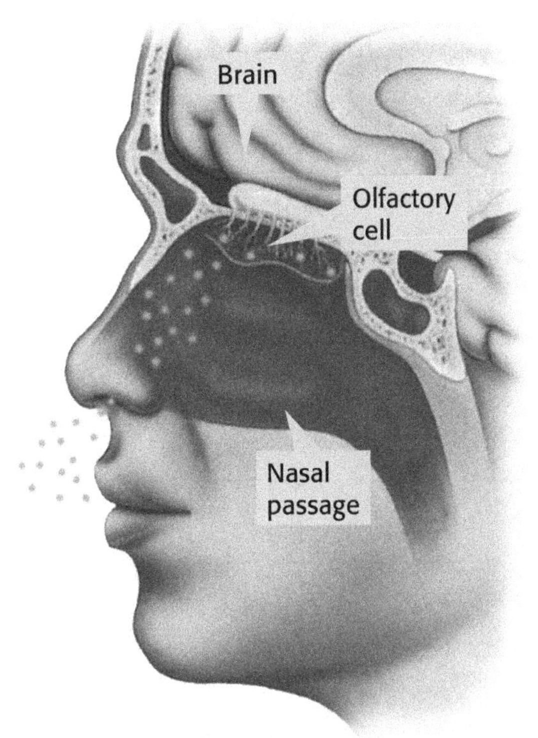

Molecules that you inhale dissolve in the moist lining of the nasal cavity and trigger an impulse. The brain interprets the impulse as the sensation of smell.

Name ______________________ Class ______________ Date ______________

Section 2 Review

NSES LS 1a, 1d, 1e, 3a

SECTION VOCABULARY

cochlea a coiled tube that is found in the inner ear and that is essential to hearing

feedback mechanism a cycle of events in which information from one step controls or affects a previous step

integumentary system the organ system that forms a protective covering on the outside of the body

reflex an involuntary and almost immediate movement in response to a stimulus

retina the light-sensitive inner layer of the eye that receives images formed by the lens and transmits them through the optic nerve to the brain

1. List What are the five senses?

____

2. Explain How do the integumentary system and nervous system work together?

____

____

____

3. Explain What are reflexes and why are they important for the body?

____

____

____

4. Explain Why is important for your eyes to have both rods and cones?

____

____

____

5. Summarize Complete the chart to summarize the major senses and sense receptors.

Sense	Receptors	What the receptors respond to
Touch	many different kinds	
Sight		
	neurons in the cochlea	
Taste		dissolved molecules
	olfactory cells	

Name ______________________ Class ______________ Date ____________

CHAPTER 25 Communication and Control

SECTION 3

The Endocrine System

BEFORE YOU READ

After you read this section, you should be able to answer these questions:

- Why is the endocrine system important?
- How do feedback systems work?
- What are common hormone imbalances?

National Science Education Standards
LS 1a, 1d, 1e

What Is the Endocrine System?

The **endocrine system** controls body functions using chemicals made by endocrine glands. A **gland** is a group of cells that makes special chemicals for your body. The chemicals made by endocrine glands are called hormones. A **hormone** is a chemical messenger. It is made in one cell or tissue and causes a change in another cell or tissue. Hormones flow through the bloodstream to all parts of the body. ☑

STUDY TIP

Describe As you read, fill in the chart at the end of the section to name the major endocrine glands and what each one does.

READING CHECK

1. Explain How do hormones move from one part of the body to another?

____

____

____

Glands and Organs of the Endocrine System

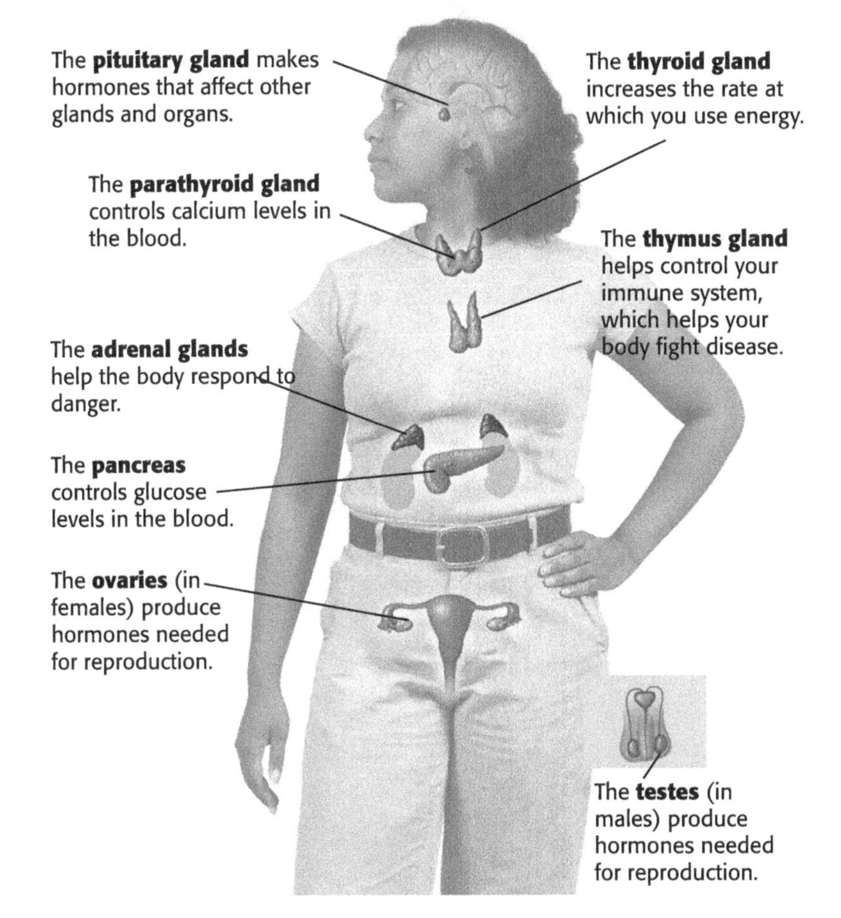

TAKE A LOOK

2. Identify What two structures produce hormones needed for reproduction?

____

____

3. Identify Which gland makes hormones that affect organs and other glands?

____

Critical Thinking

4. Infer Why do you think increasing your breathing rate helps prepare you to fight or run away?

____

ADRENAL GLANDS

Endocrine glands may affect many organs at one time. For example, the adrenal glands release the hormone *epinephrine*, sometimes called *adrenaline*. Epinephrine increases your heartbeat and breathing rate. This response is called the fight-or-flight response. When you are scared, angry, or excited, the fight-or-flight response prepares you either to fight the danger or to run from it.

How Do Feedback Systems Work?

Recall how feedback mechanisms work in the nervous system. Feedback mechanisms are cycles in which information from one step controls another step in the cycle. In the endocrine system, endocrine glands control similar feedback mechanisms.

The pancreas has specialized cells that make two different hormones, *insulin* and *glucagon*. These two hormones control the level of glucose in the blood.

TAKE A LOOK

5. Explain How could you raise your blood-glucose level without involving hormones?

____

6. Identify Which hormone tells the liver to release glucose into the blood?

____

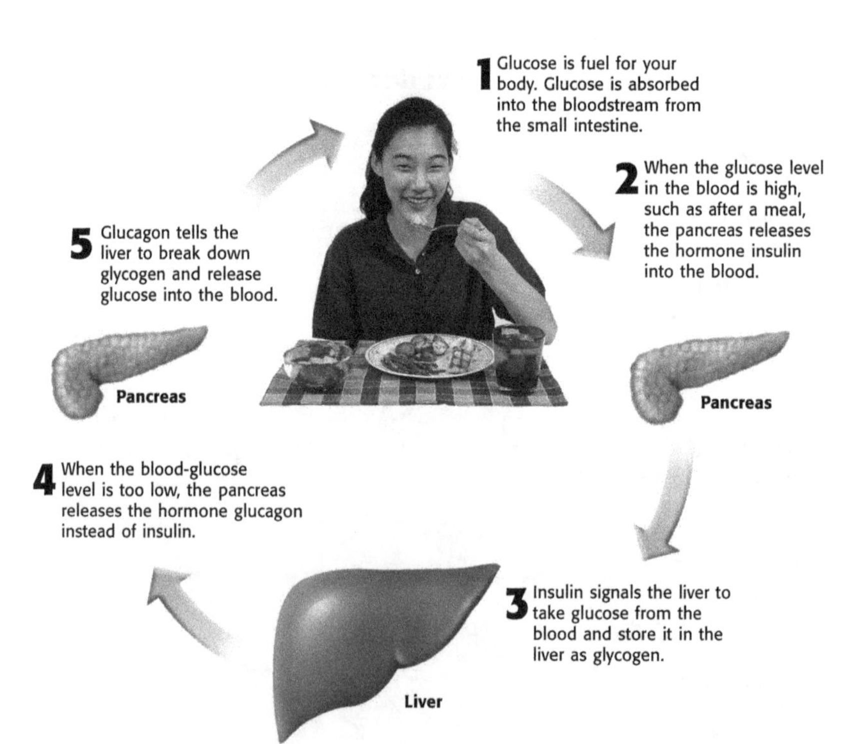

What Is a Hormone Imbalance?

Sometimes, an endocrine gland makes too much or not enough of a hormone. For example, a person's body may not make enough insulin or be able use it properly. This is a condition called *diabetes mellitus*. A person who has diabetes may need daily injections of insulin. These injections help keep his or her blood-glucose levels within safe limits. Some patients get their insulin automatically from a small machine worn on the body. ☑

READING CHECK

7. Define What is a hormone inbalance?

____

____

____

____

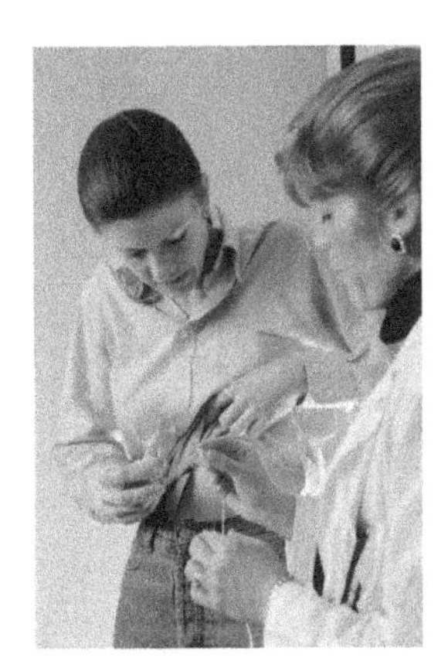

This woman has diabetes. She is wearing a device that delivers insulin to her body.

Another hormone imbalance is when a child's pituitary gland doesn't make enough growth hormone. As a result, the child does not grow as quickly or as much as he or she should. If the problem is found early in childhood, a doctor can prescribe growth hormone.

In some cases, the pituitary gland may make too much growth hormone. This may cause a child to grow taller than expected.

Endocrine gland	Function
Pituitary	
	increases the rate at which you use energy

TAKE A LOOK

8. Summarize Use this chart to help you summarize the major endocrine glands and their functions.

Name ______________________ Class ______________ Date ______________

Section 3 Review

NSES LS 1a, 1d, 1e

SECTION VOCABULARY

endocrine system a collection of glands and groups of cells that secrete hormones that regulate growth, development, and homeostasis; includes the pituitary, thyroid, parathyroid, and adrenal glands, the hypothalamus, the pineal body, and the gonads

gland a group of cells that make special chemicals for the body

hormone a substance that is made in one cell or tissue and that causes a change in another cell or tissue in a different part of the body

1. **Explain** What is the function of the endocrine system?

____

2. **Compare** How are the thymus and thyroid gland similar? How are they different?

____

____

____

3. **Identify Relationships** How do the circulatory system and the endocrine system work together?

____

____

4. **Explain** What does insulin do?

____

____

5. **Apply Concepts** Many organs in the body are part of more than one organ system. List three examples from this section of organs that are part of both the endocrine system and another organ system.

____

____

____

6. **Infer** Glucose is a source of energy. Epinephrine quickly raises your blood-glucose level when you are excited or scared. Why is epinephrine important during these times?

____

____

____

SECTION 1

Animal Reproduction

BEFORE YOU READ

After you read this section, you should be able to answer these questions:

- What is asexual reproduction?
- What is sexual reproduction?
- What is the difference between external and internal fertilization?

Why Do Organisms Reproduce?

Organisms live for different amounts of time. For example, a fruit fly lives for about 40 days. Some pine trees can live for almost 5,000 years. However, all living things will die. For a species to surivive, its members must reproduce.

STUDY TIP

Describe As you read, make a chart describing characteristics and types of asexual and sexual reproduction.

How Are Offspring Made with One Parent?

Some animals reproduce asexually. In **asexual reproduction**, one parent has offspring. These offspring are genetically identical to the parent.

Asexual reproduction can occur through budding, fragmentation, or regeneration. *Budding* happens when the offspring buds or grows out from the parent's body. The new organism then pinches off. *Fragmentation* occurs when part of an organism breaks off and then develops into a new organism. *Regeneration* occurs when part of the organism's body is lost and the organism regrows it. The lost body part can also develop into an offspring. ☑

1. List What are three types of asexual reproduction?

____

____

____

The hydra bud will separate from its parent. Buds from other organisms, such as some coral, stay attached to the parent.

The largest arm on this sea star was lost from another sea star. A new sea star has grown from this arm.

TAKE A LOOK

2. Identify What kind of asexual reproduction does this sea star in this picture show?

____

What Is Sexual Reproduction?

Most animals reproduce sexually. In **sexual reproduction**, offspring form when sex cells from more than one parent combine. The offspring will share traits from both parents. Sexual reproduction usually requires two parents—a male and a female. The female makes sex cells called **eggs**. The male makes sex cells called **sperm**. When a sperm and an egg join together, a fertilized egg, or *zygote* is made. This process is called *fertilization.* ☑

READING CHECK

3. Define What is fertilization?

____

____

____

Almost all human cells have 46 chromosomes. However, egg and sperm cells only have 23 chromosomes. These sex cells are formed by a process called *meiosis*. In humans, meiosis occurs when one cell with 46 chromosomes divides to make sex cells with 23 chromosomes. This way, when a sperm and egg join, the zygote will have the typical number of chromosomes.

How Do Offspring Get Their Parent's Traits?

Genetic information is carried by *genes*. Genes are located on *chromosomes*. At fertilization in humans, the egg and sperm both give 23 chromosomes to the zygote. This zygote will then develop into an organism that has traits from both parents. The figure below shows how genes are passed from parent to offspring.

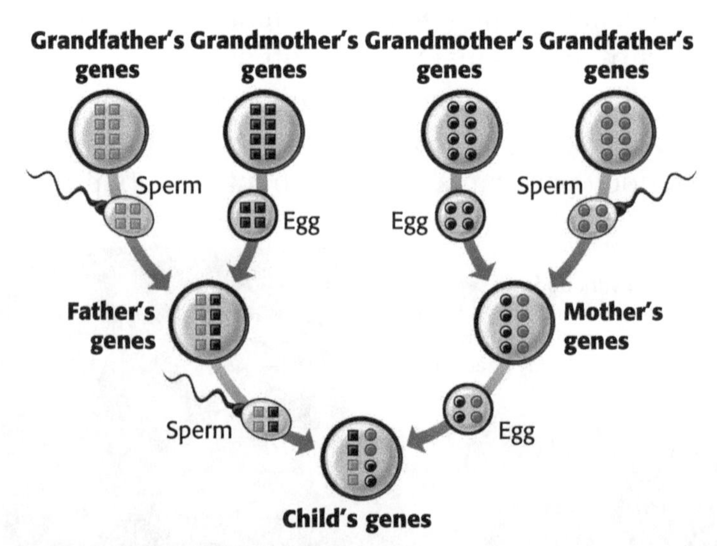

Eggs and sperm contain chromosomes. You inherited half of your chromosomes from each parent.

TAKE A LOOK

4. Identify What fraction of your genes did you inherit from each grandparent?

____

How Does Fertilization Occur?

Fertilization can occur outside or inside the female's body. **External fertilization** occurs when the sperm fertilizes the eggs outside the female's body. Some fish and amphibians, such as frogs, use external fertilization. The female releases her eggs and the male releases his sperm over the eggs. External fertilization generally takes place in wet environments so that the new zygotes do not dry out.

Internal fertilization occurs when an egg and sperm join inside a female's body. This allows the female to protect the developing zygote inside her body. Reptiles, birds, mammals, and some fishes use internal fertilization. Many mammals give birth to well-developed young. However, other animals, such as chickens, use internal fertilization to lay fertilized eggs.

Critical Thinking

5. Infer Name some environments where animals could use external fertilization.

____

____

____

The female clownfish has laid her eggs on the rock. The male will then fertilize them.

TAKE A LOOK

6. Identify Is this an example of internal or external fertlization?

____

How Do Mammals Reproduce?

All mammals reproduce sexually by internal fertilization. They also feed their young with milk. Mammals reproduce in one of the following ways. ☑

- *Monotremes* lay eggs. A platypus is a monotreme.
- *Marsupials* give birth to young that are not well developed. After birth, the young crawl into their mother's pouch to develop. A kangaroo is a marsupial.
- *Placental mammals* are fed and protected inside their mother's body before birth. These mammals are born well developed. Humans are placental mammals.

READING CHECK

7. Identify Name two ways that all mammals are alike.

____

____

____

Name ______________________ Class ______________ Date ______________

Section 1 Review

SECTION VOCABULARY

asexual reproduction reproduction that does not involve the union of sex cells and in which a single parent produces offspring that are genetically identical to the parent

egg a sex cell produced by a female

external fertilization the union of sex cells outside the bodies of the parents

internal fertilization fertilization of an egg by sperm that occurs inside the body of a female

sexual reproduction reproduction in which the sex cells from two parents unite to produce offspring that share traits from both parents

sperm the male sex cell

1. Compare How is fragmentation different from budding?

____

____

____

2. Infer Why is reproduction as important to a bristlecone pine as it is to a fruit fly?

____

____

3. Apply Concepts Why is meiosis important in sexual reproduction?

____

____

____

4. Identify What type of reproduction produces offspring that are genetically identical to the parent?

____

5. Compare What is the difference between external and internal fertilization?

____

____

6. Identify What is one advantage of internal fertilization?

____

____

7. List What are three groups of animals that use internal fertilization?

____

Name ______________________ Class ______________ Date ______________

CHAPTER 26 Reproduction and Development

SECTION 2

Human Reproduction

BEFORE YOU READ

After you read this section, you should be able to answer these questions:

- How are sperm and eggs made?
- How does fertilization occur?
- What problems can happen in the reproductive system?

National Science Education Standards
LS 1d, 1f

What Happens in the Male Reproductive System?

The male reproductive system has two functions:

- to make sperm
- to deliver sperm to the female reproductive system

To perform these functions, organs in the male reproductive system make sperm, hormones, and fluids. The **testes** (singular, *testis*) are a pair of organs that hang outside the body covered by a skin sac called the *scrotum.* They make sperm and *testosterone*, the main male sex hormone.

A male can make millions of sperm each day. Immature sperm cells divide and change shape as they travel through the testes and epididymis. The *epididymis* is a tube attached to the testes that stores sperm as they mature. ☑

Mature sperm pass into the *vas deferens*, which connects the epididymis and urethra. The *urethra* is a tube that runs from the bladder through the penis. The **penis** is the male organ that delivers sperm to the female. Before leaving the body, sperm mixes with a fluid mixture to form *semen.*

STUDY TIP

Summarize As you read, create two Process Charts. In the first, describe the path an egg takes from ovulation to fertilization. In the second, describe the path of an egg that does not get fertilized.

READING CHECK

1. Identify How many sperm can a male make in one day?

____

The Male Reproductive System

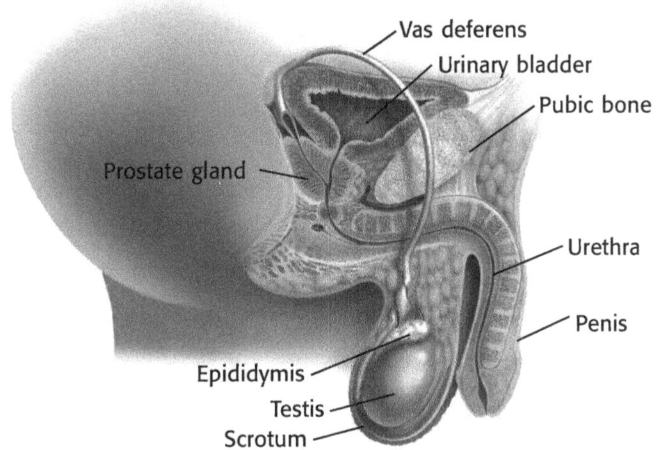

TAKE A LOOK

2. Identify On the diagram, draw an arrow pointing to the structure that makes sperm.

What Happens in the Female Reproductive System?

The female reproductive system has three functions: to produce eggs, to protect and nourish developing offspring, and to give birth. Unlike males, who produce new sperm throughout their lives, females have all their eggs when they are born.

Eggs are produced in an **ovary**. Ovaries also release the main female sex hormones: estrogen and progesterone. These hormones control the release of eggs from the ovaries and the development of female characteristics. Females generally have two ovaries.

Math Focus

3. Calculate The average woman ovulates each month from about the age of 12 to about the age of 50. How many mature eggs does she release from age 18 to age 50? Assume that she has never been pregnant.

THE EGG'S JOURNEY

During *ovulation* an ovary releases an egg. The egg passes into a *fallopian tube*. The fallopian tube leads from the ovary to the uterus. If sperm are present, fertilization usually takes place in the fallopian tube.

After fertilization, the embryo moves to the uterus and may embed in the thick lining. An embyo develops into a fetus in the **uterus**. When the baby is born, it passes from the uterus and through the vagina. The **vagina** is the canal between the outside of the body and the uterus.

The Female Reproductive System

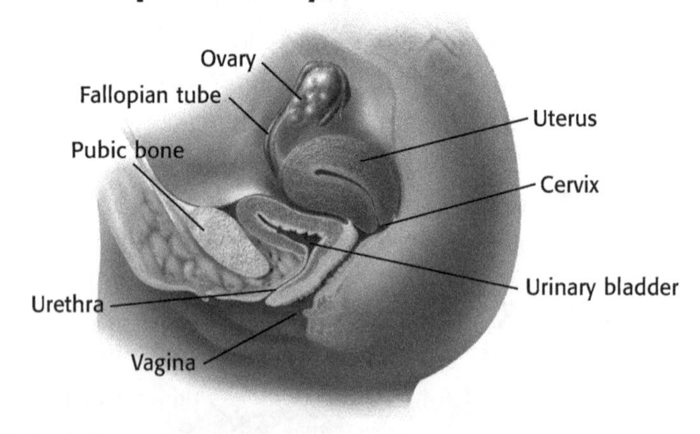

TAKE A LOOK

4. Identify On the diagram, put a circle around the structure that produces eggs. Put a square around the structure where an embryo develops.

THE MENSTRUAL CYCLE

From puberty through her late 40s or early 50s, a woman's reproductive system goes through the *menstrual cycle*. This cycle of about 28 days prepares the body for pregnancy. An ovary releases an egg at *ovulation*. This happens at around the 14th day of the cycle. If the egg is not fertilized, menstruation begins. *Menstruation* is the monthly discharge of blood and tissue from the uterus.

What Problems Can Happen in the Reproductive System?

Problems such as disease can cause the reproductive system to fail. When couples cannot have children, they are considered *infertile.* Men are infertile if they do not make enough healthy sperm. Women are infertile if they do not ovulate normally. Reproductive problems are often caused by sexually transmitted diseases and cancers.

STANDARDS CHECK

LS 1f Disease is the breakdown in structures or functions of an organism. Some diseases are the result of intrinsic failures of the system. Others are the result of damage by infection by other organisms.

5. Identify What are two common causes of reproductive problems?

____

____

SEXUALLY TRANSMITTED DISEASES

A *sexually transmitted disease* (STD) is a disease that can pass from one person to another during sexual contact. STDs are also called *sexually transmitted infections* (STIs). Sexually-active young people have the highest risk for STDs.

One example of an STD is human immunodeficiency virus (HIV), the virus that leads to AIDS. HIV destroys the immune system of the infected person. People with AIDS generally die from infections that are not fatal to people with healthy immune systems. Below is a table showing the most common STDs and how fast they are spreading in the United States.

STD	Approximate number of new cases each year
Chlamydia	3 to 10 million
Genital HPV (human papilloma virus)	5.5 million
Genital herpes	1 million
Gonorrhea	650,000
Syphilis	70,000
HIV/AIDS	40,000 to 50,000

Critical Thinking

6. Infer In women, some untreated STDs can block the fallopian tubes. How would this affect fertilization?

____

____

____

____

____

____

CANCER

Sometimes cancer happens in reproductive organs. *Cancer* is a disease in which cells grow at an uncontrolled rate. In men, the two most common cancers of the reproductive system happen in the testes and prostate gland. In women, two common reproductive cancers are cancer of the cervix and cancer of the ovaries.

Research Use your school library or the internet to research one of the STDs in the chart. What organism or virus causes it? How does it affect the body? What treatments are available? Present your findings to the class.

Name ______________________ Class ______________ Date ______________

Section 2 Review

NSES LS 1d, 1f

SECTION VOCABULARY

ovary an organ in the female reproductive system of animals that produces eggs

penis the male organ that transfers sperm to a female and that carries urine out of the body

testes the primary male reproductive organs which produce sperm and testosterone (singular, *testis*)

uterus in female placental mammals, the hollow, muscular organ in which an embryo embeds itself and develops into a fetus

vagina the female reproductive organ that connects the outside of the body to the uterus

1. **Explain** What is the purpose of the menstrual cycle?

____

____

2. **Organize** Complete the chart below to describe the functions or characteristics of structures in the female reproductive system.

Structure	Characteristic or function
	produces eggs; releases female sex hormones
uterus	
fallopian tube	
	canal that connects uterus to the outside

3. **Explain** What is fertilization and where does it occur?

____

____

4. **Apply Concepts** Fraternal twins are created when two sperm fertilize two different eggs. Paternal, or identical, twins are created when a single egg divides after fertilization. Why are fraternal twins not identical?

____

____

____

____

Name ______________________ Class ______________ Date ______________

CHAPTER 26 Reproduction and Development

SECTION 3

Growth and Development

BEFORE YOU READ

After you read this section, you should be able to answer these questions:

- What happens after an egg is fertilized?
- How does a fetus develop?
- How does a person develop after birth?

National Science Education Standards
LS 1a, 1c, 1d, 1e, 2b

How Does Fertilization Occur?

A man can release millions of sperm at once. However, only one sperm can fertilize an egg. Why are so many sperm needed?

Only a few hundred sperm survive the journey from the vagina to the uterus and into a fallopian tube. In the fallopian tube only a few sperm find and cover the egg. Once one sperm enters, or *penetrates*, the egg, it causes the egg's covering to change. This change keeps other sperm from entering the egg. When the nuclei of the sperm and egg join, the egg is fertilized.

STUDY TIP

Summarize As you read, make a a timeline that shows the stages of human development from fertilized egg to adulthood.

Critical Thinking

1. Form Hypothesis Why do you think millions of sperm are released to fertilize one egg?

____

____

____

____

____

Fertilization and Implantation

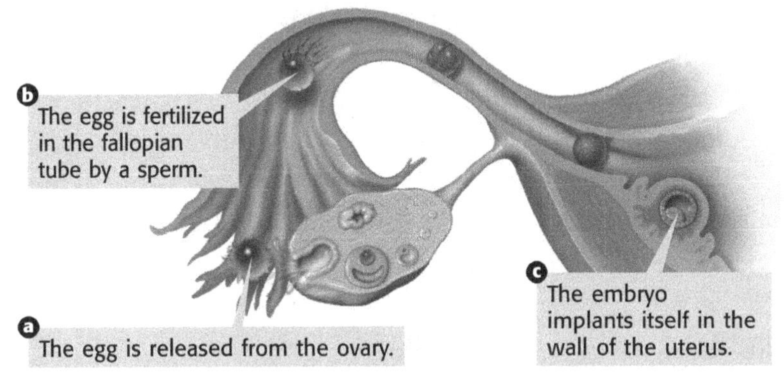

TAKE A LOOK

2. Identify Where does fertilization usually take place?

____

What Stages Does a Fertilized Egg Go Through?

At fertilization, the egg is only a single cell. At this stage, the fertilized egg is called a *zygote*. As the zygote becomes an embryo it moves down the fallopian tube, and divides many times. After about a week, the **embryo** is a ball of cells. This ball of cells implants in the uterus. ☑

As the cells of the embryo continue to divide, some cells start to *differentiate*. They develop special structures for certain jobs in the body. After week 10 of the pregnanacy, the embryo is called a **fetus**.

READING CHECK

3. Identify What is the egg called right after fertilization?

____

TAKE A LOOK

4. List List the three stages a fertilized egg goes through as it develops.

____

____

____

The fertilized egg, called a *zygote,* divides and becomes a ball of cells → The ball of cells, called an *embryo,* implants in the uterus → The cells of the embryo divide and differentiate; the embryo becomes a *fetus*

How Does an Embryo Develop?

WEEKS 1 AND 2

A woman's pregnancy starts when an egg is fertilized, and ends at birth. Pregnancy is measured from the starting date of a woman's last menstruation. This is easier than trying to determine the exact date fertilization took place. Fertilization takes place at about the end of week 2. ☑

READING CHECK

5. Explain Why is pregnancy measured from the starting date of the woman's last period?

____

____

____

____

WEEKS 3 AND 4

In week 3, the embryo moves to the uterus. As the embryo travels, it divides many times. It becomes a ball of cells that implants itself in the wall of the uterus.

WEEKS 5 TO 8

After an embryo implants in the uterus, the placenta forms. The **placenta** is an organ used by the embryo to exchange materials with the mother. The placenta has many blood vessels that carry nutrients and oxygen from the mother to the embryo. They also carry wastes from the embryo to the mother.

In week 5 of pregnancy, the **umbilical cord** forms. It connects the embryo to the placenta. A thin membrane called the amnion develops. The *amnion* surrounds the embryo and is filled with fluid. This fluid cushions and protects the embryo. ☑

READING CHECK

6. Explain What is the function of the umbilical cord?

____

____

____

WEEKS 9 TO 16

At week 9, the embryo may start to make tiny movements. The fetus grows very quickly during this stage. It doubles, then triples in size within a month. In about week 13, the fetus's face begins to look more human. During this stage the fetus's muscles also grow stronger. It can even make a fist.

WEEKS 17 TO 24

By week 17 the fetus can make faces. By week 18 the fetus starts to make movements that its mother can feel. It can also hear sounds through the mother's uterus. By week 23 the fetus makes a lot of movements. A baby born during week 24 might survive, but it would need a lot of help.

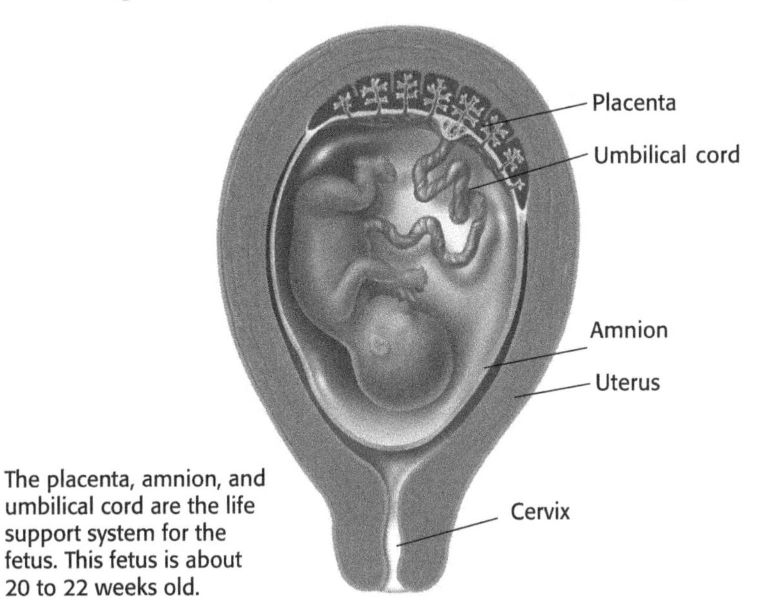

The placenta, amnion, and umbilical cord are the life support system for the fetus. This fetus is about 20 to 22 weeks old.

TAKE A LOOK

7. Identify What structure surrounds the fetus?

____

WEEKS 25 TO 36

At about 25 or 26 weeks, the fetus's lungs are well-developed. However, they are not fully mature. The fetus still gets oxygen from its mother through the placenta. The fetus will not take its first breath of air until it is born.

By the 32nd week, the fetus can open and close its eyes. Studies show that the fetus responds to light. Some scientists think fetuses at this stage show brain activity and eye movements like sleeping children or adults. These scientists think a sleeping fetus may dream. After 36 weeks, the fetus is almost ready to be born.

BIRTH

At weeks 37 to 38 the fetus is fully developed. A full pregnancy is usually 40 weeks. As birth begins, the muscles of the uterus begin to squeeze, or contract. This is called *labor*. These contractions push the fetus out of the mother's body through the vagina. ☑

Once the baby is born, the umbilical cord is tied and cut. The navel is all that will remain of the point where the umbilical cord was attached. Once the mother's body has pushed out the placenta, labor is complete.

READING CHECK

8. Define What is labor?

____

____

____

Pregnancy Timeline

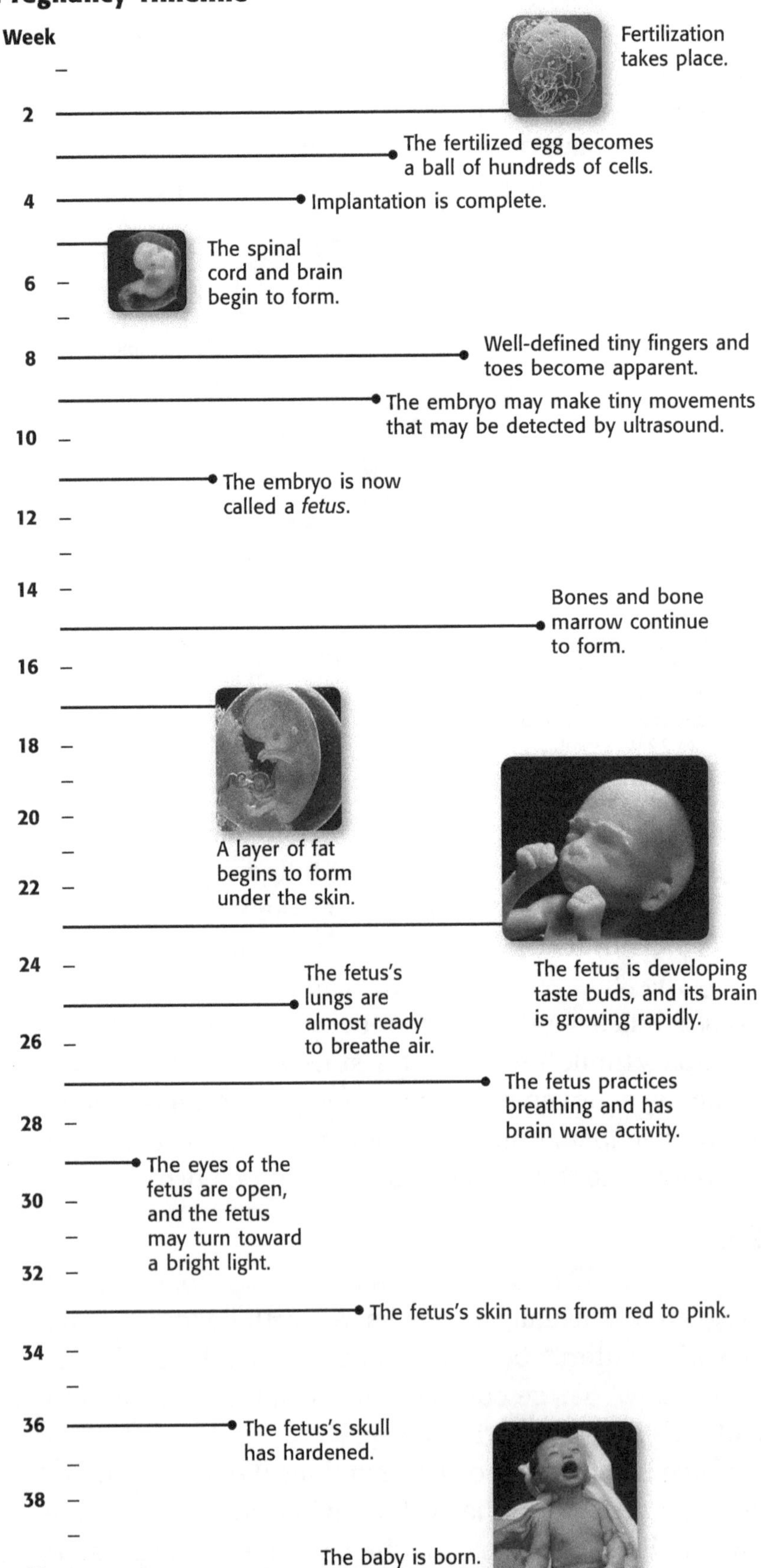

TAKE A LOOK

9. Identify By what week has the brain of the embryo started to form?

____

10. Identify At around what week does the fetus start to develop taste buds?

____

How Does a Person Grow and Change?

The human body goes through several stages of development. One noticeable difference is the change in body proportion.

Body Proportions During Stages of Human Development

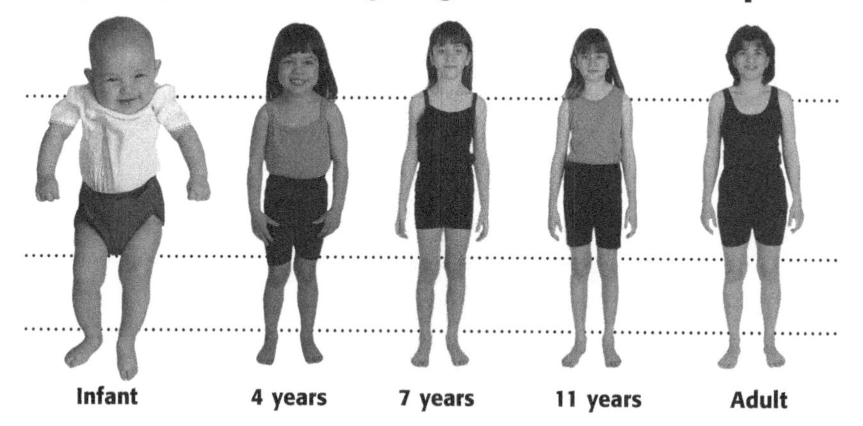

TAKE A LOOK

11. Identify At which stage or age is the head largest in proportion to the rest of the body?

____

The chart below lists the different stages of life and the characteristics of that stage.

Stage	Ages	Characteristics
Infancy	birth to age 2	• Baby teeth appear. • Nervous system and muscles develop. • Coordination improves.
Childhood	age 2 to puberty	• Permanent teeth grow. • Nerve pathways mature, and child can learn new skills. • Muscle coordination increases.
Adolescence	in females, puberty takes place between the ages of 9 and 14	**Females:** • Breasts enlarge. • Body hair appears. • Menstruation begins.
	in males, puberty takes place between the ages of 11 and 16	**Males:** • Body grows more muscular. • Voice deepens. • Facial and body hair appear.
	puberty to adulthood	• Reproductive system matures (puberty).
Adulthood	age 20+	• From ages 20 to 40, physical development is at its peak. • After age 40, hair may turn gray, skin may wrinkle, athletic ability may decrease.

Math Focus

12. Calculate Alice is 80 years old. She started puberty at age 12. Calculate the percentage of her life that she has spent in each stage of development.

____

____

____

____

TAKE A LOOK

13. Identify Which stage of development are you in?

____

Name ____________________ Class ____________ Date ____________

Section 3 Review

NSES LS 1a, 1c, 1d, 1e, 2b

SECTION VOCABULARY

embryo in humans, a developing individual from first division after fertilization through the 10th week of pregnancy

fetus a developing human from the end of the 10th week of pregnancy until birth

placenta the partly fetal and partly maternal organ by which materials are exchanged between a fetus and the mother

umbilical cord the ropelike structure through which blood vessels pass and by which a developing mammal is connected to the placenta

1. Define What is fertilization?

____

____

2. Explain Why can only one sperm enter an egg?

____

____

____

3. Explain What does it mean that cells differentiate?

____

____

4. Identify What is the function of the amnion?

____

5. Explain Why is the placenta important to a developing embryo?

____

____

____

6. Explain Why is it necessary for cells in an embryo to differentiate?

____

____

7. List What are the four stages of human development after birth?

____

____

8. Identify What is the main characteristic of adolescence?

____

Name ______________________ Class ______________ Date ______________

CHAPTER 27 Body Defenses and Disease

SECTION 1

Disease

BEFORE YOU READ

After you read this section, you should be able to answer these questions:

- What causes disease?
- How can we protect ourselves from disease?

National Science Education Standards
LS 1f

What Causes Disease?

All your life, you've probably had adults tell you to wash your hands or to cover your mouth when you sneeze. What is all the fuss about? They tell you to do these things to stop the spread of *disease*. When you have a disease, your body cannot function the way that it should.

Outline As you read, make an outline of this section. Be sure to answer the questions asked in each heading and define unfamiliar terms.

INFECTIOUS DISEASE

Many diseases, such as a cold, or the flu, can be passed from one living thing to another. This called an **infectious disease**. In other words, you can "catch" an infectious disease. Infectious diseases are caused by **pathogens**. Pathogens can be bacteria, fungi, worms, proteins, or viruses. *Viruses* are tiny particles that depend on living things to reproduce. ☑

READING CHECK

1. Identify What cause infectious diseases?

____

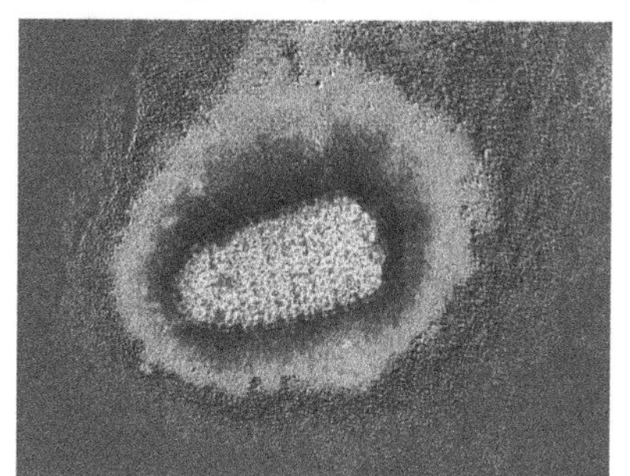

This virus causes rabies.

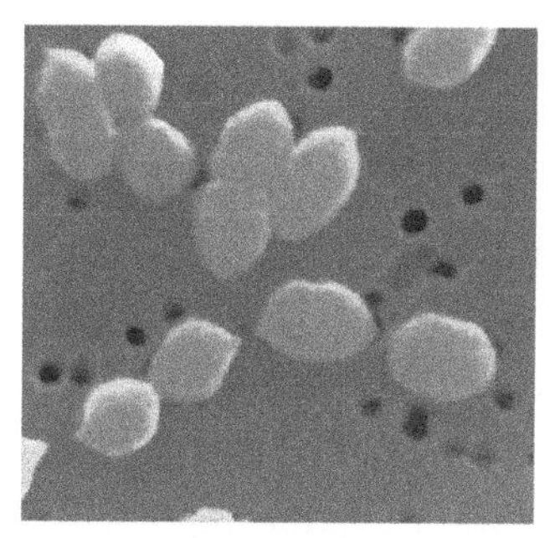

These bacteria, called *Streptococcus*, can cause strep throat.

STANDARDS CHECK

LS 1b Disease is the breakdown in structures or functions of an organism. Some diseases are the result of intrinsic failures of the system. Others are the result of damage by infection by other organisms.

2. Compare How are infectious diseases different from noninfectious diseases?

____

____

____

____

____

____

____

NONINFECTIOUS DISEASE

Some diseases, such as cancer or heart disease, cannot be passed from person to person. These are called **noninfectious diseases**. Some noninfectious diseases are caused by genes. Others can be caused by lifestyle choices, such as smoking, lack of exercise, and a high-fat diet. Avoiding harmful habits may help you avoid noninfectious diseases.

How Do Pathogens Move Between People?

Pathogens can pass from one person to another in many different ways. Being aware of how pathogens are passed can help you stay healthy.

Critical Thinking

3. Infer Why do you think it is important to cover your mouth and nose when you cough or sneeze?

____

____

____

AIR

Some pathogens travel through the air. A single sneeze can release thousands of droplets of moisture. These droplets can carry pathogens.

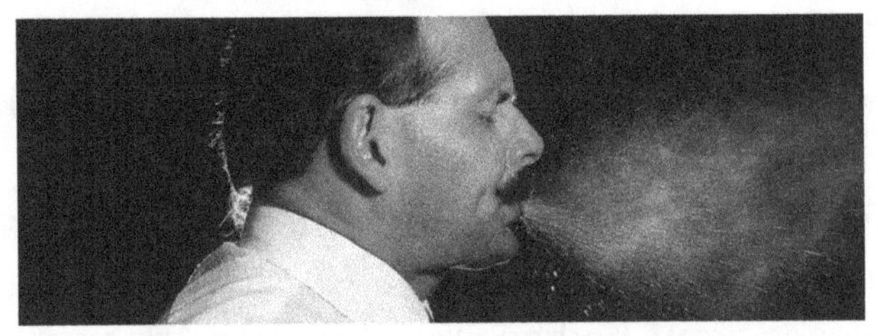

A sneeze can force thousands of droplets carrying pathogens out of your body at up to 160 km/hr. That's more than the speed limit on most highways!

Math Focus

4. Calculate You catch a cold and return to school sick. Your friends don't have immunity to the virus. On the first day, you expose five people to the virus. The next day, each of those friends passes on the virus to five more people. If this same pattern continues, how many people will be infected after 5 days?

____

OBJECTS

You probably already know that if you drink from a glass used by sick person, you can become infected too. Pathogens can also be carried on objects such as doorknobs, keyboards, combs, or towels. A sick person may leave pathogens on anything he or she touches.

PERSON TO PERSON

People can transfer pathogens directly. You can become infected by kissing, shaking hands, or touching sores on an infected person.

ANIMALS

Some pathogens are carried by animals. For example, people can get a fungus called ringworm by touching an infected cat or dog. Ticks carry bacteria that cause Lyme's disease and Rocky Mountain spotted fever.

Say It

Discuss When was the last time you were sick? How do you think you got sick? Were your friends, siblings, or other people in your class sick at the same time? Talk about these questions with a partner.

FOOD AND WATER

Food and water can contain pathogens. Most drinking water in the United States is safe. However, pathogens can enter the public water supply.

Foods such as meat, fish, and eggs that are not cooked enough can contain harmful bacteria or parasites. Because pathogens grow in food, it is important to wash all cooking surfaces and tools well.

How Can We Protect Ourselves From Disease?

Pathogens such as bacteria and viruses are everywhere. So how can we protect ourselves from them?

PASTEURIZATION

Milk and other dairy products can carry certain types of pathogens. To destroy pathogens, dairy products and other foods are treated by a process called *pasteurization.* This method was invented by the French scientist Louis Pasteur. It uses heat to kill bacteria. ☑

Juices, shellfish, and dairy products, such as milk, are all pasteurized to kill pathogens.

VACCINES

A *vaccine* is a substance that helps your body to resist a disease. The ability to resist an infectious disease is called **immunity**. Vaccines contain pathogens that have been killed or treated so they can't make you very sick. The vaccine is enough like the pathogen to help your body develop a defense against the disease.

ANTIBIOTICS

Doctors can treat bacterial infections, such as strep throat, with antibiotics. An *antibiotic* is a substance that kills bacteria or slows their growth.

Antibiotics do not affect viruses. This is because antibiotics only kill living things. Viruses are not considered to be alive because they cannot reproduce on their own. Although antibiotics do not destroy viruses, scientists are working to develop antiviral medications.

READING CHECK

5. Complete In pasteurization, bacteria are killed with

______________________.

TAKE A LOOK

6. List Name three types of foods that are pasteurized.

____

____

____

Critical Thinking

7. Compare What is the difference between a vaccine and an antibiotic?

____

____

____

____

____

____

Name ______________________ Class ______________ Date ______________

Section 1 Review

NSES LS 1f

SECTION VOCABULARY

immunity the ability to resist an infectious disease

infectious disease a disease that is caused by a pathogen and that can be spread from one individual to another

noninfectious disease a disease that cannot spread from one individual to another

pathogen a microorganism, another organism, a virus, or a protein that causes disease

1. **Explain** Why are food products, such as milk, pasteurized?

____

____

2. **Compare** Complete the table below to compare infectious and noninfectious diseases.

Type of disease	What causes it?	Can it be passed from person to person?
Infectious		
Non-infectious		

3. **Apply Concepts** Can you get a vaccine for a noninfectious disease? Explain your answer.

____

____

____

4. **List** Name five ways that you could come into contact with a pathogen.

____

____

____

5. **Apply Concepts** The common cold is caused by a virus. Should your doctor give you an antibiotic to help you fight a cold? Explain your answer.

____

____

Name ______________________ Class ______________ Date ______________

CHAPTER 27 Body Defenses and Disease

SECTION 2

Your Body's Defenses

BEFORE YOU READ

After you read this section, you should be able to answer these questions:

- How does your body keep pathogens out?
- How does your immune system destroy pathogens that do get into your body?

National Science Education Standards
LS 1b, 1d, 1e, 1f, 3a, 3b

How Does Your Body Keep Pathogens out?

Bacteria and viruses are all around us in the air, in water, and on surfaces. Your body must constantly protect itself against these pathogens. Luckily, your body has its own built-in defense system.

Keeping pathogens from entering the body is an important defense against infection. Your body has several barriers that keep pathogens from entering. These include your skin and special secretions in your eyes, nose, and throat. ☑

Your skin is made of many layers of flat cells. The outer layers are dead. Many pathogens that land on your skin cannot find a live cell to infect. These dead cells drop off your body as new skin cells grow beneath them. When the dead skin cells fall off, they carry viruses, bacteria, and other pathogens with them. Your skin also has glands that produce oil to cover the skin's surface. The oil contains chemicals that kill many pathogens.

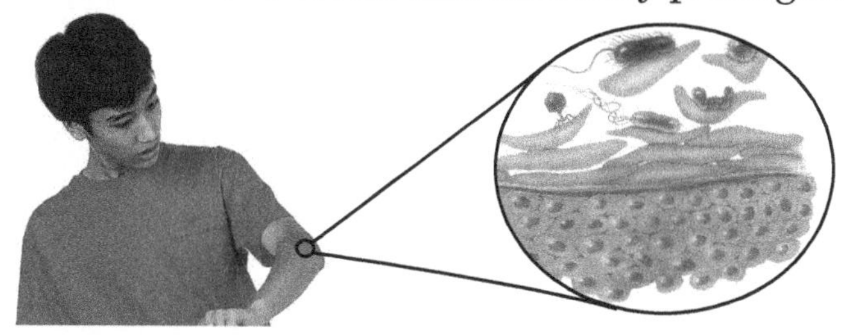

Your body loses and replaces about 1 million skin cells every 40 minutes. As skin cells fall off, pathogens fall off with them.

Pathogens also try to enter your body through your eyes, nose, or mouth. These pathogens are destroyed by chemicals called *enzymes*. Some pathogens that enter your nose are washed down your throat by a slippery fluid called *mucus*. Mucus carries pathogens to your stomach, where they are digested.

STUDY TIP

Underline As you read, underline the different defenses your body has against disease.

READING CHECK

1. List What are two of your body's barriers against infection?

____

____

____

____

Math Focus

2. Calculate About how many skin cells does your body lose and replace in 24 hours?

____

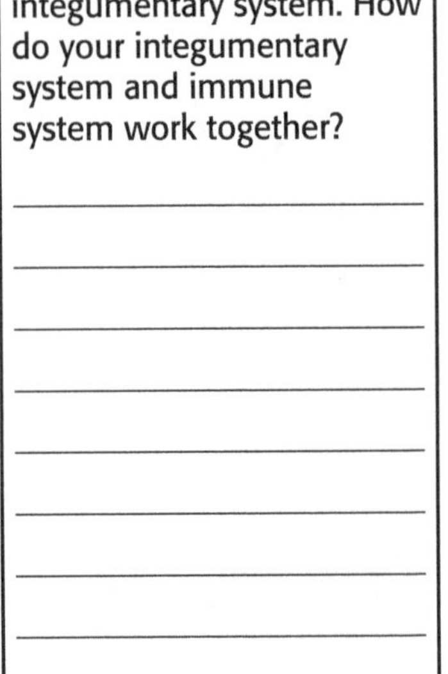

STANDARDS CHECK

LS 1e The human organism has systems for digestion, respiration, reproduction, circulation, excretion, movement, control and coordination, and protection from disease. These systems interact with one another.

3. Identify Relationships Your skin is part of the integumentary system. How do your integumentary system and immune system work together?

____

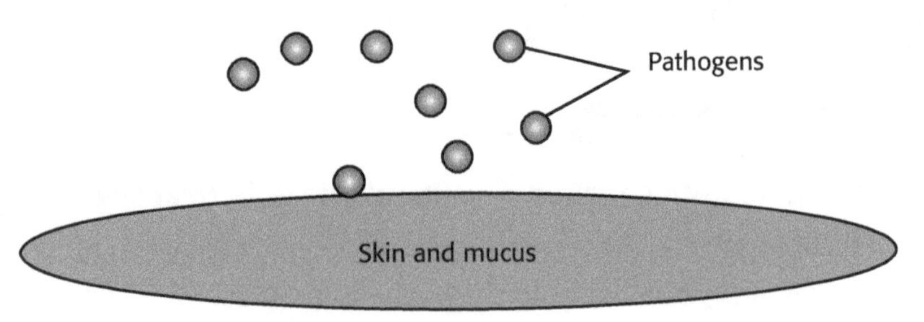

Defense Level 1: Barriers keep most pathogens from getting into your body.

What Happens If Pathogens Do Get into Your Body?

Some pathogens may get past your body's barriers. For example, pathogens can enter the body if the skin is cut or punctured. The body reacts quickly to keep more pathogens out. Blood flow to the injured area increases. Particles in the blood called *platelets* help seal the wound so no more pathogens can enter.

Blood also brings cells that belong to the immune system. The **immune system** is the body system that fights pathogens. It is not found in one specific place in your body, and it is not controlled by one organ. Instead, the immune system is a team of cells, tissues, and organs that work together to protect you from pathogens.

How Does Your Body Know a Pathogen Is an Invader?

For your body to fight pathogens, it must be able to recognize that they are invaders. How can it do this? The cells of your body have special proteins on their surfaces. Your proteins are different from everyone else's. Your proteins tell your body which cells are yours. Pathogens also have proteins on their surfaces. When pathogens get into your body, their surface proteins tell your body that the particles do not belong to you. ☑

READING CHECK

4. Explain How does your body know a pathogen does not belong in your body?

____

What Kind of Cells Attack Pathogens First?

Immune system cells called **macrophages** can surround and destroy many pathogens that enter your body. Generally, macrophages will attack any invader. If only a few pathogens enter your body at one time, macrophages can usually destroy them.

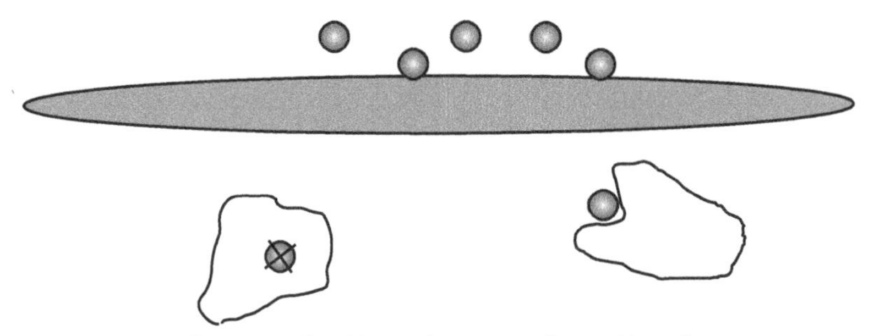

Defense Level 2: Macrophages attack most invaders.

If too many pathogens enter your body at once, the macrophages may not be able to stop all of them. Some pathogens will invade your body cells. Once this happens, another group of cells goes to work. This is known as the *immune response.* ☑

READING CHECK

5. Explain What happens when too many pathogens enter your body at one time?

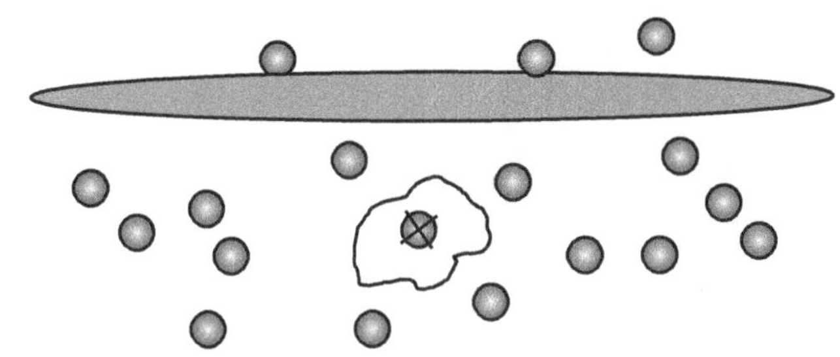

Defense Level 3: If too many pathogens enter the body, the macrophages cannot stop all of them. The pathogens will infect body cells. Other types of immune cells go to work.

What Is the Immune Response?

In the immune response, immune cells work to destroy any body cells that have been infected by a pathogen. This helps keep the pathogen from infecting more body cells.

Recall that pathogens have proteins on their surfaces that tell your body the particle does not belong to you. These proteins trigger the immune response. Anything that starts the immune response is called an *antigen.* When a pathogen invades a body cell, the pathogen's proteins are also displayed on the surface of the body cell.

T CELLS AND B CELLS

Two general kinds of immune systems cells go to work in response to antigens: T cells and B cells. While macrophages will generally attack any invader, T cells and B cells work together to attack specific invaders. These cells recognize specific invaders by their antigens. **T cells** attack infected body cells and tell other immune cells to respond to an invader. **B cells** make antibodies. ☑

READING CHECK

6. Identify What kind of immune cells produce antibodies?

ANTIBODIES

An **antibody** is a protein that attaches to a specific antigen. Antibodies tag a pathogen so that other cells can destroy it. When antibodies attach to antigens, they make the pathogens clump together. Macrophages can find these clumps easily and destroy them.

Your body can make billions of different kinds of antibodies. The shape of an antibody is very specialized. It matches an antigen like a key fits a lock. Each antibody usually attaches to just one type of antigen.

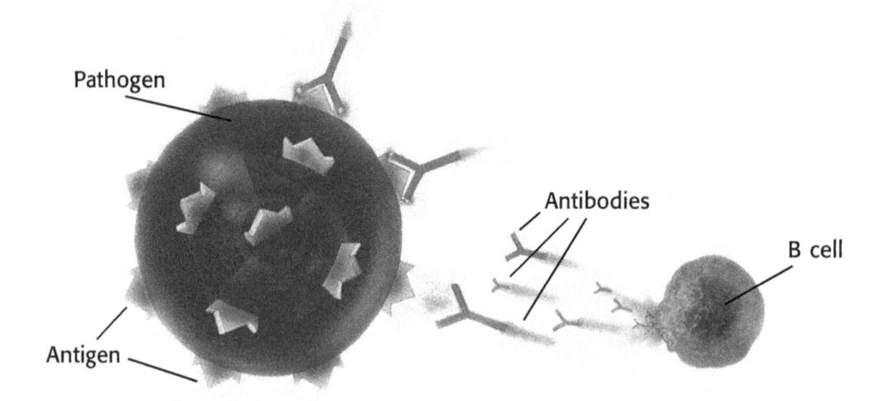

An antibody's shape is very specialized. It matches an antigen like a key fits a lock.

How Does the Immune System Fight a Virus?

The figures below show what happens when a pathogen, such as a virus, enters the body. When virus particles enter the body, one of two things may happen. Some viruses enter body cells and start to make copies of themselves. Others are attacked by macrophages.

TAKE A LOOK

7. Identify What two things can happen to a virus that enters your body?

____

____

____

____

8. Explain What happens to a macrophage after it engulfs a virus particle?

____

____

____

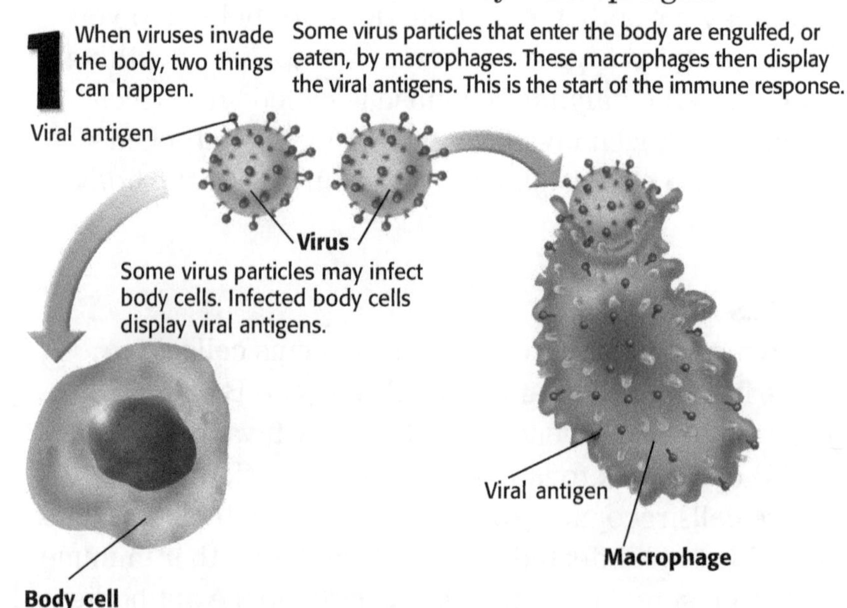

2 **Two Responses** Helper T cells have receptor proteins that recognize the shape of the viral antigen on the macrophages. These helper T cells begin two responses: a T cell response and a B cell response.

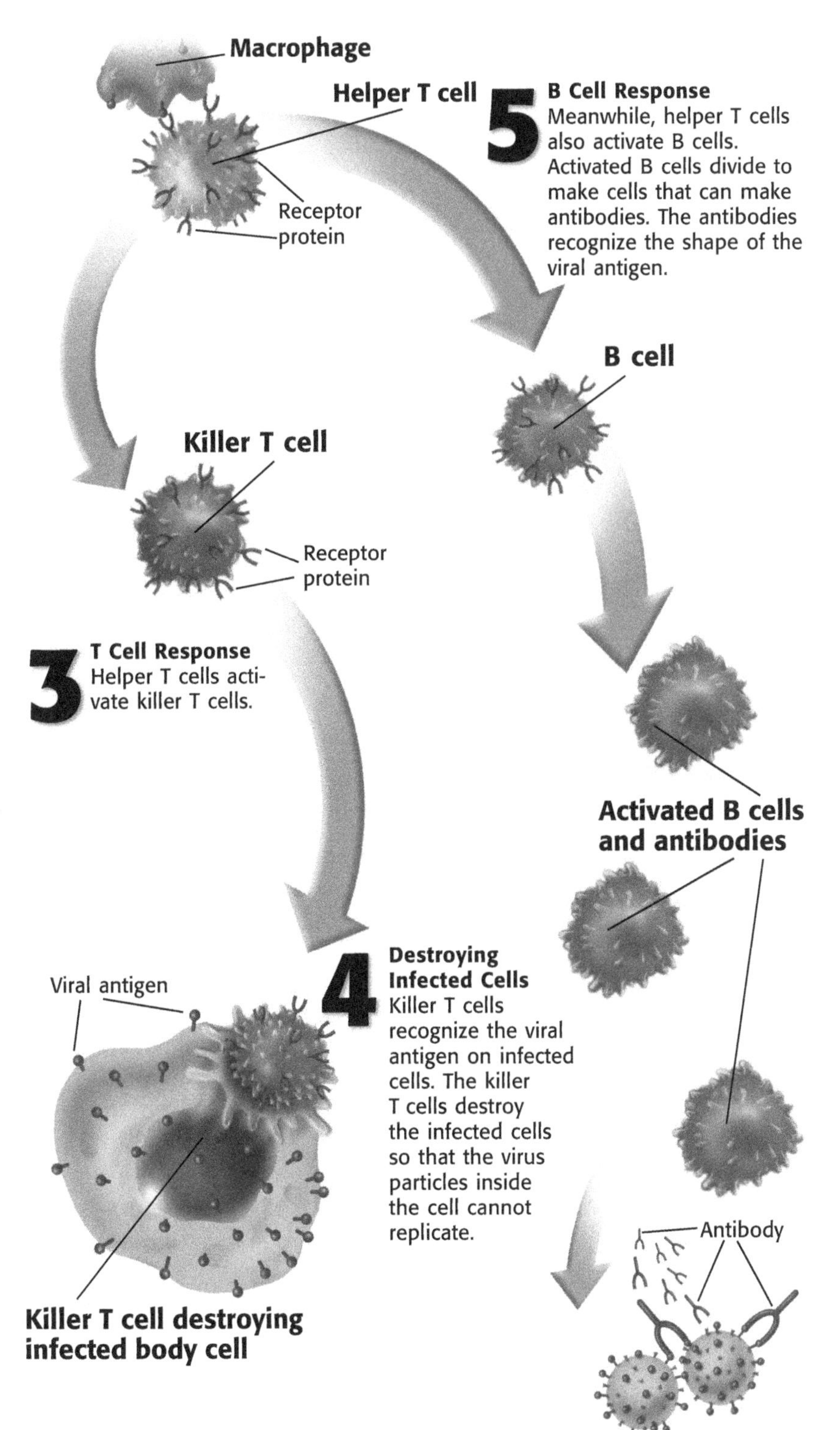

6 **Destroying Viruses** Antibodies bind to the viral antigen on the viruses. The antibodies bound to the viruses cause the viruses to clump together. Clumping marks the virus particles for destruction.

TAKE A LOOK

9. Identify What structures on helper T cells recognize viral antigens?

____

10. Explain What two things happen when a helper T cell recognizes a viral antigen?

____

____

____

____

11. Identify What is the role of killer T cells?

____

____

FEVERS

When macrophages activate helper T cells, they send a signal that tells the brain to turn up the body's temperature. In just a few minutes, your body's temperature can rise by several degrees. This rise in body temperature is a *fever*.

A fever of one or two degrees can help you get better faster. A moderate fever slows down the growth of pathogens. Fevers also help B cells and T cells reproduce faster.

Critical Thinking

12. Infer Why is it not a good idea to use medicines that stop a moderate fever?

____

____

____

____

____

____

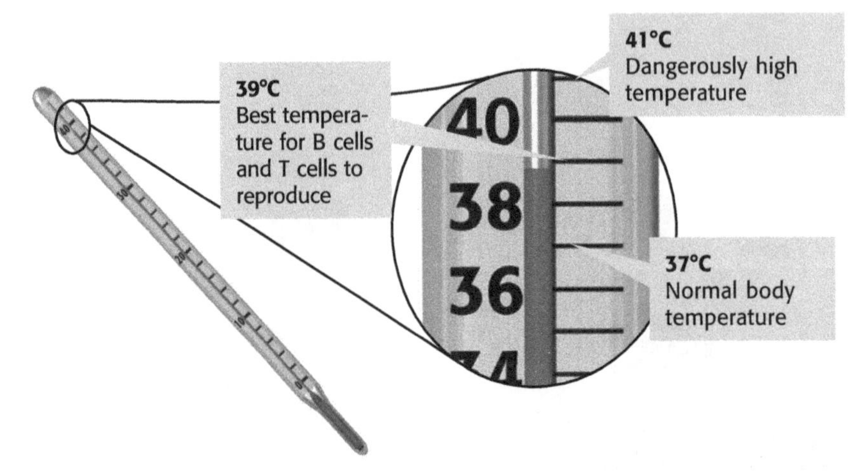

A slight fever helps immune cells reproduce. However, a fever of more than a few degrees can become dangerous.

TAKE A LOOK

13. Explain What kind of fever can become dangerous?

____

____

____

MEMORY CELLS

The first time a new pathogen enters your body, specialized B cells make antibodies that recognize the pathogen. However, this process takes about 2 weeks. That is too long to prevent an infection. The first time you are infected with a new pathogen, you usually get sick.

Your immune system can respond faster to a pathogen the second time it enters your body. After an infection, a few B cells become memory B cells. A **memory B cell** is an immune cell that can "remember" how to make the right antibody for a specific pathogen. If that pathogen enters your body a second time, memory B cells can recognize it. The immune system can make enough antibodies to fight that pathogen in just 3 or 4 days. ☑

READING CHECK

14. Define What is the role of a memory B cell?

____

____

____

____

Does the Immune System Always Work?

The immune system is not perfect. Sometimes, it can't prevent diseases. There also are conditions in which the immune system attacks the wrong cells.

ALLERGIES

Sometimes the immune system overreacts to antigens that are not harmful to the body. This inappropriate reaction is an **allergy**. Many things, including foods, medicines, plant pollen, and animals, can cause allergies.

AUTOIMMUNE DISEASES

An **autoimmune disease** happens when the immune system attacks the body's own cells. In this type of disease, immune system cells mistake body cells for pathogens. Rheumatoid arthritis, multiple sclerosis, and type 1 diabetes are all autoimmune diseases. ☑

READING CHECK

15. Explain What happens in an autoimmune disease?

____

____

____

____

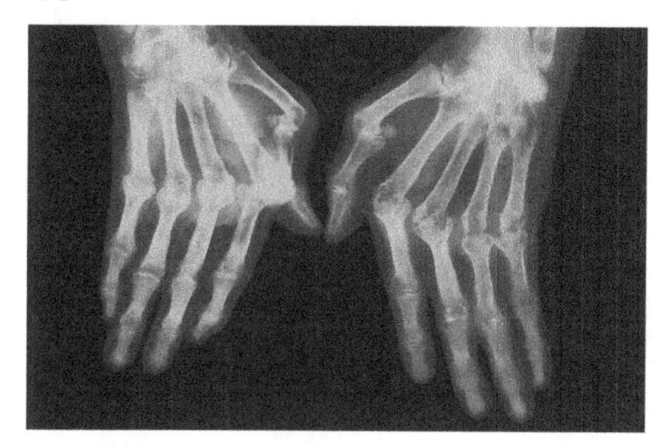

In rheumatoid arthritis, immune-system cells cause joint tissues to swell. This can cause joint deformities.

CANCER

Healthy cells divide at a regulated rate. Sometimes, a cell doesn't respond to the body's regulation, and it divides uncontrollably. Killer T cells normally kill this type of cell. But sometimes, the immune system cannot control these cells. **Cancer** is a condition in which some of the body's cells divide at an uncontrolled rate.

AIDS

The *human immunodeficiency virus* (HIV) causes *acquired immune deficiency syndrome* (AIDS). HIV is different from most viruses, which infect cells in the nose, mouth, lungs, or intestines. HIV infects the immune system itself. It uses helper T cells to make more viruses. The helper T cells are destroyed in the process. ☑

Without helper T cells, there is nothing to activate B cells and killer T cells. The immune system cannot attack HIV or any other pathogens. In fact, most people with AIDS don't die of AIDS itself. Instead, they die because their immune systems cannot fight off other diseases.

READING CHECK

16. Identify Name one way HIV is different from most other viruses.

____

____

____

____

____

Name ______________________ Class ______________ Date ______________

Section 2 Review

NSES LS 1b, 1d, 1e, 1f, 3a, 3b

SECTION VOCABULARY

allergy a reaction to a harmless or common substance by the body's immune system

antibody a protein made by B cells that binds to a specific antigen

autoimmune disease a disease in which the immune system attacks the organism's own cells

B cell a white blood cell that makes antibodies

cancer a tumor in which the cells begin dividing at an uncontrolled rate and become invasive

immune system the cells and tissues that recognize and attack foreign substances in the body

macrophage an immune system cell that engulfs pathogens and other materials

memory B cell a B cell that responds to an antigen more strongly when the body is reinfected with an antigen than it does during its first encounter with an antigen

T cell an immune system cell that coordinates the immune system and attacks many infected cells

1. **List** What are the three main kinds of cells in the immune system?

____

2. **Summarize** Complete the process chart to show what happens when a pathogen, such as a virus, enters the body.

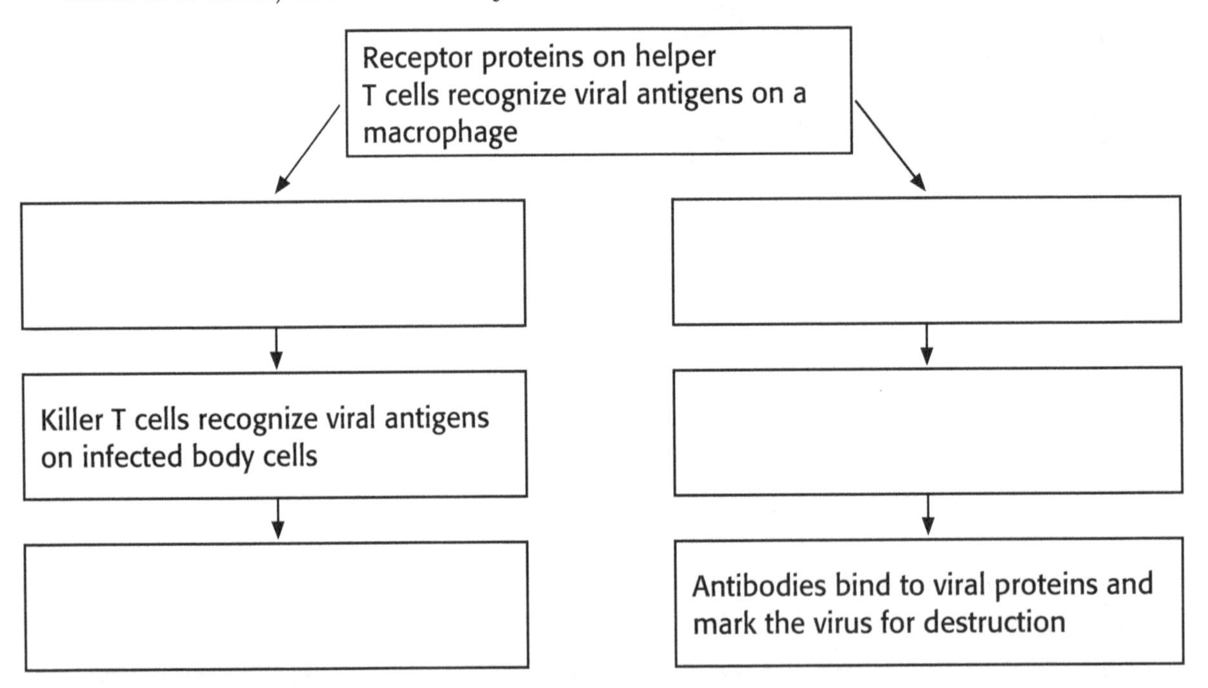

3. **Apply Concepts** Why do you think that most people only get a disease such as chickenpox once?

____

____

____

Name ______________________ Class ______________ Date ______________

CHAPTER 28 Staying Healthy

SECTION 1

Good Nutrition

BEFORE YOU READ

After you read this section, you should be able to answer these questions:

- What are nutrients?
- What are several practices for good nutrition?
- What causes malnutrition?

What Are Nutrients?

Have you ever heard someone say "You are what you eat"? This doesn't mean that you are pizza! However, the substances in pizza and other foods help build your body. These substances are called nutrients.

Nutrients provide the materials your body needs to grow and maintain itself. They are grouped into six classes: carbohydrates, proteins, fats, water, vitamins, and minerals. Carbohydrates, proteins, and fats provide energy for the body in units called *Calories* (Cal). To stay healthy, you need to take in each class of nutrients daily. ☑

STUDY TIP

List As you read, make a list of the nutrients you need for good health.

1. Complete Energy for the body is measured in units called ______________.

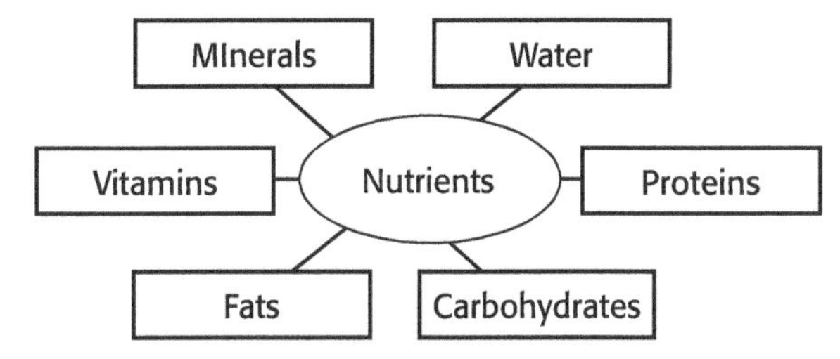

CARBOHYDRATES

Carbohydrates are your body's main source of energy. A **carbohydrate** is a compound made up of sugar molecules. There are two types of carbohydrates: simple and complex. Simple carbohydrates are made of one or a few sugar molecules linked together. You can digest them easily for quick energy.

Complex carbohydrates are made of many sugar molecules linked together. You digest complex carbohydrates more slowly than simple carbohydrates. Complex carbohydrates give you long-lasting energy.

Some complex carbohydrates are good sources of fiber. Fiber is a part of a healthy diet and is found in whole-grain foods, such as brown rice and whole-wheat bread. Many fruits and vegetables also contain fiber.

Critical Thinking

2. Apply Concepts If you wanted to have more fiber in your diet, would you choose to eat simple carbohydrates or complex carbohydrates? Explain your answer.

____

____

____

____

____

Discuss Think about what you eat in a typical day. What foods do you eat that contain protein? Talk with a partner to see if you eat the same kinds of proteins.

Math Focus

3. Calculate Percentages If you eat 2,500 Cal per day and 20% are from fat, 30% are from protein, and 50% are from carbohydrates, how many Calories of each nutrient do you eat?

____

____

____

TAKE A LOOK

4. Identify Use colored pencils to circle the foods that contain carbohydrates, fats, and proteins. Use red for carbohydrates, blue for fats, and green for proteins. You may circle a food more than once.

PROTEINS

Proteins are nutrients your body uses to build and repair itself. Your body can make the proteins it needs. However, it must get building blocks called *amino acids* from the proteins you eat. Foods such as poultry, fish, milk, eggs, nuts, and beans are good sources of protein.

FATS

Fats are nutrients that store energy. Your body also uses fats to store vitamins and make hormones. Fats help keep your skin healthy and insulate your body.

There are two types of fats: saturated and unsaturated. *Saturated fats* are found in meat, dairy products, coconut oil, and palm oil. Saturated fats raise blood cholesterol levels. Although *cholesterol* is found naturally in the body, high levels can increase the risk of heart disease.

Unsaturated fats and foods high in fiber may help reduce blood cholesterol levels. Your body cannot make unsaturated fats. They come from foods such as vegetable oils and fish.

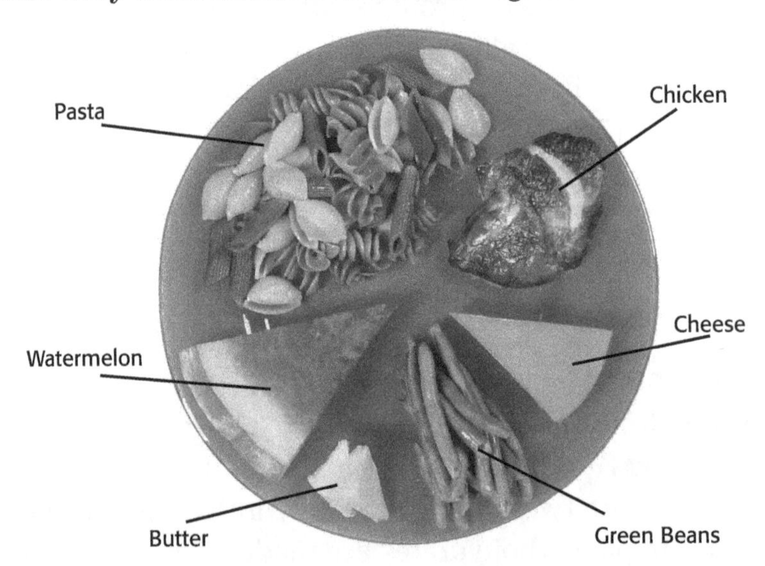

WATER

We could not live without water. Water moves substances through our bodies, keeps our temperatures stable, and keeps body tissues moist.

You cannot survive for more than a few days without water. In fact, your body is about 70% water. Many health professionals think you should drink at least eight glasses of water a day. You also get water from other liquids you drink and the foods you eat. Fresh fruits and vegetables, juices, soups, and milk are good sources of water.

VITAMINS AND MINERALS

Vitamins are compounds that your body needs to control many body functions. Some of these functions are listed in the table below. Your body cannot make most vitamins, so you need to get them from food.

Vitamin	What it does	Where you get it
A	keeps skin and eyes healthy; builds strong teeth and bones	yellow and orange fruits and vegetables, leafy greens, meats, and milk
B	helps body use carbohydrates; helps blood, nerves, and heart function	meats, whole grains, beans, peas, nuts, and seafood
C	strengthens tissues; helps the body absorb iron and fight disease	citrus fruits, leafy greens, broccoli, peppers, and cabbage
D	builds strong bones and teeth; helps the body use calcium and phosphorus	sunlight, enriched milk and soymilk, eggs, and fish
E	protects red blood cells from destruction; keeps skin healthy	oils, fats, eggs, whole grains, wheat germ, liver, and leafy greens
K	helps your blood to clot	leafy greens, tomatoes, and potatoes

TAKE A LOOK

5. Identify Which vitamins can you get from leafy greens?

____

6. List Which foods could you eat to help give you strong bones and teeth?

____

____

Minerals are chemical elements that are essential for good health. You need six minerals in large amounts: calcium, chloride, magnesium, phosphorus, potassium, and sodium. Your body needs other minerals, such as fluorine, iodine, iron, and zinc, in small amounts. Some minerals and their functions are listed in the table below.

Mineral	What it does
Calcium	makes teeth and bones strong
Magnesium and sodium	help the body use proteins
Potassium	helps regulate heartbeat and makes muscles move
Iron	helps make blood cells

TAKE A LOOK

7. Identify Which mineral do you need to make your muscles move?

____

How Can You Get All the Nutrients You Need?

For good health, you need a variety of nutrients from different foods. The next figure can help you find out how much of each type of food you should eat every day. The six stripes represent five food groups. The width of each stripe shows how much food from that group you should eat.

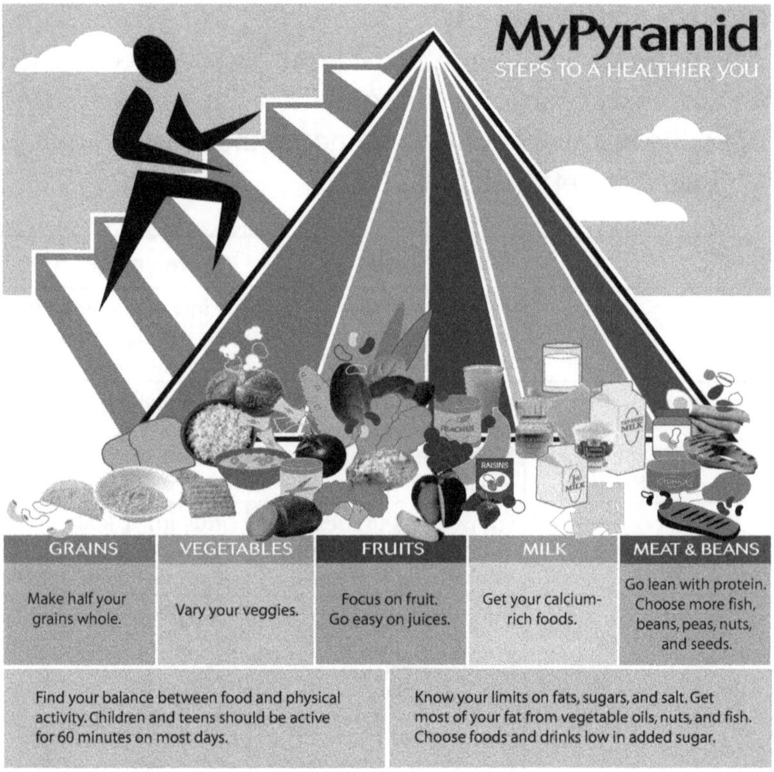

MyPyramid
STEPS TO A HEALTHIER YOU

GRAINS	VEGETABLES	FRUITS	MILK	MEAT & BEANS
Make half your grains whole.	Vary your veggies.	Focus on fruit. Go easy on juices.	Get your calcium-rich foods.	Go lean with protein. Choose more fish, beans, peas, nuts, and seeds.

Find your balance between food and physical activity. Children and teens should be active for 60 minutes on most days.

Know your limits on fats, sugars, and salt. Get most of your fat from vegetable oils, nuts, and fish. Choose foods and drinks low in added sugar.

TAKE A LOOK

8. Identify Which food group should you eat the most foods from each day?

____

9. Apply Ideas Design a healthy lunch that includes one food from each group. What would you eat?

____

____

____

____

What Happens If You Do Not Get the Nutrients You Need?

Malnutrition occurs when someone does not eat enough of the nutrients the body needs. It can happen when you do not eat enough food. It can also happen when you eat too much food that is high in fat and low in other nutrients. Malnutrition can affect how you look. It can also keep your body from fighting illness or repairing itself quickly.

Critical Thinking

10. Infer How could eating only junk food, such as candy and fast food, cause malnutrition?

____

____

____

____

____

EATING DISORDERS

Conditions called *eating disorders* can cause malnutrition. People with *anorexia nervosa* starve themselves by not eating. They have a strong fear of gaining weight. People with *bulimia nervosa* eat very large amounts of food at once. They will then try to get rid of the food quickly. They may do this by vomiting or using diuretics and laxatives.

Anorexia can lead to severe malnutrition. Bulimia can damage the teeth and nervous system. Both anorexia and bulimia can cause problems such as weak bones, low blood pressure, and heart problems. They can even lead to death. ☑

READING CHECK

11. List What are three problems that both anorexia and bulimia can cause?

____

____

____

OBESITY

Obesity is having a high percentage of body fat. Obesity increases the risk of certain health problems. These include high blood pressure, heart disease, and diabetes. People suffering from obesity may not be eating foods that give them the correct balance of nutrients. Exercising very little can also lead to obesity.

Eating a balanced diet and exercising regularly can help reduce obesity. However, obesity can be caused by other things. For example, people who are obese often have obese parents. Scientists are studying the link between obesity and heredity. ☑

READING CHECK

12. Identify Name two things that can help reduce obesity.

____

____

What Do Food Labels Tell You?

Reading food labels can help you make healthy eating choices. Nutrition facts show how much of each nutrient is in one serving of a food. Packaged foods, such as cereal, peanut butter, and pasta, all have these labels. The label below is from a can of chicken noodle soup.

Nutrition Facts

Serving Size 1/2 cup (120 ml) — **Serving information**
Servings per Container 2.5

Amount per Serving	**Prepared**
Calories	70
Calories from Fat	25

Number of Calories per serving

	% Daily Value
Total Fat 2.5 g	4%
Saturated Fat 1 g	5%
Trans Fat 0 g	
Cholesterol 15 mg	5%
Sodium 960 mg	40%
Total Carbohydrate 8 g	3%
Dietary Fiber less than 1 g	4%
Sugars 1 g	
Protein 3 g	
Vitamin A	15%
Vitamin C	0%
Calcium	0%
Iron	4%

Percentage of daily values

*Percent Daily Values are based on a 2,000 Calorie diet. Your daily values may be higher or lower depending on your Calorie needs:

	Calories	2,000	2,500
Total Fat	Less than	65g	80g
Sat Fat	Less than	20g	25g
Cholesterol	Less than	300mg	300mg
Sodium	Less than	2,400mg	2,400mg
Total Carbohydrate		300g	375g
Dietary Fiber		25g	30g
Protein		50g	60g

Math Focus

13. Calculate How many Calories are in the entire can of soup?

____

TAKE A LOOK

14. Evaluate Is this soup a good source of vitamin C? Explain your answer.

____

____

____

Name ______________________ Class ______________ Date ______________

Section 1 Review

SECTION VOCABULARY

carbohydrate a class of energy-giving nutrients that includes sugars, starches, and fiber

fat an energy-storage nutrient that helps the body store some vitamins

malnutrition a disorder of nutrition that results when a person does not consume enough of each of the nutrients that are needed by the human body.

mineral a class of nutrients that are chemical elements that are needed for certain body processes

nutrient a substance in food that provides energy or helps form body tissues and that is necessary for life and growth

protein a molecule that is made up of amino acids and that is needed build and repair body structures and to regulate processes in the body

vitamin a class of nutrients that contain carbon and that are needed in small amounts to maintain health and allow growth

1. **Apply Concepts** Name some of the nutrients that can be found in a glass of milk.

____

2. **Infer** Which type of fats are healthier—saturated or unsaturated?

____

____

3. **Compare** How do anorexia nervosa and bulimia nervosa differ?

____

____

____

4. **Infer** Is it possible for someone to suffer from both obesity and malnutrition? Explain your answer.

____

____

____

5. **Explain** How would you use a Nutrition Facts label to choose a food that is high in calcium?

____

____

____

6. **List** What are the five main food groups?

____

____

Name ______________________ Class ______________ Date ______________

CHAPTER 28 Staying Healthy

SECTION 2

Risks of Alcohol and Other Drugs

BEFORE YOU READ

After you read this section, you should be able to answer these questions:

- What is a drug?
- What are some of the problems caused by drug abuse?

What Is a Drug?

Sometimes when people talk about drugs, they are talking about helpful substances, such as medicines. Other times, we hear about illegal drugs and their dangers. So what is a drug?

A **drug** is any chemical substance that causes a physical or psychological change in your body. Some drugs enter the body through the skin. Other drugs are swallowed, inhaled, or injected.

Underline This section may have many words that are unfamiliar. As you read, underline these words and find out what they mean. Write their definitions in the margin.

All of these products contain drugs.

Discuss Does it surprise you that all of the items in the figure contain drugs? What kinds of things do you usually think about when you hear the word "drug"? Share your ideas in small groups.

You can buy some drugs at the grocery store, while others must be prescribed by a doctor. Some drugs are illegal to buy, sell, or have. When used correctly, legal drugs can help your body heal. When used improperly, however, even legal drugs can be harmful.

How Can Drugs Be Harmful?

Addiction is dependence on a substance. Many drugs cause addiction. They are *addictive*. A person with an addiction finds it very hard to stop taking a drug.

The body can form a *physical dependence*, or need for a drug. If the body doesn't get a drug that it is physically dependent on, it can show withdrawal symptoms. These can include nausea, vomiting, pain, and tremors or shakes.

Some people also form *psychological dependence* on a drug. This means they have strong cravings for the drug. ☑

READING CHECK

1. Explain Why does a person experience withdrawal symptoms?

____

____

____

____

What Are Some Types of Drugs?

HERBAL MEDICINES

Some types of plants called *herbs* contain chemicals with healing properties. Information about these *herbal medicines* has been handed down for centuries. Even though they come from plants, herbs are drugs and should be used carefully. The United States Food and Drug Administration (FDA) does not regulate herbal medicines or teas. This means it cannot guarantee that these products are safe to use. ☑

READING CHECK

2. Explain What does it mean that the FDA doesn't regulate herbal medicines?

____

____

____

PRESCRIPTION AND OVER-THE-COUNTER DRUGS

Many drugs need a prescription from a doctor. A *prescription* describes the drug, how to use it, and how much to take. Below are some important drug safety tips you should always follow.

Drug Safety Tips

- Never take another person's prescription medicine.
- Read the label before each use. Always follow the instructions on the label and those provided by your doctor or pharmacist.
- Do not take more or less medication than prescribed.
- Consult a doctor if you have any side effects.
- Throw away leftover and out-of-date medicines.

Some drugs, such as aspirin, can be bought without a prescription. These are called *over-the-counter (OTC) drugs*. You should still read and follow directions carefully for OTC drugs. Both prescription and OTC drugs can cause side effects. *Side effects* are uncomfortable symptoms, such as headache, nausea, or sleepiness.

TOBACCO

Cigarettes contain tobacco. They are addictive, and smoking has serious health effects. Tobacco contains a chemical called nicotine. **Nicotine** increases heart rate and blood pressure, and it is extremely addictive.

Smoking increases the chances of lung cancer. It has also been linked to other cancers, bronchitis, and heart disease. Experts estimate that there are more than 430,000 deaths related to smoking each year in the United States. Secondhand smoke also can be harmful.

Critical Thinking

3. Infer Why can someone become addicted to cigarettes?

____

____

____

____

Effects of Smoking

▼ Healthy lung tissue of a nonsmoker

▼ Damaged lung tissue of a smoker

Cilia in your lungs clean the air you breathe and keep debris out of your lungs. Smoking damages the cilia so that they cannot do their job.

TAKE A LOOK

4. Identify What is the function of cilia in your lungs?

____

____

____

Smokeless, or chewing, tobacco is also addictive and can cause health problems. Nicotine is absorbed through the lining of the mouth. Smokeless tobacco increases the risk of several cancers, including mouth and throat cancer. It also causes gum disease and yellowing of the teeth.

ALCOHOL

In most of the United States, using alcohol is illegal for people under age 21. Alcohol slows down the central nervous system and can cause memory loss. Excessive use of alcohol can damage the liver, pancreas, brain, nerves, and cardiovascular system. It can even cause death. ☑

Alcohol affects decision-making and can lead you to take unhealthy risks. Some people suffer from **alcoholism**, which means that they are physically and psychologically dependent on alcohol. Alcoholism is considered a disease. Scientists think certain genes can make some people more likely to develop alcoholism.

READING CHECK

5. Identify What is the legal age for using alcohol in most of the United States?

____

MARIJUANA

Marijuana is a drug that comes from a plant called hemp. It is illegal in most places. Marijuana affects different people in different ways. It may increase anxiety or cause feelings of paranoia. Marijuana slows down your ability to react, clouds your thinking, and makes you less coordinated. ☑

READING CHECK

6. List What are three effects of marijuana?

____

____

____

____

COCAINE

Cocaine and its purified form, crack, are made from the coca plant. Both drugs are illegal and highly addictive. Users can become addicted to them in a very short time. Cocaine can make a person feel very excited, but later it can make them feel anxious and depressed. Both drugs increase heart rate and blood pressure and can cause heart attacks.

Name ______________________ Class ______________ Date ______________

NARCOTICS AND DESIGNER DRUGS

Drugs made from the opium plant are called **narcotics**. Some narcotics are used to treat severe pain. These narcotics are illegal unless a doctor prescribes them.

Some narcotics, such as heroin, are always illegal. Users usually inject heroin, and they might share needles. This puts heroin users at a high risk of diseases such as hepatitis and AIDS. Heroin users can also die of an overdose. ☑

READING CHECK

7. Explain Why are heroin users at a high risk for developing diseases such as hepatitis and AIDS?

____

____

____

____

Other kinds of illegal drugs include inhalants, barbiturates, amphetamines, and designer drugs. Designer drugs are made by making small changes to other drugs. Ecstasy, or "X," is a designer drug. Over time, the drug can cause lesions or holes to develop in a user's brain. Ecstasy users are also more likely to develop depression.

HALLUCINOGENS

Users of hallucinogens have *hallucinations*. This means that they see and hear things that are not real. Drugs such as LSD and PCP are powerful, illegal hallucinogens. Sniffing glue or solvents, such as paint thinner, can also cause hallucinations and serious brain damage. ☑

READING CHECK

8. Define What is a hallucination?

____

____

____

TAKE A LOOK

9. Complete Use the text to complete the chart on the harmful effects of certain drugs.

Type of drug	Legal or illegal?	Harmful effects
Herbal medicines		not fully known
Prescription	legal when used properly	
Tobacco	legal over age 18 in most of U.S.	
Alcohol		
Marijuana		
Cocaine		
Narcotics		
Hallucinogens		

What Is Drug Abuse?

Drug use and drug abuse are not the same thing. Someone can use drugs to prevent or help a medical problem. The drug user gets these drugs legally and uses them properly. People who abuse drugs usually do not take the drugs for a medical problem. Instead, they may take drugs to help them feel good, escape from problems, or fit in. Drug abusers often get drugs illegally. ☑

READING CHECK

10. List Give three reasons a person may abuse drugs.

______

______

______

______

How Does Drug Abuse Start, and How Can It Stop?

Young people often start to use drugs because of peer pressure. Teenagers may drink, smoke, or try marijuana to make friends. Many teenagers begin using illegal drugs to feel like part of a group. However, drug abuse can lead to many problems with friends, family, school, and money. These problems often lead to depression and social isolation.

Many people who start using drugs do not recognize the dangers. Incorrect information about drugs can be found in many places. Some of the most common myths about drugs are shown below.

Drug Myths

Myth	**"It's only alcohol, not drugs."**
Reality	Alcohol is a mood-altering and mind-altering drug. It affects the central nervous system and is addictive.
Myth	**"I won't get hooked on one or two cigarettes a day."**
Reality	Addiction is not related to the amount of a drug used. Some people become addicted after using a drug once or twice.
Myth	**"I can quit any time I want."**
Reality	Addicts may quit and return to drug usage many times. Their inability to stay drug-free shows how powerful the addiction is.

TAKE A LOOK

11. Explain Can someone become addicted to a drug only after using a lot of it or using it for a long time? Explain your answer.

______

______

______

______

The first step to quitting drugs is to admit to drug abuse and to decide to stop. The addicted person should get proper medical treatment. Getting off drugs can be very difficult. However, people who stop abusing drugs lead happier and healthier lives.

Name ______________________ Class ____________ Date ____________

Section 2 Review

SECTION VOCABULARY

addiction a dependence on a substance, such as alcohol or another drug

alcoholism a disorder in which a person repeatedly drinks alcoholic beverages in an amount that interferes with the person's health and activities

drug any substance that causes a change in a person's physical or psychological state

narcotic a drug that is derived from opium and that relieves pain and induces sleep

nicotine a toxic, addictive chemical that is found in tobacco and that is one of the major contributors to the harmful effects of smoking

1. **Compare** What is the difference between physical and psychological dependence?

____

____

____

2. **Explain** Can prescription and over-the-counter drugs be harmful? Explain your answer.

____

____

3. **Define** What are side effects? Give several examples.

____

____

____

4. **Analyze** Is smokeless tobacco safer to use than cigarettes? Explain your answer.

____

____

____

5. **List** Name five drugs that are always illegal.

____

6. **Explain** Why are drug use and drug abuse not the same thing?

____

____

____

____

____

SECTION 3

Healthy Habits

BEFORE YOU READ

After you read this section, you should be able to answer these questions:

- What are some habits you can practice to stay healthy?
- What are some ways to prevent accidents and injuries?

Why Should You Take Care of Your Health?

Do you like playing sports or acting in plays? How does your health affect your favorite activities? The better your health is, the better you can perform. To keep yourself healthy, you need to practice healthy habits every day.

STUDY TIP

Underline As you read this section, underline all the things you can do to take care of your health.

What Is Hygiene?

Hygiene is the science of preserving and protecting your health. Practicing good hygiene is not hard. One of the easiest things you can do to stay healthy is to wash your hands. This helps prevent the spread of disease. You should always wash your hands after using the bathroom and before and after handling food. Taking care of your skin, hair, and teeth is also important for good hygiene. ☑

READING CHECK

1. Identify What is one of the easiest things you can do to prevent the spread of disease?

____

Why Is Posture Important?

Standing up straight not only makes you look better, but it can make you feel better. Posture is important to your health. Bad posture strains your muscles and makes breathing difficult. To have good posture, imagine a line passing through your ear, shoulder, hip, knee, and ankle when you stand. You can have good posture at a desk by pulling your chair forward and keeping your feet on the floor. ☑

READING CHECK

2. Identify What problems can bad posture cause?

____

____

____

When you have good posture, your ear, shoulder, hip, knee, and ankle are in a straight line.

Bad posture strains your muscles and ligaments and can make breathing difficult.

Why Is Exercise Important?

Aerobic exercise at least three times a week is essential to good health. **Aerobic exercise** is vigorous, constant exercise of the whole body for 20 minutes or more. ☑

READING CHECK

3. Identify How long should aerobic exercise last?

____

Types of aerobic exercise include:

- walking
- dancing
- running
- swimming
- biking
- playing sports, such as soccer, basketball, or tennis

Discuss Do you play any sports, or perform other aerobic activities? How do you feel after you exercise? Tell the class about your favorite form of exercise.

Aerobic exercise increases your heart rate. This helps your body take in more oxygen. Aerobic exercise protects your physical and mental health in the following ways:

- strengthens the heart, bones, and lungs
- burns Calories
- helps your body conserve some nutrients
- helps your digestive system work well
- gives you more energy and stamina

How Much Sleep Do You Need?

Believe it or not, teenagers actually need more sleep than younger children. Do you ever fall asleep in class, like the girl in the figure below? If so, you may not be getting enough sleep. Scientists say teenagers need about 9.5 hours of sleep each night. At night, the body goes through several cycles of deeper and deeper sleep. If you do not sleep long enough, you will not enter the deepest, most restful period of sleep. ☑

READING CHECK

4. Identify About how many hours of sleep each night do scientists say teenagers need?

____

If you fall asleep easily during the day, you are probably not getting enough sleep.

What Is Stress?

Have you ever been excited about your team's upcoming soccer game? Have you ever been upset or angry because you got a bad grade on a test? The game and the test were causing you stress. **Stress** is the physical and mental response to pressure, such as being worried, anxious, or excited.

Some stress is a normal part of life. Stress tells your body to prepare for difficult or dangerous situations. However, sometimes you may have no outlet for the stress, and it builds up. Too much stress can be harmful to your health. It can also make it hard for you to do everything you need to do each day.

Some signs of too much stress include:

- headaches
- upset stomach
- inability to fall asleep at night
- nervous habits, such as nail biting
- becoming irritable or resentful

Say It

Describe Have you ever felt stressed? What do you think caused your stress? How did it make you feel? Describe this to a partner.

Critical Thinking

5. Analyze How can stress be both good and bad?

____

____

____

____

____

____

TAKE A LOOK

6. Identify Circle all the things in this picture that could cause stress.

What Are Some Ways to Cope with Stress?

Different things stress different people. Once you learn what things cause you stress, you can find ways to deal with it. If you cannot remove the cause of stress, here are some ideas for handling stress. ☑

- Share your problems. Talk them over with a parent, friend, teacher, or school counselor.
- Make a list of all the things you would like to get done. Rank these things in order of how important they are. Do the most important thing first.
- Exercise regularly.
- Get enough sleep
- Pet or play with a friendly animal.
- Spend some quiet time alone.
- Practice deep breathing or other relaxation techniques.

READING CHECK

7. Explain What do you need to do before you can deal with stress?

____

____

What Are Some Ways to Prevent Injuries?

Practicing healthy habits includes preventing injuries. Accidents can happen any time. It is impossible to prevent all accidents. However, you can decrease your risk of injury by following some basic safety rules.

OUTDOOR SAFETY

If you enjoy outdoor activities, such as biking, hiking, or boating, some rules you should follow are:

- never hike or camp alone
- dress for the weather
- tell someone where you are going and when you will return
- learn how to swim, and never swim alone
- wear a life jacket when you are in a boat
- if a storm approaches, get out of the water and seek shelter
- always wear a helmet when you are biking, and obey the rules of the road

You should practice other common sense habits as well. For example, you should always wear a seat belt in a moving vehicle. Never ride in a car with someone who has been drinking alcohol or using illegal drugs.

Critical Thinking

8. Infer Why do you think it is important to tell somone where you are going and when you will return?

____

____

____

____

____

____

SAFETY AT HOME

Many accidents can be avoided. The figure below shows ways you can help prevent accidents in your home.

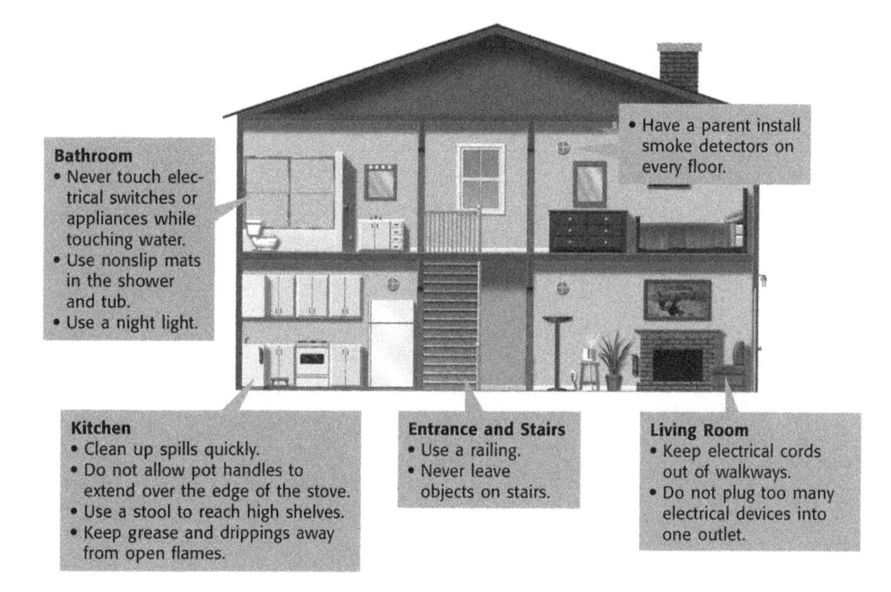

TAKE A LOOK

9. Apply Concepts List three things from this picture that you can do to protect your home from fires.

____

____

____

____

____

What Should You Do If an Accident Happens?

Accidents can happen, so you should be prepared. If you see someone in danger, such as a person choking, you should call for medical help right away. In most communities, you can dial 911 on the telephone. This will connect you with emergency personnel who will send help. Once you dial 911, you should:

- stay calm
- speak slowly and clearly
- give the complete address, phone number, and description of where you are
- describe the accident and tell how many people are injured and what kind of injuries they have
- ask what to do and listen carefully to the instructions
- let the other person hang up the phone first to be sure there are no more questions or instructions for you ☑

READING CHECK

10. Explain When you make a 911 call, why should you let the other person hang up first?

____

____

____

____

Name ______________________ Class ______________ Date ____________

Section 3 Review

SECTION VOCABULARY

aerobic exercise physical exercise intended to increase the activity of the heart and lungs to promote the body's use of oxygen	**hygiene** the science of health and ways to preserve health **stress** a physical or mental response to pressure

1. List Name four ways aerobic exercise can help you stay healthy.

____

____

2. Summarize Complete the organizer below to identify some healthy habits.

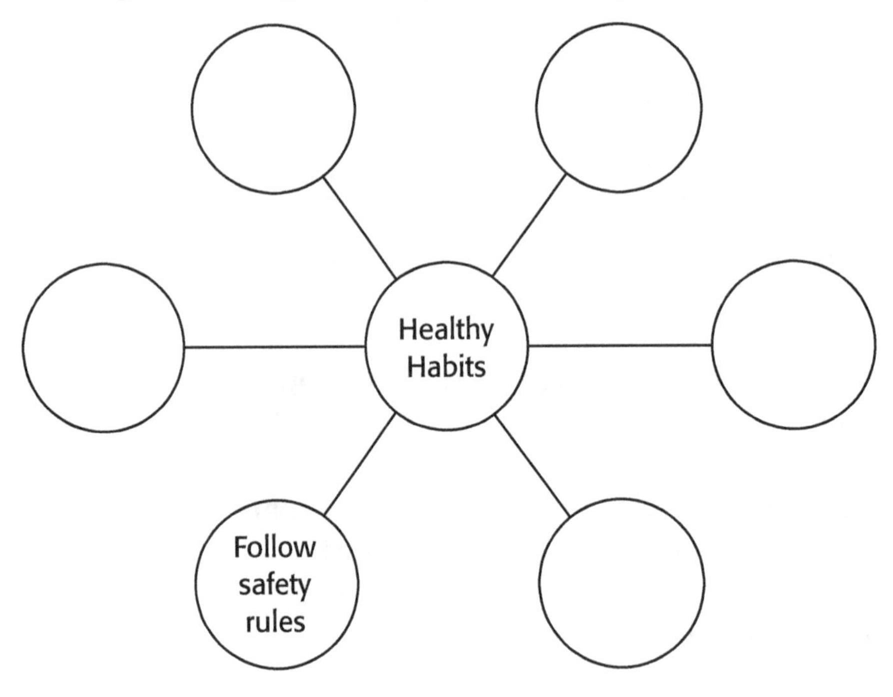

3. List Give three symptoms of too much stress.

____

____

4. Explain Why do you miss the most restful hours of sleep if you don't sleep long enough?

____

____

5. Describe How can you have good posture while sitting at a desk?

____

____

Photo Credits

Abbreviations used: c-center, b-bottom, t-top, l-left, r-right, bkgd-background.

Page 1 Sam Dudgeon/HRW; 2 Hank Morgan/Photo Researchers; 3 (t) NASA; (c) Gerry Gropp; (b) Chip Simons Photography; 6 Sam Dudgeon/HRW; 7 Sam Dudgeon/HRW; 11 Sam Dudgeon/HRW; 13 Royalty Free/CORBIS; 17 (c) CENCO; (inset) Robert Brons/Biological Photo Service; 18 (tl) Sinclair Stammers/Science Photo Library/Photo Researchers; (inset) Microworks/Phototake; (tr) Pascal Goetgheluck/Photo Researchers; (inset) Visuals Unlimited/Karl Aufderheide; 23 (l) Wolfgang Kaehler/Liaison International/Getty News Images; (inset) Visuals Unlimited; 24 (tl) David M. Dennis/Tom Stack and Associates; (tr) David M. Dennis/Tom Stack and Associates; 25 (tl) James M. McCann/Photo Researchers; (tr) Visuals Unlimited/Stanley Flegler ; 27 Wolfgang Bayer; 28 Alex Kerstitch/Visuals Unlimited; 29 (tl) William J. Hebert/Getty Images; (tr) SuperStock; (b) Kevin Schafer/Peter Arnold; 33 (l) Kevin Collins/Visuals Unlimited; (inset) Leonard Lessin/Peter Arnold; 34 (tl) M.I. Walker/Photo Researchers; (tr) Steve Allen/Photo Researchers; (bl) Michael Abbey/Visuals Unlimited; (br) Biodisc/Visuals Unlimited; 37 (t) Wolfgang Baumeister/Science Photo Library/Photo Researchers; (b) Biophoto Associates/Photo Researchers; 51 (bl) Sam Dudgeon/HRW; (br) Sam Dudgeon/HRW; 56 (br) John Langford/HRW; 59 (b) CNRI/Science Photo Library/Photo Researchers ; 60 Biophoto Associates/Photo Researchers; 61 (tl) Ed Reschke/Peter Arnold; (tr) Ed Reschke/Peter Arnold; (cl) Ed Reschke/Peter Arnold; (cr) Ed Reschke/Peter Arnold; (bl) Biology Media/Photo Researchers; (br) Biology Media/Photo Researchers; (b) Visuals Unlimited; 63 (b) National Geographic Image Collection/Ned M. Seidler; 69 (b) Joe McDonald/Visuals Unlimited; 75 (b) CNRI/Phototake NYC; (inset) Biophoto Associates/Photo Researchers; 80 (t) Rob vanNostrand; 87 (inset) J.R. Paulson & U.K. Laemmli/University of Geneva; 89 (c) David M. Phillips/Visuals Unlimited; 93 (tr) Jackie Lewin/Royal Free Hospital/Science Photo Library/Photo Researchers; (br) Jackie Lewin/Royal Free Hospital/Science Photo Library/Photo Researchers; 95 (l) James Beveridge/Visuals Unlimited; (c) Doug Wechsler/Animals Animals; (r) Gail Shumway/Getty Images; 96 (bl) Ken Lucas; (br) John Cancalosi/ Tom Stack & Associates; 98 (t) Courtesy of Research Casting International and Dr. J. G. M. Thewissen; (tc) Courtesy of Research Casting International and Dr. J. G. M. Thewissen; (bc) Philip Gingerich/Courtesy of the Museum of Paleontology, The University of Michigan; (b) Courtesy of Betsy Webb, Pratt Museum, Homer, Alaska; 99 (camel) Ron Kimball/Ron Kimball Stock; (pig) Carol & Ann Purcell/CORBIS; (bison) SuperStock; (hippo) Martin B. Withers/CORBIS; (whale) Martin Ruegner/Alamy Photos; (whale) James D. Watt/Alamy Photos; 103 (b) Carolyn A. McKeone/Photo Researchers; 107 (b) Getty Images; 123 (br) Daniel J. Cox/Getty Images; 125 John Reader/Science Photo Library/Photo Researchers; 126 John Gurche; 127 Volker Steger/Nordstar-4 Million Years, of Man/Science Photo Library/Photo Researchers; 135 Biophoto Associates/Photo Researchers; 136 (t) Dr. Tony Brian & David Parker/Science Photo Library/Photo Researchers; (b) Sherrie Jones/Photo Researchers; 138 (t) Phil A. Dotson/Photo Researchers; (b) © Royalty Free/CORBIS; 139 (t) © Royalty-Free/CORBIS; 143 (tl) Andrew Syred/Photo Researchers; (tr) David M. Phillips/Visuals Unlimited; (b) CNRI/Science Photo Library/Photo Researchers; 144 (c) David M. Phillips/Visuals Unlimited; (inset) SuperStock; 145 (c) SuperStock; (inset) Dr. Kari Lounatmaa/Science Photo Library/Photo Researchers; 148 Peter Van Steen/HRW; 149 (t) Carmela Leszczynski/Animals Animals/Earth Scenes; (b) Aaron Haupt/Photo Researchers; 151 E.O.S./Gelderblom/Photo Researchers; 152 (tl) Hans Gelderblom/Visuals Unlimited; (tr) K.G. Murti/Visuals Unlimited; (bl) Dr. O. Bradfute/Peter Arnold; (br) Oliver Meckes/MPI-Tubingen/Photo Researchers; 155 (tl) David Phillips/Visuals Unlimited; (tr) Matt Meadows/Peter Arnold; (bl) Breck P. Kent; (br) Michael Abbey/Photo Researchers; 156 Dr. Hilda Canter-Lund; 157 (t) Eric Grave/Science Source/Photo Researchers; 159 Robert Brons/Biological Photo Service; 160 Manfred Kage/Peter Arnold; 163 (t) P. Parks/OSF/Animals Animals; (b) Manfred Kage/Peter Arnold; 165 Matt Meadows/Peter Arnold; 167 (l) Runk/Schoenberger/Grant Heilman; (r) Stan Flegler/Visuals Unlimited; 169 A. Davies/Bruce Coleman; 170 (t) Andrew Syred/Science Photo Library/Photo Researchers; (b) J. Forsdyke/Gene Cox/Science Photo Library/Photo Researchers; 171 (t) Bill Beatty; (b) Michael Fogden/DRK; 172 (t) David M. Dennis/Tom Stack; (b) Walter H. Hodge/Peter Arnold; 173 (tl,tr) Stephen & Sylvia Duran Sharnoff/National Geographic Society Image Collection; (bl) Stan Osolinski; 177 (l) Bruce Coleman; (r) Runk/Schoenberger/Grant Heilman; 187 T. Branch/Photo Researchers; 189 (tl) Dwight R. Kuhn; (tr) Runk/Schoenberger/Grant Heilman; (b) Nigel Cattlin/Holt Studios International/Photo Researchers; 191 (br) Stephen J. Krasemann/Photo Researchers; 192 Tom Bean; 193 Everett Johnson/Jupiterimages; 195 George Bernard/Science Photo Library/Photo Researchers; 203 (bl) Jerome Wexler/Photo Researchers; (bc) Paul Hein/Unicorn Stock Photos; (br) George Bernard/Earth Scenes; 205 (b) Cathlyn Melloan/Getty Images; 206 (l) R. F. Evert; (r) R. F. Evert; 209 (cl) David B. Fleetham/Getty Images; 210 (cl) Visuals Unlimited/Fred Hossler; 213 (cl) Michael & Patricia Fogden/CORBIS; 214 Tim Davis/Getty Images; 215 Gerard Lacz/Peter Arnold; 216 Ralph A. Clevenger/CORBIS; 217 Brian Kenney; 219 Peter Weimann/Animals Animals/Earth Scenes; 220 Johnny Johnson/Animals Animals/Earth Scenes; 221 (t) © Ron Kimball/Ron Kimball Stock; (b) Richard R. Hansen/Photo Researchers; 227 (tl) Biophoto Associates/Science Source/Photo Researchers; (tr) Lee Foster/Getty Images; (bl) David Fleetham/Getty Images; (br) Randy Morse/Tom Stack & Associates; 228 CNRI/Science Photo Library/Photo Researchers; 229 (t) Visuals Unlimited/R. Calentine; (b) Visuals Unlimited/A. M. Siegelman; 231 (inset) Nigel Cattlin/Holt Studios International/Photo Researchers; (b) Visuals Unlimited/David M. Phillips; 233 Milton Rand/Tom Stack & Associates; 235 Leroy Simon/Visuals Unlimited; 236 (t) M.H. Sharp/Photo Researchers; (b) CNRI/Science Photo Library/Photo Researchers; 237 (tl) Visuals Unlimited/A. Kerstitch; (tr) Dr. E. R. Degginger/Color-Pic; (b) Daniel Gotshall/Visuals Unlimited; (inset) M.I. Walker/Photo Researchers; 238 Leroy Simon/Visuals Unlimited; 241 (l) Cabisco/Visuals Unlimited; (r) Paul McCormick/Getty Images; 243 (tl) Andrew J. Martinez/Photo Researchers; (tr) Robert Dunne/Photo Researchers; (cl) Flip Nicklin/Minden Pictures; (cr) Marty Snyderman/Visuals Unlimited; (bl) Chesher/Photo Researchers; (br) Daniel W. Gotshall/Visuals Unlimited; 245 (tl) Randy Morse/Tom Stack; (tr) G.I. Bernard OSF/Animals Animals; 247 James Watt/Animals Animals/Earth Scenes; 249 (t) Steinhart Aquarium/Photo Researchers; (c) Hans Reinhard/Bruce Coleman; (bl) Norbert Wu; (br) Martin Barraud/Getty Images; (bc) Index Stock Imagery; 250 (l) Steinhart Aquarium/Tom McHugh/Photo Researchers; (r) Ron & Valerie Taylor/Bruce Coleman; (b) Bruce Coleman; 253 David M. Dennis/Tom Stack & Associates; 255 M.P.L. Fogden/Bruce Coleman; 256 (tl) Richard Thom/Visuals

Unlimited; (tr) Stephen Dalton/NHPA; (bl) Leonard Lee Rue/Photo Researchers; (br) Breck P. Kent; 257 Felix Labhardt/Getty Images; 259 Danilo B. Donadoni/Bruce Coleman; 260 (l) Stanley Breeden/DRK Photo; (r) Visuals Unlimited/Joe McDonald; 261 (l) Mike Severns/Getty Images; (r) Dr. Carl H. Ernst; 262 (tl) Wayne Lynch/DRK Photo; (tr) Kevin Schafer/Peter Arnold; (bl) Animals Animals/Earth Scenes; (br) Uhlenhut, Klaus/Animals Animals/Earth Scenes; 263 Kevin Schafer/CORBIS; 269 (t) D. Cavagnaro/DRK Photo; (b) Hal H. Harrison/Grant Heilman; 271 (l) Kevin Schafer/Getty Images; (c) Gavriel Jecan/Getty Images; (r) APL/J. Carnemolla/Westlight; 272 (t) Tui De Roy/Minden Pictures; (c) S. Nielsen/DRK Photo; (b) Wayne Lankinen/Bruce Coleman; 273 (tr) Stephen J. Krasemann/DRK Photo; (tl) Frans Lanting/Minden Pictures; (bl) Greg Vaughn/Getty Images; (br) Fritz Polking/Bruce Coleman; 275 (tl) Gerard Lacz/Animals Animals; (tr) Nigel Dennis/Photo Researchers; (b) Tim Davis/Photo Researchers; 276 Hans Reinhard/Bruce Coleman; 277 (l) Tom Tietz/Getty Images; (r) Sylvain Cordier/Photo Researchers; 279 (l) Wayne Lynch/DRK Photo; (r) John D. Cunningham/Visuals Unlimited; 280 (tl) D. R. Kuhn/Bruce Coleman; (tr) Gail Shumway/Getty Images; (bl) Gerry Ellis/Minden Pictures; (br) Frans Lanting/Minden Pictures; 281 (tl) David Cavagnaro/Peter Arnold; (tr) John Cancalosi; (bl) Art Wolfe/Getty Images; (br) Merlin D. Tuttle/Bat Conservation International; 282 (tl) Gail Shumway/Getty Images; (tr) Arthur C. Smith III/Grant Heilman; (bl) Art Wolfe/Getty Images; (br) Manoi Shah/Getty Images; 283 (tl) Gail Shumway; (tr) Roberto Arakaki/International Stock/ImageState; (b) Scott Daniel Peterson/Liaison/Getty News Images; 284 (tl) Flip Nicklin/Minden Pictures; (tr) Pete Atkinson/NHPA; (b) Tom & Therisa Stack; 285 (t) Inga Spence/Tom Stack; (bl) J. & P. Wegner/Animals Animals; (r) Martin Harvey/NHPA; 286 Joe McDonald/Bruce Coleman; 287 Pavel German/NHPA; (r) Edwin & Peggy Bauer/Bruce Coleman; 288 Dave Watts/Nature Picture Library; 289 (tl) Art Wolfe/Getty Images; (tr) Jean-Paul Ferrero/AUSCAPE; (b) Hans Reinhard/Bruce Coleman; 301 Ross Hamilton/Getty Images; 302 (t) Jack Wilburn/Animals Animals/Earth Scenes; (b) Art Wolfe; 303 Stanley Breeden/DRK Photo; 304 (t) Tom J. Ulrich/Visuals Unlimited; (b) Jeff Lepore/Photo Researchers; 305 Thomas Kitchin/Natural Selection; 306 (t) Robert & Linda Mitchell Photography; (b) Ed Robinson/Tom Stack & Associates; 307 (t) Gay Bumgarner/Getty Images; (b) Rick & Nora Bowers/Visuals Unlimited; 313 (l) Diana L. Stratton/Tom Stack & Associates; (r) Stan Osolinski; 326 (c) Stuart Westmorland/Getty Images; (inset) Manfred Kage/Peter Arnold; 329 Jeff Hunter/Getty Images; 335 Larry Lefever/Grant Heilman Photography; 338 REUTERS/Jonathan Searle /NewsCom; 339 Rex Ziak/Getty Images; 341 (l) Peter Van Steen/HRW; (r) Peter Van Steen/HRW; (c) Peter Van Steen/HRW; 342 (t) Stephen J. Krasemann/DRK Photo; (b) Photo Disc/Getty Images; 343 (c) Novopix / Luc Novovitch; (inset) Singeli Agnew/Taos News; 344 Martin Bond/Science Photo Library/Photo Researchers; 345 K. W. Fink/Bruce Coleman; 346 Stephen J. Krasemann/DRK Photo; 408 Sam Dudgeon/HRW; 409 Will & Deni McIntyre/Photo Researchers; 411 (r) Cabisco/Visuals Unlimited; (l) Innerspace Visions Photography; 413 Getty Images; 422 (tr) David Phillips/Photo Researchers; (tl) Petit Format/Nestle/Science Source/Photo Researchers; (c) Photo Lennart Nilsson/Albert Bonniers Forlag AB, A Child Is Born, Dell Publishing Company; (b) Keith/Custom Medical Stock Photo; 423 Peter Van Steen/HRW; 425 (l) Tektoff-RM/CNRI/Science Photo Library/Photo Researchers; (r) CNRI/Science Photo Library/Photo Researchers; 426 Kent Wood/Photo Researchers; 427 Peter Van Steen/HRW; 429 Peter Van Steen/HRW; 435 Clinical Radiology Dept., Salisbury District Hospital/Science Photo Library/Photo Researchers; 438 (cl) Sam Dudgeon/HRW; (cl) Peter Van Steen/HRW; 445 (l) E. Dirksen/Photo Researchers; (r) Dr. Andrew P. Evans/Indiana University; 449 Sam Dudgeon/HRW; 450 Peter Van Steen/HRW; 451 Sam Dudgeon/HRW.

4500773488
July 2, 2019
Printed in the U.S.A